NOVEL NANOMATERIALS

NOVEL NANOMATERIALS

Jennifer Powers

Kruger Brentt
Publishers
2024

Kruger Brentt Publishers UK. LTD.
Company Number 9728962

Regd. Office: 68 St Margarets Road, Edgware, Middlesex HA8 9UU

ISBN: 978-1-78715-269-4

For information on all our publications visit our website at http://krugerbrentt.com/

Preface

Nanomaterials have emerged as promising tools in the fields of biomedical and biotechnological applications, revolutionizing the landscape of medicine and healthcare. Their unique properties at the nanoscale, such as high surface area, tunable surface chemistry, and quantum effects, enable them to interact with biological systems in unprecedented ways. In the realm of biomedicine, nanomaterials have found applications in drug delivery, imaging, diagnostics, and therapeutics. Nanoparticles, for instance, can be engineered to carry therapeutic agents to specific target sites within the body, enhancing drug efficacy while minimizing side effects. Moreover, nanomaterials like quantum dots and magnetic nanoparticles have demonstrated exceptional potential in medical imaging, allowing for precise and real-time visualization of biological structures. The multifunctionality of these materials opens avenues for personalized medicine, where treatments can be tailored based on individual patient characteristics.

In the biotechnological domain, nanomaterials play a pivotal role in advancing diagnostic techniques, biosensing, and tissue engineering. Nanoscale sensors and probes enable highly sensitive and specific detection of biomolecules, paving the way for early diagnosis of diseases. Additionally, nanomaterials contribute to the development of biosensors that can rapidly detect pathogens or monitor specific physiological parameters. In tissue engineering, nanomaterials act as scaffolds to support cell growth and tissue regeneration. They provide a platform for controlled release of growth factors and other bioactive molecules, promoting cell adhesion, proliferation, and differentiation. The integration of nanotechnology into biotechnological applications not only enhances the precision and efficiency of diagnostic tools but also facilitates the creation of innovative solutions for tissue repair and regeneration. Despite the immense potential of nanomaterials in biomedical and biotechnological applications, challenges such as biocompatibility, toxicity, and long-term safety need to be addressed. Researchers are continually exploring new materials and fabrication techniques to overcome these hurdles and unlock the full potential of nanotechnology in improving healthcare and advancing biotechnological processes. The interdisciplinary nature of this field, combining expertise from materials science, biology, medicine, and engineering, underscores the collaborative efforts required to harness the transformative power of nanomaterials for the benefit of human health and biotechnological innovation.

Eyeing ocular drug delivery, "Novel Nano-Formulations and Nanomedicines" highlights promising applications in ocular health. The global challenge of infectious diseases takes center stage in "Incorporating Nanotechnology for the Diagnosis and Treatment of Sars-Cov-2," proposing a strategic approach for pandemic prevention. "Nanoparticles of Cerium Oxide" bridges the realms of nanotechnology and biomedicine, showcasing diverse synthesis techniques with biomedical implications. "Polycaprolactone Composite Nanofibrous Material" introduces an alternative material for drug extraction, demonstrating the versatility of nanotechnology in chromatography. Enhancing the efficiency of microalga harvesting is the focus of "Improving the Magnetic Harvesting Efficiency of Microalga Chlorella Vulgaris," emphasizing sustainable practices. Finally, "Covalent Immobilization of Papain on Films Comprising Chitosan–Clay Nanocomposites" explores applications in synthetic and real white wine, showcasing the diverse range of fields impacted by nanotechnology. It is a valuable resource for anyone interested in the intersection of nanotechnology with biomedical and biotechnological fields. Graduate and postgraduate students studying nanotechnology, biotechnology, biomedical sciences, or related fields can use the book as a comprehensive reference for coursework.

We are grateful to all those persons as well as various books, manuals, periodicals, newsletters, magazines, journals etc. that helped in the preparation of this book. In spite of the best efforts, it is possible that some errors may have occurred into the compilation and editing of the book. Further queries, constructive suggestions and criticisms for the further improvement of the book are always welcome and shall be thankfully acknowledged.

Jennifer Powers

Contents

01 PROPELLING A MAGNETIC MICRODROPLET THROUGH THE USE OF AN ELECTROMAGNETIC ACTUATION SYSTEM

1. INTRODUCTION: AN OVERVIEW

In the burgeoning field of microfluidics, the precise manipulation of small volumes of liquid, particularly in the form of microdroplets, has become a focal point for numerous applications ranging from healthcare to advanced materials. One intriguing avenue within this realm involves the utilization of magnetic forces for the propulsion and control of microscale droplets. The remote manipulation of nano and microparticles attracts researchers from various scientific fields. Several technologies were adopted, whereby magnetism and composite materials seem to be the most appropriate tools for this task. The applications of magnetic particles in biomedical and clinical research fields are several. They are used for drug delivery in tumors [1–4], infective diseases [5], thrombosis [6], localized administration of drugs [7], stem-cell therapies [8–10], and gene delivery [11,12]. To manipulate magnetic particles, permanent magnets or electromagnets can be used. Permanent magnets propagate high-gradient magnetic fields able to retain magnetic particles in relatively small distances (up to 5 cm); however, the propagated field intensity cannot be adjusted. Therefore, their clinical applications are limited, e.g., for retaining magnetically responsive materials at small distances. A second alternative of magnetic manipulation relies on the use of electromagnets which enable the possibility to fine-tune or switch on/off the generated magnetic field. Many attempts were made to manipulate magnetic agents for clinical purposes such as the Octomag [13] or the Mag-µBot system [14]. Nevertheless, all these systems can only attract magnetically labeled items because of the paramagnetic nature of magnetic nano and microparticles. Furthermore, an additional obstacle in driving pharmaceutical agents in deep tissues and organs is human body size [15]. In order to reach deep targets while preserving the ability to manipulate pharmaceutical agents, it is necessary to magnetically push nano- and/or microparticles (magnetic injection) deep into the human body. In an early study using permanent magnets with a specific configuration, Sarwar and his colleagues successfully managed to push magnetic particles into rat inner ears, keeping the apparatus at a corresponding "human head" working distance [16,17]. Nevertheless, the system proposed by Sarwar has the major drawback of moving parts. To achieve the pushing effect, all magnetic apparatus must be moved, implying a complex design and a laborious use of the entire device. Electromagnetic driving devices designed and developed by Probst et al. and Bijarchi et al. were also proposed recently [18,19]. However, these electromagnetic devices with adjustable magnetic gradients cannot manipulate magnetic microdroplets at human body working distances.

In this work, we present the design of an actuation system able to manipulate (deflect and push) magnetic particles at relatively big distances from its cores without the need for moving parts. Indeed, the pushing effect is obtained by fine-tuning the electric current feeding a specifically configurated electromagnet array. The optimum design and working conditions (angle, current for each coil, distance, etc.) were optimized by using computer simulations; subsequently, the magnetic system was constructed and tested successfully. This design paves the way for driving pharmaceutical agents in deep areas of the human body. The fact that electromagnets are used offers a promising and fully controllable tool for biomedical applications to push, deflect, or pull magnetic particles at relatively long distances from the magnetic core.

2. MATERIALS AND METHODS

2.1 Magnetic System

The distance between the electromagnet and the magnetic particles to be driven is a key parameter in the equation governing any magnetic system that is intended to be used for the remote control of magnetic nano- and/or microparticles [20]. Indeed, the magnetic field intensity must be increased as the distance increases from the magnets. To design our system, we took advantage of the basic principal that, thanks to a specific electromagnet configuration, the magnetic field lines sum together at one specific point due to the linearity of Maxwell's equations. On this specific point, magnetic lines cancel each other, creating a magnetic field minimum. The magnetic field strength will increase outward from this minimum and create outward forces. The apparatus that was constructed following this reasoning and presented here consists of four coils, two steel cores, four power supplies, and a control system.

To achieve an increasing magnetic field gradient which pushes magnetic particles, the magnetic lines have to cancel themselves (meet at 180°) at a desirable point and, after that, the magnetic field vectors should meet again with an angle different from 180°. We used the finite element method through ANSYS 17 (Ansys, Canonsburg, PA, USA, licensed to National Technical University of Athens) to test different electromagnet arrangements with various angles and various current combinations to optimize the magnetic line alignments and obtain a cancellation node, where only one configuration was selected to be experimentally validated and is presented hereafter. The magnetic core dimensions were60 × 60 × 170 mm3, made from electrical steel SA1008 (Sidenor steel industry, Maroussi, Greece), and theinner and outer coils were wound in 350 and 540 turns, respectively, with 2-mm-diameter copper wire. At each core, two coils were wound (Figure 1). Magnetic cores were surrounded by wooden holders and connected by an adjustable wooden bar to regulate the angle between electromagnetic elements.

2.2 Control System

The control system was designed and developed in order to control the characteristics of the supply signal and feed each coil separately with pulsing direct current (DC) (1 s duty cycle, 1 s pulse width). In order to accomplish this, the control system was based on the pulse-width modulation (PWM) technique, according to which the modulation of the width (duration) of a pulsed signal results in the amplitude adjustment of the output voltage. As a result, the designed control system can be used to adjust the current, as well as the pulse width (duty cycle) supplied to each coil. The schematic of the designed circuit is provided in Figure S1 (Supplementary Materials). It consists of four channels, each of which can be connected to a different coil. Each channel is controlled by a current mode controller (TS3843B). The output of each controller can be adjusted by a potentiometer, which is located at the upper panel of the control system. A fifth knob is used in order to control the pulse duration, through a selection among 12 different capacitor networks, which are connected to the controller through a Schmitt trigger inverter (74HC14). Finally, the adjusted controller's output is fed to the gate of a power metal–oxide-semiconductor field-effect transistor (MOSFET) (IRFP3306PbF), which allows the current to flow to the connected coil.

The designed circuit was fabricated on a printed circuit board (PCB), having the required terminals for the connection of the coils. Four power supplies with a maximum output power of 600 W (24 V, 25 A) each were used in order to provide sufficient power to the coils.

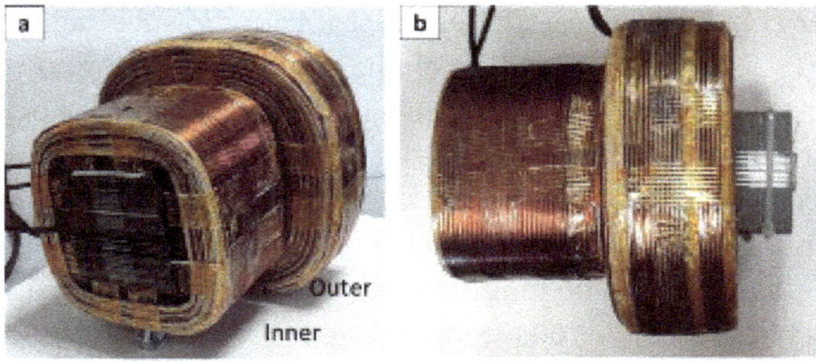

Figure 1. Electromagnet unit arrangement with the two coils (outer and inner) and the electrical steel core: (a) front view; (b) top view.

2.3 Nanoparticles

For the experiments, a microdroplet of super-paramagnetic nanoparticles was used. The product (fluidMAG-Lipid, article number 4119-5) was purchased from Chemicell GmbH, Berlin, Germany. The nanoparticle core was magnetite-functionalized with a matrix phosphatidylcholine, resulting in a hydrodynamic diameter of 100 nm. The surfactant (matrix phosphatidylcholine) is hydrophilic; therefore, nanoparticles were dissolved in water. For the experiment, oleic acid was used in order to force the nanoparticles to form a microdroplet. The nanoparticle buffer was sterile water, and a micropipette with adjustable volume between 0–10 μL was used to transfer the microdroplet of nanoparticles in the oleic acid.

2.4. Microscope

Images and videos were recorded using a simple microscope made of plastic elements to avoid interference with the magnetic field (Bresser junior DM400 digital microscope, Bresser, Germany).

3. RESULTS

3.1. Magnetic Actuator Modeling and Construction

The electromagnetic system, as well as the control system regulating the current at each coil, was constructed at our laboratory facilities. The validation of the design was done using a Hall sensor in order to compare the generated field with the values from the simulations. Measurements and simulations were taken for 18 A at the outer coils and 6 A in the inner coils. In Figure 2, the whole apparatus is shown together with the simulation results. Experimental measurements are compared and a very small deviation between them can be observed. In Figure 2b, the magnetic field intensity that propagated at a straight line from the middle of the cores is depicted for 18 A at the outer coils and 6 A in the inner coils. At a distance of 7.5 cm from the cores, for the specific values of current in the coils, the magnetic field intensity is minimized due to the opposite direction of the vectors at this point.

3.2. Validation of the Pushing Effect

We performed preliminary tests in order to verify the efficiency of the system and the reliability of the simulations. A paramagnetic sphere of 4-mm radius was placed on an aluminum rail at 9 cm, nearby the cancellation node (7.5 cm) but are the point where outward force was maximum (Figure 3a) according to the simulation results as seen in Figure 2. Subsequently, by feeding the coils with 18 A (outer coils) and 6 A (inner coils), the sphere received a magnetic push that moved it by about 10 cm from the initial point (Figure 3b). The results of this experiment confirmed the accuracy of our simulations, as well as the measurements with the Hall sensors. However, since the aim of our study was to demonstrate the possibility to magnetically push small particles in the human body, we set up a second round of experiments.

3.3. Remote Control of Magnetic Micro droplet

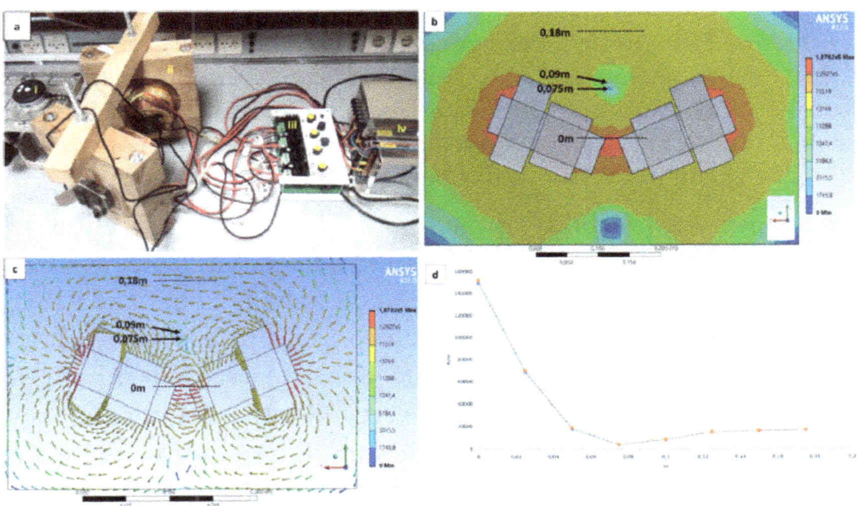

Figure 2. The whole apparatus is depicted in (a): (i) the microscope used to monitor the movement of the magnetic droplet; (ii) electromagnets; (iii) the control system; (iv) the power supplies. (b,c) Simulations for the magnetic field intensity propagated by the electromagnets are shown, as well as the vectors of the field. Arrows in (b,c) indicate the cancellation area created at given distance from magnet cores and the initial position of the magnetic sphere and microdroplet. (d) The magnetic field intensity measured by a Hall sensor (orange colour) in comparison with the results of the simulation from ANSYS (blue colour). The measurements are taken on the centerline starting from the last point of the cores (dashed lines). The scale bars are in meters.

For this experiment we used a microdroplet with a volume of 10 μL of water-soluble ferrofluid material placed in 100% oleic acid (Figure 4). The petri disc containing the microsphere of ferrofluid nanoparticles and oleic acid was put at a specific distance (9 cm) from the cores in order to align the microdroplet with the point where outward force was maximum (Figure 4), and electromagnets were activated as described previously. Due to the high viscosity of the oleic acid and the size of the specimen, the movement of the microdroplet was restricted only to a few mm as seen in Figure 4. Nevertheless, this significant magnetic push received by the microdroplet demonstrates that it is possible to remotely control paramagnetic material.

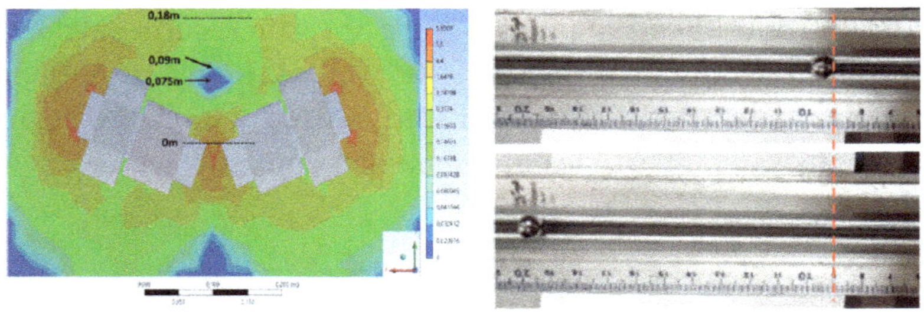

Figure 3. (a) Map of magnetic forces generated around the electromagnets; units reported in the simulation are arbitrary and are intended to highlight areas where the force is almost zero (dark blue), while the other colors indicate forces >0. The scale bar is in meters. (b) A paramagnetic sphere pushed from 9 to 19 cm on an aluminum rail using the actuator system.

Having assured the reliability of the simulations, the next step was to test the deflecting efficiency. To do so, the ferrofluid microdroplet, a cluster of nanoparticles with average size of 100 nm, was put at the same starting point, and the coils were fed with specific currents (Figure 5 and Video S1, Supplementary Materials). In particular, coil 1 and coil 2 of the left array (Figure 5a,b) were fed with 15.5 A and 4 A, whereas coil 3 and coil 4 of the right array were fed with 4 A and 13 A, respectively. This deferential power supplied to the four coils allowed us to create a magnetic flux that moved the droplet toward the X-axis (Figure 5c). For the deflecting step that allowed us to move the droplet toward the Y-axis (Figure 5d), coil 1 and coil 2 of the left array (Figure 5e) were fed with 17.5 A and 4 A, whereas coil 3 and coil 4 of the right array were fed with 4 A and 11 A, respectively. The i_erential current supplied to each coil in these experiments allowed remotely controlling a microdroplet of magnetic material; the experimental data were confirmed by the simulations that predicted a hange in the direction of the magnetic field vectors (Figure 5b,f) as a consequence of power supplied to the coils. The velocity of the ferrofluid droplet was also calculated to be approximately 135 μm/s.

The driving of the magnetic droplet was based on the optical feedback provided by a microscope connected to the computer. However, it is possible to predesign a route and program the actuation system accordingly, in order to drive the microdroplet at a specific route, without intervening during the procedure. The programming of the actuation system will be presented in a future work.

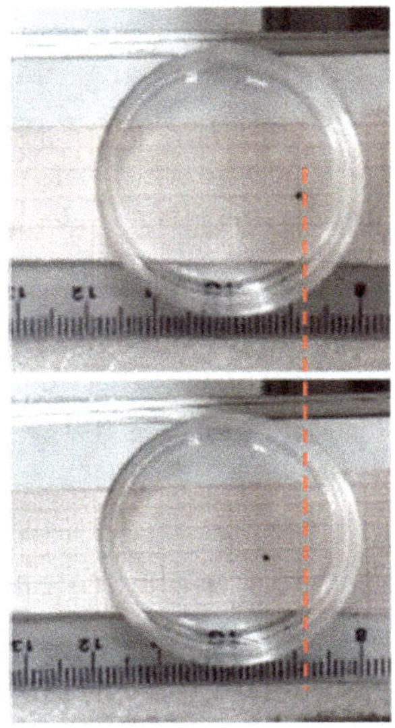

Figure 4. Ferrofluid microdroplet being pushed in oleic acid.

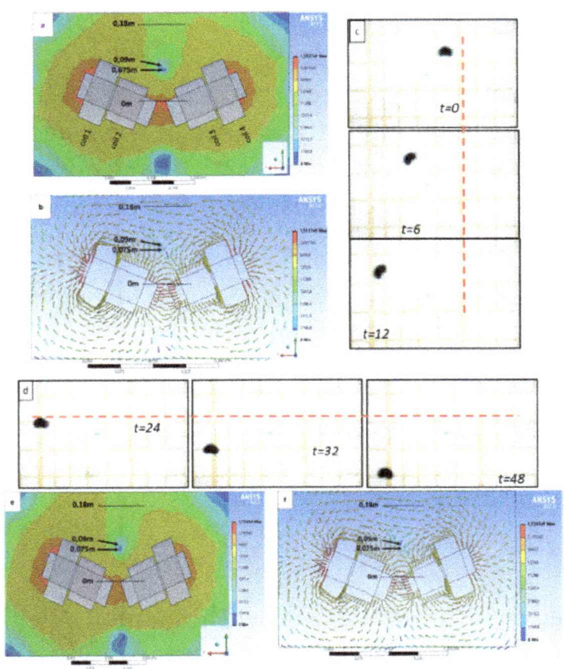

Figure 5. (a, b) Simulations of the magnetic field intensity and the vectors of the field for currents used to push forward the droplet. (c) The positions at t = 0, 6, and 12 s. (d) The position of the droplet for t = 24, 32, and (e, f) Simulations of the magnetic field intensity and the vectors of the to deflect the droplet. Note the change of direction of the vector in the cancellation node (arrows) in (b, f) due to different current supply to the four coils. The measurements are taken on the centerline starting from the last point of the cores (dashed lines). The scale bars are in meters.

4. DISCUSSION

To achieve pushing and deflecting effects of any magnetic material, a controllable cancellation node in a given bidimensional space must be realized. In fact, at this unique point, the magnetic lines at the minimum field point meet each other with opposite vector directions but with the same absolute field intensity values. Since the magnetic fields do not cancel at other points, this point is a unique location of a locally minimum (zero) magnetic field strength. The forces increase from low to high magnetic field; therefore, in the region beyond the cancelation node, the generated forces will push particles away from the magnetic system. An ideal magnetic actuator system should offer the ability to adjust the position of the local minimum point and, therefore, to remotely drive the magnetic objects by regulating the power provided to the electromagnets, and not by moving the magnetic elements.

The magnetic force applied by a non-homogeneous magnetic field on a particle suspended in a fluid is given by the following equation:

$$F_m = \mu_o\left(\left(m_p - m_f\right)\cdot\nabla\right)H_{ap}, \qquad (1)$$

where μo is the permeability of free space, r is the Hamilton operator, Hap is the value of the magnetic field intensity applied at the place where the particle is located, mp is the magnetic moment of the particle, and mf is the magnetic moment of the liquid where the particle suspends. Xp and Xf are the susceptibilities of the particle and the fluid, respectively.

$$m_p - m_f = V_p \frac{3\left(\chi_p - \chi_f\right)}{\left[\left(\chi_p - \chi_f\right) + 3\left(\chi_f + 1\right)\right]} H_{ap}. \qquad (2)$$

In our case, the Xf of oleic acid is ~0; thus, the equation can be written as

$$F_m = \mu_o V_p \frac{3\chi_p}{\left(\chi_p + 3\right)}\left(H_{ap}\cdot\nabla\right)H_{ap}. \qquad (3)$$

The force applied by a magnetic field with magnetic field intensity Hap at a magnetic particle of radius X(m) and magnetic susceptibility X is

$$F_m = \frac{4\pi\alpha^3}{3}\frac{\mu_o\chi}{\left(1+\frac{\chi}{3}\right)}H_{ap}\frac{dH_{ap}}{dx} = \frac{2\pi\alpha^3}{3}\frac{\mu_o\chi}{\left(1+\frac{\chi}{3}\right)}\nabla\left(\left|H_{ap}\right|^2\right). \qquad (4)$$

Based on literature data, the magnetic susceptibility of the magnetic microdroplet is 20 [17] and that of the human body is 10.6–10.4 [21]; thus, the efficiency of the magnetic field on the magnetic microdroplet is not affected remarkably by human tissue. The magnetic force applied depends strongly on the particle size. Very small objects with size on the order of nm can be manipulated by using cooling systems, which allow providing higher currents feeding the coils. Moreover, the magnetic forces can be increased by choosing particles with anisotropic shape [22] or particles with higher magnetic susceptibility. Furthermore, it must be noted that the magnetic force increases according to the increase in the magnetic field gradient. A steeper gradient results in a bigger force on a given microdroplet. Manipulation of magnetic ferrofluid attracted the interest of various scientific fields due to the multidisciplinary nature of applications. Nacev et al. and Torres-Diaz et al. [23,24] described various approaches that were done in order to manipulate ferrofluid; however, they reported the control of ferrofluid at small distances and without the ability to push outward. Other attempts were made by Zhang et al. and Hassan et al. [25,26] to manipulate ferrofluid using uniform magnetic fields which are easier to acquire, but the feasibility of this manipulation applied to human body still presents drawbacks due to system complexity and operational distances. The system presented in this work overcomes many of these issues.

A highly important field of interest is anti-tumor therapy. Indeed, magnetic drug delivery could enable a better treatment efficiency, while minimizing side effects of the chemotherapy. By functionalizing drug nanocarriers with therapeutic molecules able to interact specifically with tumor cells, it is possible to deliver drugs exclusively to the tumor tissue [27]. Another important application is targeted regenerative medicine and, more specifically, magnetically driven stem cells. Stem cells play a special role in renewing and repairing damaged tissues and organs. However, a big obstacle in such therapies is the inability to retain stem cells to the desired location. To achieve this goal, magnetic nanoparticles could be inserted in the stem cells, conferring in this way magnetic properties that allow immobilizing them via an externally applied magnetic field. Furthermore, with magnetically modified cells, it would be possible to avert the problem of the short-term cell engraftment in organs or tissue after the injection. One application of stem cells that is becoming increasingly interesting with significant opportunities in the field of neurosurgery is the treatment of spinal cord injuries. Spinal cord injuries (SCIs) are defined as complicated path physiological damages caused by injuries to the spinal column, and there are over 700,000 cases per year worldwide [28]. SCI path physiology can be divided into two distinct phases. Primary injury is the initial shearing or compression of the spinal cord tissue. The mechanical force of the primary injury can lead to hemorrhage, disruption of cell membrane integrity, and ion and neurotransmitter imbalance that immediately affects neural function. Secondary injury is related to the progressive inflammatory, ischemic, and apoptotic cascade that follows the initial mechanical assault [29]. The target of stem-cell therapies for SCI is to minimize the spread of secondary injury, improve the function of remaining cell populations, and engage regeneration of neuronal and glial populations. All these therapeutic effects could be achieved via a remote control of magnetically responsive stem cells and an apparatus like that proposed in the present work.

Another interesting field of application of magnetically responsive cells is cellular transplantation, which can be augmented with a combination of growth factors, scaolds, or other biomaterials that improve cell survival, engraftment, and dierentiation. Intraspinal application of these therapies leads to engraftment of transplanted cells, which may promote neural regeneration through several proposed mechanisms [30]. The engraftment on specific locations can be effectively increased by using magnetically modified stem cells. Furthermore, various studies were conducted on the delivery of magnetically modified stem cells at targets such as the myocardium or at the spinal canal, but the targeting in all these studies was primeval, since no study used a sophisticated method to guide and, most importantly, to inject the magnetic stem cells to the location needed [31]. Implementation of our devices could offer a solution for those pathologies that, at the moment, remain without any feasible operative option.

5. CONCLUSIONS

In this work, we presented an electromagnet system of four coils arranged in two units with a particular configuration. Laboratory experiments and modelling were conducted to prove the effectiveness of the proposed system. We showed, for the first time, an apparatus able to push and deflect magnetic particles just by acting on the currents provided to the four coils without moving any system part. The reason why two units of two coils per core were utilized is essentially due to a better control of the magnetic object during the deflection movement. Indeed, by regulating the power supplied to each coil forming a single unit, the magnetic force generated can be precisely modulated. The pushing effect could be realized with only two coils; however, in order to deflect and pull the microdroplet, the four-coil configuration is necessary. Implementation of our device could offer a solution for pathologies that can be treated with drug target therapies but, at the moment, remain without any feasible operative option.

6. CONCISE SUMMARY

Treatment of certain diseases requires the administration of drugs at specific areas of tissues and/or organs to increase therapy effectiveness and avoid side effects that may harm the rest of the body. Drug targeting is a research field that uses various techniques to administrate therapies at specific areas of the body, including magnetic systems able to drive nano "vehicles", as well as magnetically labelled molecules, in human body fluids and tissues. Most available actuation systems can only attract magnetic elements in a relatively small workspace, limiting drug target applications to superficial tissues, and leaving no alternative cases where deep targeting is necessary. In this paper, we propose an electromagnetic actuation system able to push and deflect magnetic particles at distance of ~10 cm, enabling the manipulation of magnetic nano and microparticles, as well as administration of drugs in tissues, which are not eligible for localized drug targeting with state-of-the-art systems. Laboratory experiments and modelling were conducted to prove the effectiveness of the proposed system. By further implementing our device, areas of the human body that previously were impossible to treat with magnetically labelled materials such as drugs, cells, and small molecules can now be accessible using the described system.

REFERENCES

1. Lübbe, A.S.; Alexiou, C.; Bergemann, C. Clinical Applications of Magnetic Drug Targeting. J. Surg. Res. 2001,95, 200–206. [CrossRef] [PubMed]
2. Wu, M.; Huang, S. Magnetic Nanoparticles in Cancer Diagnosis, Drug Delivery and Treatment. Mol. Clin. Oncol. 2017, 7, 738–746. [CrossRef] [PubMed]
3. Chertok, B.; Moffat, B.A.; David, A.E.; Yu, F.; Bergemann, C.; Ross, B.D.; Yang, V.C. Iron Oxide Nanoparticles as a Drug Delivery Vehicle for MRI Monitored Magnetic Targeting of Brain Tumors. Biomaterials 2008, 29, 487–496. [CrossRef] [PubMed]
4. Toro-Cordova, A.; Flores-Cruz, M.; Santoyo-Salazar, J.; Carrillo-Nava, E.; Jurado, R.; Figueroa-Rodriguez, A.P.; Lopez-Sanchez, P.; Medina, A.L.; Garcia-Lopez, P. Liposomes Loaded with Cisplatin and Magnetic Nanoparticles: Physicochemical Characterization, Pharmacokinetics, and In-vitro Efficacy. Molecules 2018, 23, 2272. [CrossRef]
5. Carinelli, A.; Martí, M.; Alegret, S.; Pividori, M.I. Biomarker detection of global infectious diseases based on magnetic particles. New Biotechnol. 2015, 32, 521–532. [CrossRef]
6. Li, Q.; Liu, X.; Chang, M.; Lu, Z. Thrombolysis Enhancing by Magnetic Manipulation of Fe3O4 Nanoparticles. Materials 2018, 11, 2313. [CrossRef]
7. McBain, S.; Yiu, H.H.P.; Dobson, J. Magnetic Nanoparticles for Gene and Drug Delivery. Int. J. Nanomed.2008, 3, 169–180.
8. Snider, E.J.; Kubelick, K.P.; Tweed, K.; Kim, R.K.; Li, Y.; Gao, K.; Read, A.T.; Emelianov, S.; Ethier, C.R. Improving Stem Cell Delivery to the Trabecular Meshwork Using Magnetic Nanoparticles. Sci. Rep. 2018, 8, 12251. [CrossRef]
9. Labusca, L.; Herea, D.D.; Mashayekhi, K. Stem Cells as Delivery Vehicles for Regenerative Medicine-Challenges and Perspectives. World J. Stem Cells 2018, 10, 43–56. [CrossRef]

10. Cores, J.; Caranasos, G.T.; Cheng, K. Magnetically Targeted Stem Cell Delivery for Regenerative Medicine.J. Funct. Biomater. 2015, 6, 526–546. [CrossRef]

11. Kami, D.; Takeda, S.; Itakura, Y.; Gojo, S.; Watanabe, M.; Toyoda, M. Application of Magnetic Nanoparticles to Gene Delivery. Int. J. Mol. Sci. 2011, 12, 3705–3722. [CrossRef] [PubMed]

12. Majidi, S.; Zeinali Sehrig, F.; Samiei, M.; Milani, M.; Abbasi, E.; Dadashzadeh, K.; Akbarzadeh, A. Magnetic Nanoparticles: Applications in Gene Delivery and Gene Therapy. Artif. Cells Nanomed. Biotechnol. 2016, 44, 1186–1193. [CrossRef] [PubMed]

13. Kummer, M.P.; Abbott, J.J.; Kratochvil, B.E.; Borer, R.; Sengul, A.; Nelson, B.J. OctoMag: An Electromagnetic System for 5-DOF Wireless Micromanipulation. In Proceedings of the 2010 IEEE International Conference on Robotics and Automation, Anchorage, AK, USA, 3–7 May 2010; pp. 1610–1616. [CrossRef]

14. Nelson, B.J. Making Microrobots Move. In Proceedings of the 2010 7th International Conference Informatics Control, Automation and Robotics (ICINCO), Funchal, Madeira, Portugal, 15–18 June 2010; Volume 1, pp. 13–18.

15. Niu, F.; Li, J.; Ma, W.; Yang, J.; Sun, D. Development of an Enhanced Electromagnetic Actuation System with Enlarged Workspace. IEEE/ASME Trans. Mechatron. 2017, 22, 2265–2276. [CrossRef]

16. Sarwar, A.; Lee, R.; Depireux, D.A.; Shapiro, B. Magnetic Injection of Nanoparticles into Rat Inner Ears at a Human Head Working Distance. IEEE Trans. Magn. 2013, 49, 440–452. [CrossRef]

17. Shapiro, B.; Dormer, K.; Rutel, I.B. A Two-Magnet System to Push Therapeutic Nanoparticles. Aip Conf. Proc.2010, 1311, 77–88. [CrossRef]

18. Probst, R.; Lin, J.; Komaee, A.; Nacev, A.; Cummins, Z.; Shapiro, B. Planar steering of a single ferrofluid drop by optimal minimum power dynamic feedback control of four electromagnets at a distance. J. Magn. Magn.Mater. 2011, 323, 885–896. [CrossRef

19. Bijarchi, M.A.; Favakeh, A.; Sedighi, E.; Shafii, M.B. Ferrofluid droplet manipulation using an adjustable alternating magnetic field. Sens. Actuators A Phys. 2020, 301, 1. [CrossRef]

20. Sarwar, A.; Nemirovski, A.; Shapiro, B. Optimal Halbach Permanent Magnet Designs for Maximally Pulling and Pushing Nanoparticles. J. Magn. Magn. Mater. 2012, 324, 742–754. [CrossRef]

21. Di Luzio, S.; Obletter, G.; Comani, S.; Del Gratta, C.; Romani, G.L. Magnetic Mapping of DC Fields Related to Tissue Susceptibility in the Human Body BT—Advances in Biomagnetism; Williamson, S.J., Hoke, M., Stroink, G., Kotani, M., Eds.; Springer: Boston, MA, USA, 1989; pp. 505–508. [CrossRef]

22. Lim, J.K.; Yeap, S.P.; Chee, H.L.; Toh, P.Y.; Low, S.C. Magnetophoresis of iron oxide nanoparticles at low field gradient: The role of shape anisotropy. J. Colloid Interface Sci. 2014, 170–177. [CrossRef]

23. Nacev, A.; Komaee, A.; Sarwar, A.; Probst, R.; Kim, S.H.; Emmert-Back, M.; Shapiro, B. Towards Control of Magnetic Fluids in Patients: Directing Therapeutic Nanoparticles to Disease Locations. IEEE Control Syst. 2012, 32, 32–74. [CrossRef]

24. Torres-Diaz, I.; Rinaldi, C. Recent progress in ferrofluid research: Novel applications of magnetically controllable and tunable fluids. Soft Matter 2014, 43, 8584–8602. [CrossRef] [PubMed]

25. Zhang, J.; Hassan, M.R.; Rallabandi, B.; Wang, C. Migration of ferrofluid droplets in shear flow under a uniform magnetic field. Soft Matter 2019, 15, 2439–2446. [CrossRef] [PubMed]

26. Hassan, M.R.; Zhang, J.; Wang, C. Deformation of a ferrofluid droplet in simple shear flows under uniform magnetic fields. Phys. Fluids 2018, 30, 092002. [CrossRef]

27. Kheirkhah, P.; Denyer, S.; Bhimani, A.D.; Arnone, G.D.; Esfahani, D.R.; Aguilar, T.; Zakrzewski, J.; Venugopal, I.; Habib, N.; Gallia, G.L.; et al. Magnetic Drug Targeting: A Novel Treatment for Intramedullary Spinal Cord Tumors. Sci. Rep. 2018, 8, 11417. [CrossRef] [PubMed]

28. Kumar, R.; Lim, J.; Mekary, R.A.; Rattani, A.; Dewan, M.C.; Sharif, S.Y.; Osorio-Fonseca, E.; Park, K.B. Traumatic Spinal Injury: Global Epidemiology and Worldwide Volume. World Neurosurg. 2018, 113, e345–e363. [CrossRef] [PubMed]

29. Sekhon, L.H.S.; Fehlings, M.G. Epidemiology, Demographics, and Pathophysiology of Acute Spinal Cord Injury. Spine 2001, 26, S24. [CrossRef]

30. Jin, M.C.; Medress, Z.A.; Azad, T.D.; Doulames, V.M.; Veeravagu, A. Stem Cell Therapies for Acute Spinal Cord Injury in Humans: A Review. Neurosurg. Focus Foc. 2019, 46, E10. [CrossRef]

31. Santoso, M.R.; Yang, P.C. Magnetic Nanoparticles for Targeting and Imaging of Stem Cells in Myocardial Infarction. Stem Cells Int. 2016, 2016. [CrossRef]

02 OPTIMIZING PROCESSES AND EXPLORING INDUSTRIAL APPLICATIONS OF CUBIC NANOPARTICLES IN MAGNETIC HYPERTHERMIA

1. INTRODUCTION: AN OVERVIEW

In the realm of nanotechnology, the pursuit of innovative strategies for medical applications has led to the development of cubic nanoparticles designed for magnetic hyperthermia. This cutting-edge approach harnesses the unique properties of nanoscale cubic structures to induce controlled heating in targeted areas, holding immense promise for revolutionizing cancer treatment and other therapeutic interventions. This article delves into the optimization processes involved in crafting these cubic nanoparticles for magnetic hyperthermia, exploring the intricacies of their synthesis and highlighting the potential for industrial implementation. Through a comprehensive examination of this emerging technology, we aim to unravel the transformative impact it could have on the landscape of medical treatments and its role in shaping the future of nanoparticle-based therapies. According to the World Health Organization (WHO), cancer is the main cause of death in the world, having been responsible for 9.6 million of deaths in 2018. In the next two decades, the increase in the number of cases is expected to be 60%, according to a new WHO report [1]. Thus, research in oncology is extremely important to develop therapies to treat this disease and increase the expectation and quality of life of patients. Currently, the main treatments are based on surgery, radiotherapy, and chemotherapy. Thus, these techniques have concentrated most of the research and development till now. Nevertheless, the known side effects of these treatments have not been completely eliminated yet. Due to all this, the search for more efficient techniques and more specific action, with fewer unwanted effects, is important in promoting a better quality of life for patients. Various types of complementary therapies exist and are continuously being developed. In particular, the so-called hyperthermia therapies, which take advantage of the fact that tumor cells have a lower thermal tolerance than healthy cells [2]. In fact, hyperthermia is a therapeutic procedure in which tissues/cells are heated above normal physiological ranges (between 41 and 46°C) in order to kill tumor or sensitize cells to increase the efficiency of standard therapies.

A more recent modality is magnetic hyperthermia (MHT) [3–6], where the increase in temperature occurs by applying alternating magnetic fields to a magnetic material with specific characteristics. In fact, the use of Magnetic Nanoparticles (MNPs) in medicine allows to treat hard-to-reach regions of the body. Chemical manipulation at a nanoscale size has conferred the ability to conjugate biomolecules, such as antibodies, with magnetic particles, for a more effective therapy or to achieve a specific goal, making MNPs ideal vehicles for new therapies based on localized and selective heat dissipation. Thus, MNPs can simultaneously combine various therapeutic functionalities, such as drug carriers, contrast agents in magnetic resonance imaging, or magnetic heating agents. Although MHT has been incorporated into clinical practice to treat a relatively wide range of types of cancer (prostate, esophagus, brain, etc.), today, it has shown to be advantageous in the treatment of sarcomas, carcinomas, and lymphomas.

The ability of the magnetic nanoparticles to generate heat when subjected to alternating magnetic fields is the basis of MHT. In fact, when exposed to an alternating magnetic field, MNPs produce heat through two main mechanisms [7]: hysteresis loss and Néel and Brown relaxation loss. Hysteresis losses occur in large MNPs that have multiple domains, while relaxation losses occur mainly in single domain MNPs.

Magnetite (iron oxide) particles are well-known for their biocompatibility, possible mono-dispersity and easiness of synthesis [7–9]. Besides, mono-domain particles are usually easy to obtain. Shape and size are important characteristics and it has been proven that critical size of magnetite to form multi-domain structure is between 76 and 80 nm for cubic-shaped particles and 128 nm for spherical particles [10]. Super-magnetic behaviour is known to exist, at least, below 50–20 nm for magnetite particles, depending on their shape and other characteristics [11–13]. Besides, SAR values (Specific Absorption Rate) for cubic- shaped particles have been shown to be higher than the ones presented by spherical-shaped particles [14,15]. The correlation between particles microstructure and MHT response has been demonstrated (e.g., [16–18]).

The main methods to produce MNPs are coprecipitation, thermal decomposition, microemulsion, sol-gel, hydrothermal synthesis, and synthesis with polyols [8,9]. These methods have been used to prepare particles with an appropriate composition and size distribution, according to the application in mind. For example, for environmental non- specific treatments, coprecipitation is usually the preferred method [8,19–23], while for biomedical applications where mono-dispersed particles are usually required, less simple methods, like thermal decomposition and hydrothermal, are preferred [8]. A summary of the main methods and their characteristics is presented in Table 1.

Table 1. Comparison between selected magnetic nano-particle manufacturing methods.

Coprecipitation Very simple	Very simple	20–90	Minutes	Water	During/after reaction	Relatively narrow	Not good	High
Microemulsion Complicated	Complicated	20–50	Hours	Organic agents	During reaction	Relatively	Good	Low
Thermal decomposition Complicated	Complicated	100–320	Hours- days	Organic agents	During reaction	narrow Very narrow	Very good	High
Hydrothermal Simple	Simple	200–250	Hours- days	Water-ethanol	During reaction	Very narrow	Very good	Medium
Poliol	Simple	25-Boiling Point	Hours	Ethylene PEG	During reaction	Narrow	Very Good	High

In the case of magnetic hyperthermia, magnetic nanoparticles must present good size distribution, good shape control, and a good SAR generation capability.

Most of the published literature is concerned with laboratorial methods to produce magnetic nanoparticles for hyperthermia applications, but, so far, to the best of our knowledge, none has studied the economic and technological viability of the scale-up of the method and its possible application at large-scale industrial production (which is an important step towards its widespread production and availability, especially after the recent license of a real medical clinical application of magnetic hyperthermia). This will also limit the number of producing methods that may be selected, as the method must be easily scalable (a fundamental characteristic for a production plant).

The present work is concerned with the maximization of the magnetic hyperthermia capabilities of the particles, without increasing the costs to an unbearable level. Thus, the goal is to produce cubic magnetic nanoparticles with sizes lower than the mono-domain critical value and, if possible, the super-magnetism critical value. The optimization of the main variables (ramp-up temperature, stirring speed, type of solvent, proportions of raw materials) for the selected method of production is one of the goals. The other goal is to study the economic and technological viability of the production, in order to set the path for future industrial implementation to support clinical widespread applications.

2. MATERIALS AND METHODS

2.1 Reagents

In this study was used: iron (III) acetylacetonate (99%) from Sigma Aldrich (St. Louis, MO, USA), oleic acid (90%) from Panreac (Barcelona, Spain), benzyl ether (99%) from Sigma Aldrich (Madrid, Spain), 4-biphenylcarboxylic acid (99%) from Sigma Aldrich (Madrid, Spain), toluene (99.8%) from Panreac (Barcelona, Spain), hexane (99%) from Panreac (Barcelona, Spain), chloroform (99.9%) from Panreac (Barcelona, Spain), and nitrogen (99%) from Air Liquide (Paris, France). All solutions were made with distilled water.

2.2 General Experimental Methodology

2.2.1 Synthesis of Cubic Magnetic Nanoparticles

For the synthesis of the magnetic nanoparticles, the method developed by Kim et al. in 2008 [24] was followed, which gives rise to MNPs with cubic morphology and with sizes close to the critical size of the super-paramagnetic state. In a typical synthesis of magnetite nanocubes, 2 mmol of iron acetylacetonate (Fe(acac)3)—Fe(C5H7O2)3,4.5 mmol of oleic acid (C18H34), and 52.5 mmol of benzyl ether - C14H14O-(or 2 mmol of4-biphenylcarboxylic acid - C13H10O2—depending on the size and shape) were used. The initial mixture is first degassed with nitrogen gas. Then, temperature is risen up to 290 °C at a temperature gradient ramp-up of about 20°C/min, under energetic stirring. Once the final temperature is reached, the reaction remains at 290 °C for 30 min. At the end of the process, the sample is cooled and a mixture of toluene and hexane in a 4:1 volumetric ratio is added, and the sample centrifuged at 1700 rpm. Finally, chloroform is used to clean the particles. The main reaction that occurs at these conditions is:

$$aFe(C_5H_7O_2)_3 + bC_{18}H_{34} + cC_{13}H_{10}O_2/dC_{14}H_{14}O \,!\, eFe_3O_4 + \text{Subproducts} \quad (1)$$

The steps of the reaction process are represented in Figure 1.

2.2.2 Particles' Characterization

Particles were characterized by determining their magnetic properties, size, and shape and percentage of magnetite. Hence, to certify the percentage of obtained magnetite, X-ray studies were made (X-ray diffraction (XRD) Bruker D8 Advance, Karlsruhe, Germany); to determine the morphology and size of the particles, Scanning Electron Microscopy (SEM) (JEOL JSM-840, Madrid, Spain) was performed; finally, to access magnetic susceptibility, samples were analyzed with a Kappabridge KLY-4 susceptometer (a semiautomatic auto balance inductivity bridge, Porto, Portugal).

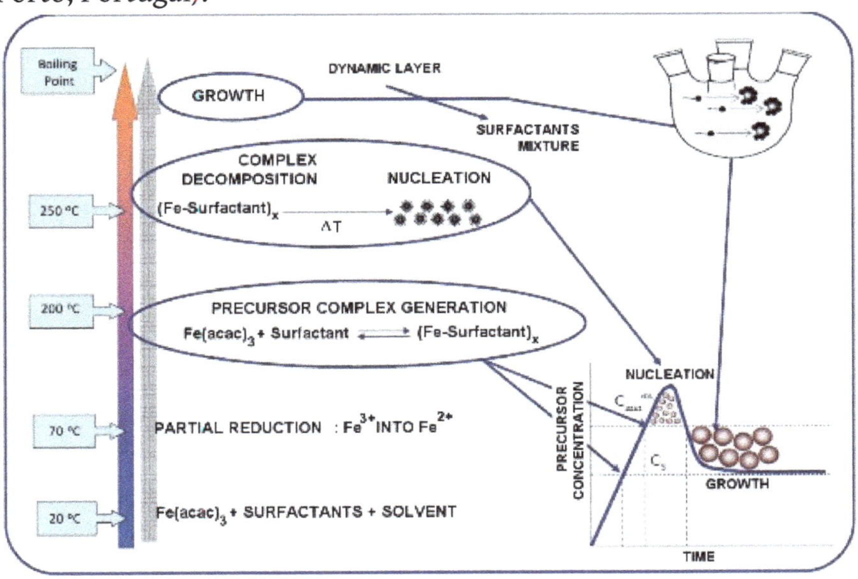

Figure 1. Reaction mechanism—based on [25].

2.2.3. Influence of the Main Variables

For all these experiments, the default options are: iron acetylacetonate (III)—2 mmol; oleic acid—4.5 mmol; 4-biphenylcarboxylic acid—2 mmol; stirring—200 rpm; reaction temperature—290 °C; temperature ramp-up: 20 °C/min; centrifugal speed—1700 rpm. The analyzed variables were:

- Effect of mechanical stirring (four values were used: 150, 170, 200, and 220 rpm).
- Effect of temperature ramp-up (the increasing temperature gradients studied were: 15, 20, 25, and 30 °C/min).
- Effect of the type of solvent (the solvents studied were: (a) 4-biphenylcarboxylic acid—2 mmol; (b) benzyl ether—52.5 mmol; (c) both solvents: 4-biphenylcarboxylic acid—2 mmol—and benzyl ether—52.5 mmol).
- Effect of reagents ratio (the ratios studied were: doubling oleic acid amount, doubling 4-biphenylcarboxylic acid amount, and doubling all reagents).
- Degassing time (the degassing time with nitrogen was reduced by half).

Table 2 shows the values used in each experiment. It is important to notice that experiment 7 corresponds to the predetermined conditions with similar results to those already presented in literature, thus no results are presented here.

Table 2. Experiments done for optimization of the synthesis process (when the value of a specific parameter differs from the default, the value of the parameter is highlighted in grey).

1	2	4.5	2	--	290	20	150	1700	Normal
2	2	4.5	2	--	290	20	170	1700	Normal
3	2	4.5	2	--	290	20	220	1700	Normal
4	2	4.5	2	-	290	25	220	2000	Normal
5	2	4.5	2	--	290	15	220	2000	Normal
6	2	4.5	2	--	290	30	220	2000	Normal
7	2	4.5	--	52.5	290	15	200	2500	Normal
8	2	4.5	--	52.5	290	20	200	2500	Normal
9	2	4.5	--	52.5	290	25	200	2500	Normal
10	2	4.5	2	--	290	20	220	2500	Normal
11 (*)	4	9	4	--	290	25	200	2500	Normal
12 (**)	2	4.5	2	52.5	290	25	200	2500	Normal
13	2	4.5	2	--	290	30	200	2500	Half
14	2	4.5	2	52.5	290	25	200	2500	Normal
15	2	9	2	--	290	20	200	2500	Normal
16	2	4.5	4	--	290	30	200	2500	Normal

(*): It is important to notice that experiment 11 is not an upscale version of experiment 14. In fact, the only parameters that were doubled in experiment 11 were the reagents and one of the solvents, and not all the remaining parameters (to perform an upscale all the other parameters should be also modified accordingly—as done in the upscale and economic analysis presented at the final section of the paper).

2.2.4. Economic Analysis and Industrial Scale-Up Study

The following methodological steps were applied for the economic analysis and industrial scale-up [8]: (a) determine the optimized process for magnetic hyperthermia; (b) perform a market study to determine the demand of the product (magnetic nanoparticles) for hyperthermia applications; (c) determine the size of the plant; (d) analyze all the process steps, choose all the process units and perform all the mass and energy balances required in the plant; (e) detailed scale-up design of the main equipment; and (f) analyze the global economic impact and profitability.

2.3. Instrumentation

The main instrumentation used was: Three-way Reactor (Nahita, Madrid), Orbital stirrer (Nahita, Madrid), Thermal Heater (Nahita, Madrid), Centrifuge (Fischer, Barcelona), besides regular laboratory glass material.

3. RESULTS AND DISCUSSION

In order to determine the best set of values for the main producing parameters, the obtained particles need to be analyzed and compared according to their main characteristics. In this context, it is important to analyze: magnetic properties (magnetic susceptibility), phase composition or phase distribution ratio (XRD), particle size, and morphology (SEM).

3.1 Magnetic Properties

In Table 3 are shown the results obtained for the magnetic properties of the final product corresponding to each experiment.

Table 3. Magnetic properties of the produced nanoparticles.

Experiment Number	1	2	3	4	5	6	7	8
Magnetic susceptibility (SI)	2.51×10^{-4}	5.29×10^{-4}	2.84×10^{-4}	1.56×10^{-4}	1.51×10^{-4}	2.56×10^{-4}	9.71×10^{-5}	2.00×10^{-4}
Experiment number	9	10	11	12 and 14	13	15	16	
Magnetic susceptibility (SI)	2.13×10^{-4}	9.25×10^{-5}	4.64×10^{-4}	3.12×10^{-4}	2.44×10^{-4}	—	—	

As it may be seen, the large majority of the obtained nanoparticles present the same order of magnitude for the magnetic susceptibly. However, particles obtained in experiments 2, 11, 12, and 14 present the highest values, while the lowest values were obtained for experiments 7 and 10.

3.2 Composition

Concerning their structure, X-ray analysis were made on all particles obtained from the set of the performed experiments. In Figure 2 is presented one of the diffraction patterns obtained for these. They all are almost coincident and present the same peaks, so they may be represented by the one depicted in Figure 2 (except for experiment 15 and 16 for which no analysis was possible to be done due to the nature of the obtained products). The typical magnetite peaks are shown in red, according to Powder Diffraction File PDF2 dataset #82- 1533 (Powder Diffraction File, International Centre for Diffraction Data, Newtown Square, PA, USA). No additional peaks are found, so samples are essentially pure magnetite.

In addition to the qualitative analysis, an estimation of the crystallite size was calculated by using the Scherrer equation (Table 4). The main peak (at ca. 35.5°) was used for the calculations. Lanthanum hexaboride (NIST-660b) was used as crystalline standard for instrumental broadening.

Table 4. Estimation of the crystallite size (calculated by using the Scherrer equation).

Experiment Number	1	2	3	4	5	6	7	8
D (nm)	73	50	44	39	32	37	88	92
Experiment number	9	10	11	12 and 14	13	15	16	
D (nm)	121	36	54	32	50	—	—	

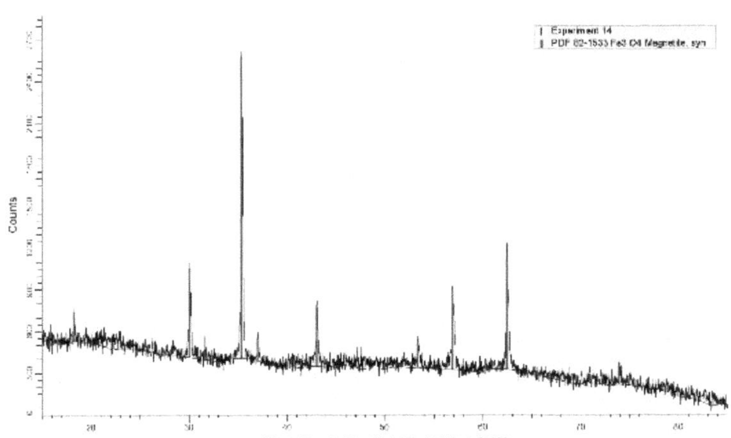

Figure 2. X-ray of experiment 14.

3.3 Particle Size and Morphology

In Figure 3 are presented the secondary electron SEM images of the particles obtained in experiments 1-14, while in Table 5 are presented their size ranges percentages, extracted from the corresponding histograms shown in Supplementary Materials. For some particles, high-quality images were not possible to obtain, mostly due to charging, even after gold sputter coating. In spite of this setback, relevant data were obtained, as summarized in Tables 5 and 6. Some general comments may be emphasized. Typically, when poorly shaped particles are obtained, some tendency to large agglomeration is present, maybe suggesting some kind of growth interference (e.g., Figure 3a,d). When moderately shaped particles are attained, there is a tendency for euhedral forms, almost cubic, though occasional octahedra may be present (e.g., some triangles in Figure 3k), but no rhombic-dodecahedra have been detected. At last, when well-shaped forms are obtained, there is an undoubted preference for cubes and parallelepipeds, in accordance to the results of Kim et al. 2008 [24]. Some stacking is observed (e.g., Figure 3g-i), but it has to be underlined that twinning is not detected in any well-shaped sample.

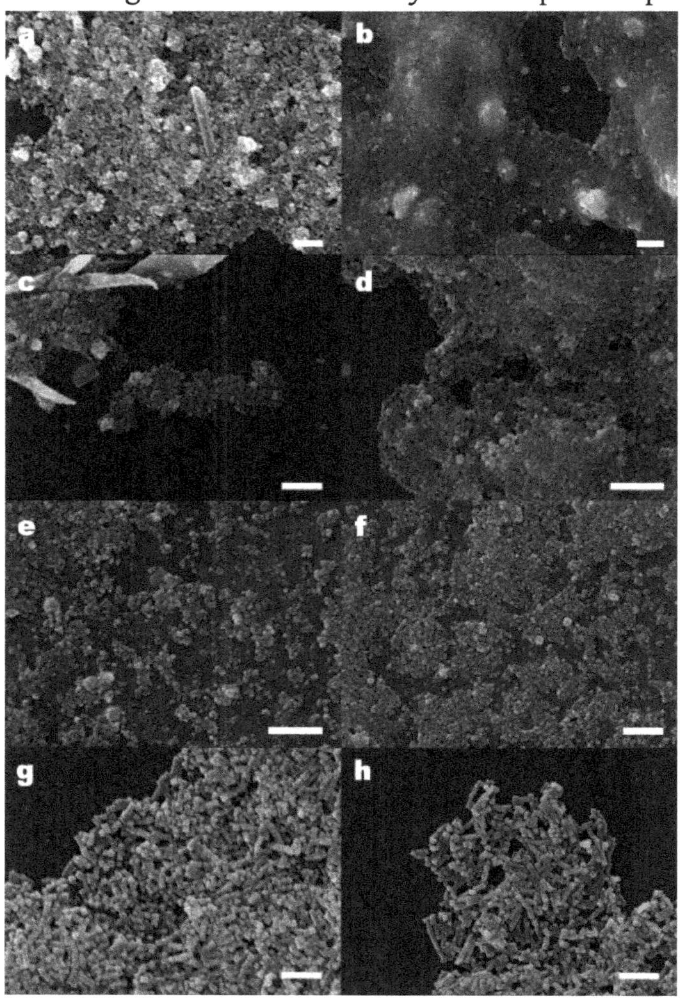

Figure 3. Cont.

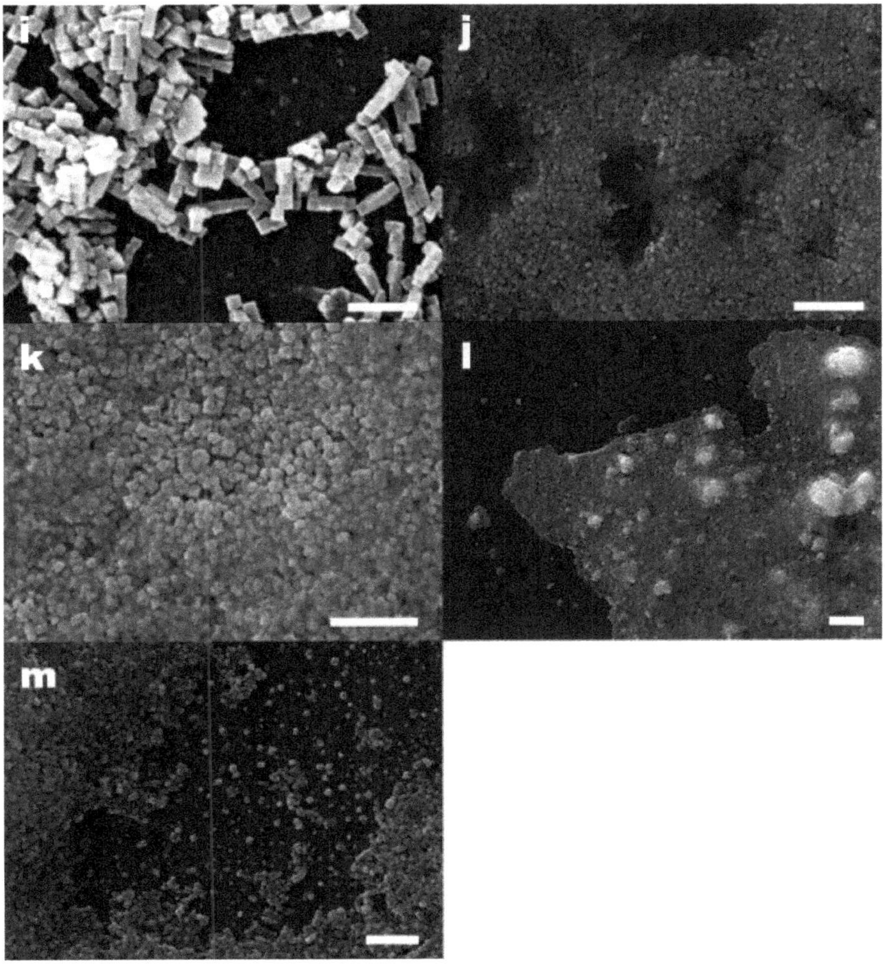

Figure 3. SEM of the nanomagnetite particles obtained in experiments: (a) 1; (b) 2; (c) 3; (d) 4; (e) 5; (f) 6; (g) 7; (h) 8; (i) 9; (j) 10; (k) 11; (l) 13; (m) 12 and 14. White bar length = 1 μm.

Table 5. Size ranges percentages of each experiment product. The size ranges showing highest percentage are highlighted for each case: light grey— between 10–19%; medium gray- between 20–29%; dark grey— more than 30%.

Exp. Number	10–20 (nm)	20–28 (nm)	28–36 (nm)	36–44 (nm)	44–52 (nm)	52–60 (nm)	60–68 (nm)	68–76 (nm)	76–84 (nm)	84–92 (nm)	92–100 (nm)	>100 (nm)
1	0%	0%	0%	0%	0%	0%	1%	2%	3%	2%	12%	80%
2	0%	0%	0%	0%	2%	8%	8%	9%	14%	14%	13%	32%
3	0%	0%	4%	15%	24%	27%	15%	12%	1%	0%	2%	0%
4	0%	5%	12%	28%	26%	18%	7%	2%	0%	2%	0%	0%
5	0%	3%	12%	25%	26%	14%	9%	6%	3%	1%	2%	0%
6	0%	6%	24%	25%	17%	17%	5%	2%	2%	2%	0%	0%
7	0%	0%	0%	0%	1%	4%	4%	11%	12%	12%	10%	46%
8	0%	0%	0%	0%	0%	1%	1%	3%	3%	3%	7%	82%
9	0%	0%	0%	0%	0%	0%	0%	0%	1%	2%	4%	93%
10	0%	11%	9%	19%	30%	18%	9%	4%	0%	0%	0%	0%
11	0%	4%	13%	18%	16%	20%	10%	13%	2%	3%	2%	0%
12, 14	6%	26%	32%	16%	9%	5%	4%	1%	1%	0%	0%	0%
13	0%	4%	6%	6%	15%	20%	24%	12%	5%	5%	4%	3%
15	—	—	—	—	—	—	—	—	—	—		
16	—	—	—	—	—	—	—	—	—	—		

Table 6. Main conclusions from the obtained results.

1	Undefined	92–100 nm: 12% >100 nm: 80%	Regular	The low resolution does not allow to determine their precise morphology. Size much larger than desired. Presents some large agglomerations.
2	Undefined	76–84 nm: 14% 84–92 nm: 14% 92–100 nm: 13%	Slightly Higher	The low resolution does not allow to determine their precise morphology. Size much larger than desired. Presents some large agglomerations.
3	Generally irregular. Some cubic-shaped particles are present.	36–44 nm: 15% 44–52 nm: 24% 52–60 nm: 27% 60–68 nm: 15% 68–76 nm: 12%	Regular	Acceptable morphology. Size larger than desired. Some large agglomerations are present.
4	Irregular	28–36 nm: 12% 36–44 nm: 28% 44–52 nm: 26% 52–60 nm: 18%	Regular	Poorly defined morphology. Size slightly larger than desired. Small agglomerations of medium size are seen.

5	Irregular	28–36 nm: 12% 36–44 nm: 25% 44–52 nm: 26% 52–60 nm: 14%	Regular	Poorly defined morphology. Size slightly larger than desired. Small agglomerations of medium size are seen.
6	Generally irregular. Some cubic-shaped particles are present.	28–36 nm: 24% 36–44 nm: 25% 44–52 nm: 17% 52–60 nm: 17%	Regular	Acceptable morphology. Size slightly larger than desired. Some agglomerations of medium size are seen.
7	Cubic/cobblestone particles	68–76 nm: 11% 76–84 nm: 12% 84–92 nm: 12% 92–100 nm: 10% >100 nm: 46%	Slightly Lower	Ideal morphology. Size much larger than desired. There are no agglomerations.
8	Cubic/cobblestone particles	>100 nm: 82%	Regular	Ideal morphology. Size much larger than desired. There are no agglomerations.
9	Cubic/cobblestone particles	>100 nm: 93%	Regular	Ideal morphology. Size much larger than desired. There are no agglomerations.
10	Generally irregular. Some cubic-shaped particles are present.	20–28 nm: 11% 28–36 nm: 9% 36–44 nm: 19% 44–52 nm: 30% 52–60 nm: 18%	Slightly Lower	Acceptable morphology. Size slightly larger than desired. Large agglomerations are not visible.
11	Generally irregular. Some cubic-shaped particles are present.	28–36 nm: 13% 36–44 nm: 18% 44–52 nm: 16% 52–60 nm: 20% 60–68 nm: 10% 68–76 nm: 13%	Slightly Higher	Acceptable morphology. Size slightly larger than desired. There are several medium-sized agglomerations

13	Undefined	44–52 nm: 15% 52–60 nm: 20% 60–68 nm: 24% 68–76 nm: 12%	Regular	The low resolution does not allow to determine their precise morphology. Size much larger than required. Presents some large agglomerations.
12 and 14	Generally cubic-shaped particles are present. Some irregular shape also appears.	20–28 nm: 26% 28–36 nm: 32% 36–44 nm: 16%	Slightly Higher	Very good morphology. Very good size. There are no agglomerations.
15	No particles obtained	-----	-----	-----
16	No particles obtained	-----	-----	-----

3.4 Discussion of Results

In Table 6 is presented a summary of the main conclusions, considering all the results presented previously.

From Table 3, it may be concluded that except for the case of experiments 15 and 16 (where no particles are obtained), the particles present the same order of magnitude concerning magnetic properties (being experiments 2, 11, 12, and 14 the ones that present particles with higher magnetic properties and 7 and 10 the ones that present particles with lower magnetic properties, by a ratio of 3–4). Clearly, the process is well behaved for the majority of parameters, does not pose serious threat to the attainment of magnetite crystals, and does not have a strong influence on their anisotropy and the alignment of magnetic moments. The exceptions are experiments 15 and 16, corresponding to the increase of the surfactant (oleic acid) and one of the tested solvents (4-biphenylcarboxylic acid), respectively.

Considering the attainment of the cubic shape, only experiments 3 (highest stirring), 6 (highest ramp-up temperature scale gradient), 10 (highest centrifuging speed), and 11 (doubling of all reagents quantities) are acceptable, and experiments 7, 8, 9 (all using benzyl ether as solvent), 12, and 14 (using both solvents—benzyl ether and 4-biphenylcarboxylic acid, highest centrifuging speed and high temperature ramp-up gradient) are ideal. Com- paring the products of these last experiments, considering the size of the particles, it may be concluded that particles obtained in experiments 7, 8, and 9 present a much larger size than desired (using benzyl ether as solvent alone is good for the shape, but not for obtaining small sizes); in experiments 3, 6, 10, and 11 the obtained particles present slightly larger size than desired and present agglomerations, while in experiments 12 and 14, obtained particles present the desired size and no agglomerations are observed.

Therefore, steps followed to obtain particles in experiments 12 and 14 stand-out as the optimized particle production method, as these experiments present final cubic-shaped nanoparticles with high magnetic properties, with the desired size and no agglomerations. Hence, using both solvents at the same time (benzyl ether and 4-biphenylcarboxylic acid) and maintaining the selected base preferences for the other reagents and stirring speed (while increasing the centrifuging speed) seems to be the best option to obtain the desired magnetic nanoparticles.

3.4.1 Parameters Influence

By analyzing each parameter, it may be concluded that:

- increasing stirring speed is beneficial (experiments 1, 2, and 3),
- increase in temperature ramp-up does not affect the results (experiments 4, 5, 6, 7, 8, 9, and 10),
- substituting the solvent (4-biphenylcarboxylic acid by benzyl ether) improves the percentage of particles presenting a cubic form, but increases particle sizes (experiments 7, 8, and 9),
- doubling the quantities of oleic acid or 4-biphenylcarboxylic acid leads to a null production of particles (experiments 15 and 16),
- doubling the quantity of the two reactives and one of the tested solvents is not detrimental (experiment 11),
- reducing the degasification time is detrimental (experiment 13),
- using both solvents at the same time (4-biphenylcarboxylic acid and benzyl ether) improves the process (experiments 12 and 14).

3.4.2 Ideal Value of the Parameters

From all the above, it may be concluded that the ideal conditions are: iron acetylacetonate (III)—2 mmol; oleic acid—4.5 mmol; 4-biphenylcarboxylic acid—2 mmol; benzyl ether—52.5 mmol; stirring—200 rpm; reaction temperature—290 °C; temperature ramp- up—25 °C/min; centrifugal speed 2500 rpm.

4. POTENTIAL INDUSTRIAL IMPLEMENTATION

To determine the potential industrial implementation of the cubic magnetic nanoparticle manufacturing process, the methodology described in [8] is followed: first, determine the potential international demand of the product (magnetic nanoparticles for magnetic hyperthermia applications), then, determine the plant size, analyze all the process steps and chose all the process units (based on the mass and energy balances), perform a detailed scale-up design of the main equipment, and finally analyze the economic impact and profitability. By following this methodology it was possible to determine the technological and economic viability of the industrial implementation of the optimized process for production of cubic-shaped magnetite nanoparticles based on the optimized values found in Section 3. In what follows, only the main results are present (in Supplementary Material are given all the details concerning these calculations).

4.1 Size of the Plant

The minimum size of the plant in order to be profitable is 3850 kg/year. The plant will be designed for an annual income of EUR 203,200,000, at a price of EUR 30,000/kg magnetic nanoparticles. The annual production is equivalent to 6773 kg/year.

4. 2 Process Engineering Design

In Figure 4 is depicted the process diagram of the proposed industrial plant and in Supplementary Material the characteristics of each stream.

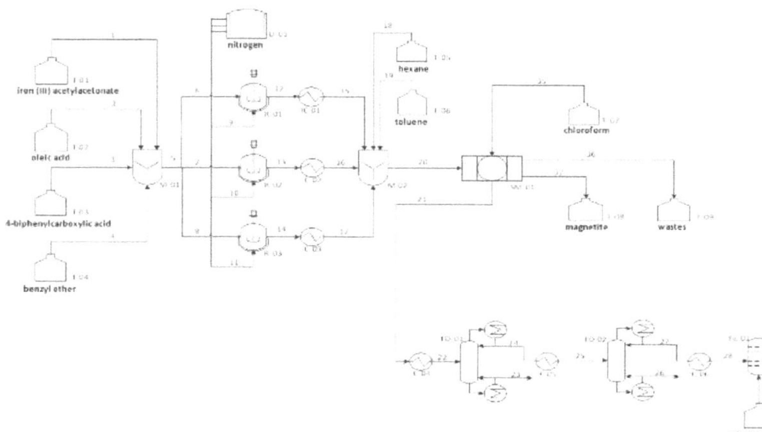

Equipment List	
Equipment	Nomenclature
T	Storage Tank
M	Mixer
D	Degasifier
IC	Heat Exchanger
R	Reactor
SM	Magnetic Separator
TD	Destilation Tower
TE	L-L Extraction Tower

Figure 4. Process Engineering Flowsheet of the Plant to Produce Cubic-Shaped Magnetite Nanoparticles.

The process begins in the M-01 mixer where the main components (stored in the T-01, T-02, T-03, and T-04 tanks) are mixed in their stoichiometric proportions. The main raw materials are: iron (III) acetylacetonate, which will act as the precursor agent, oleic acid, which will be the surfactant and intervenes in the ligand exchange reaction with iron acetylacetonate, and 4-biphenylcarboxylic acid and benzyl ether, which will act as solvents. When the raw materials have been mixed in the M-01, the mixture is taken to the R-01 reactor, as well as to reactors R-02 and R-03. In the reactors, the degassing is done with nitrogen delivered by the degasser D-01. The degassed mixture is then heated through a jacket heater up to 290 ◦C (563 K), with a constant temperature ramp-up rate of 25 ◦C/min and a stirring of 800 rpm. When the mixture reaches 290 ◦C, it remains constant for 30 min. Then, the sample is fed to the heat exchanger IC01, where it is cooled to room temperature. Subsequently, the mixture is introduced in the mixing container

M-02 and a mixture of hexane and toluene is added, from tanks T-05 and T-06, with a ratio of 1:4, to facilitate the subsequent separation. The entire mixture is then passed through a magnetic separator SM-01, which is responsible for retaining the MNPs synthesized in the reactor. Lastly, the MNPs are cleaned with chloroform and stored in tank T-07. Regarding the remaining compounds, they are discarded by the magnetic separator through stream 21, including benzyl ether, hexane, toluene, and the other by-products (acetone), which are recycled to reduce raw material costs. For such an effect, distillation towers and a liquid-liquid extraction tower are used.

4.3 Global Economic Impact and Profitability

In Table 7 is presented a summary of the overall economic balance of the plant.

Table 7. Overall economic balance of industrial implementation of MNPs production for magnetic hyperthermia applications.

1.1 Raw Materials	48,298,030	1.1 Instrumentation	827,043
1.2 Direct Human Labor	829,870	1.2 Initial Setup	473,602
1.3 Indirect Human Labor	207,358	1.3 Piping and Valves	372,270
1.4 General Services	580,909	1.4 Measuring and Control	165,409
1.5 Supplies	80,096	1.5 Heat Isolation	57,893
1.6 Maintenance	234,179	1.6 Electrical Installation	124,057
1.7 Laboratory	165,974	1.7 Land and Structures	1,122,893
1.8 Board and Technical Staff	224,564	1.8 Auxiliary Facilities	330,817
1.9 Amortization	82,704	1.9 Project and Design	361,923
1.10 Taxes and Insurances	160,192	1.10 Constructor Hiring	208,439
TOTAL COST OF FABRICATION	50,863,876	1.11 Unexpected Expenses	521,098
2.1 Commercial Expenses	10,172,775	1.12 Preliminary Studies	2,803,343
2.2 Management	131,826	1.13 Preliminary Startup	640,764
2.3 Financial Expenses	5,809,661	-	-
2.4 Research	240,286	TOTAL IMMOBILIZED	8,009,550
2.5 Technical Services	2,032,000	CIRCULATING CAPITAL	21,038,753
TOTAL COST OF MANAGEMENT	18,386,548	TOTAL INVESTED CAPITAL	29,048,303
TOTAL PRODUCTION COSTS	69,250,424	TOTAL INCOME	203,200,000

As it may be seen from Table 7 and the calculations in the Supplementary Material, a maximum net benefit of EUR 100,462,182/year and a margin of profits over costs of 145% may be attained.

4.4 Discussion of Results—Industrial Adaptation

From the obtained results, the economic viability of the industrial implementation of a plant producing magnetic nanoparticles for hyperthermia applications seems evident. Nonetheless, it is important to notice that market demand was based on the current and foreseen demand for biomedical applications. This means that the produced particles must also be able to be used in other biomedical applications in order to reach the expected selling amount and price. The characteristics of the particles seem appropriate for this goal also (although in some cases it will probably require a further process stage for fictionalization, but the economic margin is enough to absorb this step and maintain economic viability).

In what concerns the technological viability, the designed process seems appropriate, also considering all the mass and heat balances that were performed and the process units applied. The recovery and recycling of the raw materials after reaction is considered, as well as an appropriate recovery and purification step concerning the produced magnetic nanoparticles.

5. CONCLUSIONS

Hyperthermia applications of magnetic nanoparticles is currently a highly intensive field of research. Cubic-shaped magnetite particles seem to be the most promising type of particles for magnetic hyperthermia. Nonetheless, only a few studies present proper ways to obtain this type of particle, no optimization of the processes has been done, and no proper study of the influence of the main operating parameters is available in the current literature. In this work, a method for the production of cubic-shaped magnetite nanoparticles was applied and the influence of the solvent, precursor and surfactant quantities and ratio, type of solvent, stirring speed, centrifuge speed, temperature ramp-up gradient, reaction temperature, and degassing time was studied. It was concluded that increasing stirring speed is beneficial, an increase in temperature ramp-up does not affect the results, substituting the solvent (4-biphenylcarboxylic acid by benzyl ether) improves cubic form presence but increases particle size, doubling the quantities of oleic acid or 4-biphenylcarboxylic acid leads to a null production of particles, doubling the quantity of the two reactives and one of the solvents is not detrimental, reducing the degasification time is detrimental, and using both solvents at the same time (4-biphenylcarboxylic acid and benzyl ether) improves the process. The best values for the studied variables are: iron acetylacetonate (III)—2 mmol; oleic acid—4.5 mmol; 4-biphenylcarboxylic acid—2 mmol; benzyl ether—52.5 mmol; stirring—200 rpm; reaction temperature—290 °C; temperature ramp-up—25 °C/min; centrifugal speed—2500 rpm.

Then, the industrial implementation of a process to produce these magnetite nanoparticles based on the optimized values reached in the research stage was fully designed and studied in order to study the potential industrial implementation viability. Full process engineering, including energy and mass balances, was also conducted for the plant. The designed plant has proven to be viable economically and technically, with a maximum net benefit of EUR 100,462,182/year. Since the profit is much larger than regular bank benefits, investors may consider attractive the construction and start-up of the designed industrial plant, even if several of the parameters have to be adjusted (e.g., adjustment of the sell price of the particles, estimation of some costs, etc.). It is expected that this work will serve as a demonstration of the connection between science and practical applications, and to allow scientific improvement of other processes bearing in mind their possible industrial implementation.

6. CONCISE SUMMARY

Cubic nanoparticles are referred to as the best shaped particles for magnetic hyperthermia applications. In this work, the best set of values for obtaining optimized shape and size of magnetic particles (namely: reagents quantities and proportions, type of solvents, temperature, etc.) is determined. A full industrial implementation study is also performed, including production system design and technical and economic viability.

REFERENCES

1. Wild, C.; Weiderpass, E.; Steward, B. World Cancer Report: Cancer Research for Cancer Prevention; International Agency for Research on Cancer, World Health Organization: Geneva, Switzerland, 2020.
2. Bruvera, J.I. Disipación de Potencia Por Nanopartículas Magnéticas Expuestas a Campos de Radiofrecuencia Para Terapia Oncológica Por Hipertermia; Universidad Nacional de La Plata: La Plata, Argentina, 2015.
3. Horny, M.-C.; Gamby, J.; Dupuis, V.; Siaugue, J.-M. Magnetic Hyperthermia on γ-Fe2O3@SiO2 Core-Shell Nanoparticles for mi-RNA 122 Detection. Nanomaterials 2021, 11, 149. [CrossRef]
4. Caizer, C. Optimization Study on Specific Loss Power in Super-paramagnetic Hyperthermia with Magnetite Nanoparticles for High Efficiency in Alternative Cancer Therapy. Nanomaterials 2021, 11, 40. [CrossRef]
5. Veloso, S.R.S.; Silva, J.F.G.; Hilliou, L.; Moura, C.; Coutinho, P.J.G.; Martins, J.A.; Testa-Anta, M.; Salgueiriño, V.; Correa- Duarte, M.A.; Ferreira, P.M.T.; et al. Impact of Citrate and Lipid-Functionalized Magnetic Nanoparticles in Dehydropeptide Supramolecular Magnetogels: Properties, Design and Drug Release. Nanomaterials 2021, 11, 16. [CrossRef] [PubMed]
6. Strbak, O.; Antal, I.; Khmara, I.; Koneracka, M.; Kubovcikova, M.; Zavisova, V.; Molcan, M.; Jurikova, A.; Hnilicova, P.; Gombos, J.; et al. Influence of Dextran Molecular Weight on the Physical Properties of Magnetic Nanoparticles for Hyperthermia and MRI Applications. Nanomaterials 2020, 10, 2468. [CrossRef]

7. Wu, K.; Wanga, J.-P. Magnetic hyperthermia performance of magnetite nanoparticle assemblies under different driving fields. AIP Adv. 2017, 7, 056327. [CrossRef]
8. Augusto, P.A.; Castelo-Grande, T.; Vargas, D.; Pascual, A.; Hernández, L.; Estevez, A.M.; Barbosa, D. Upscale Design, Process Development, and Economic Analysis of Industrial Plants for Nanomagnetic Particle Production for Environmental and Biomedical Use. Materials 2020, 13, 2477. [CrossRef] [PubMed]
9. Augusto, P.A.; Castelo-Grande, T.; Vargas, D.; Hernández, L.; Merchán, L.; Estevez, A.M.; Gómez, J.; Compaña, J.M.; Barbosa, D. Water Decontamination with Magnetic Particles by Adsorption and Chemical Degradation. Influence of the Manufacturing Parameters. Materials 2020, 13, 2219. [CrossRef] [PubMed]
10. Li, Q.; Kartikowati, C.W.; Horie, S.; Ogi, T.; Iwaki, T.; Okuyama, K. Correlation between particle size/domain structure and magnetic properties of highly crystalline Fe3O4 nanoparticles. Sci. Rep. 2017, 7, 9894. [CrossRef]
11. Butler, R.F.; Banerjee, S.K. Theoretical single-domain grain size range in magnetite and titanomagnetite. J. Geophys. Res. 1975, 80, 4049–4058. [CrossRef]
12. Muxworthy, A.R.; Williams, W. Critical superparamagnetic/single-domain grain sizes in interacting magnetite particles: Implications for magnetosome crystals. J. R. Soc. Interface 2009, 6, 1207–1212. [CrossRef]
13. Shah, R.R.; Davis, T.P.; Glover, A.L.; Nikles, D.E.; Brazel, C.S. Impact of magnetic field parameters and iron oxide nanoparticle properties on heat generation for use in magnetic hyperthermia. J. Magn. Magn. Mater. 2015, 387, 96–106. [CrossRef]
14. Khurshid, H.; Alonso, J.; Nemati, Z.; Phan, M.H.; Mukherjee, P.; Fdez-Gubieda, M.L.; Barandiarán, J.M.; Srikanth, H. Anisotropy effects in magnetic hyperthermia: A comparison between spherical and cubic exchange-coupled FeO/Fe3O4 nanoparticles. J. Appl. Phys. 2015, 117, 17A337. [CrossRef]
15. Avugadda, S.K.; Wickramasinghe, S.; Niculaes, D.; Ju, M.; Lak, A.; Silvestri, N.; Nitti, S.; Roy, I.; Samia, A.C.S.; Pellegrino, T. Uncovering the Magnetic Particle Imaging and Magnetic Resonance Imaging Features of Iron Oxide Nanocube Clusters. Nanomaterials 2021, 11, 62. [CrossRef] [PubMed]
16. Abenojar, E.C.; Wickramasinghe, S.; Bas-Concepcion, J.; Samia, A.C.S. Structural effects on the magnetic hyperthermia properties of iron oxide nanoparticles. Prog. Nat. Sci. Mater. Int. 2016, 26, 440–448. [CrossRef]
17. Martinez-Boubeta, C.; Simeonidis, K.; Makridis, A.; Angelakeris, M.; Iglesias, O.; Guardia, P.; Cabot, A.; Yedra, L.; Estradé, S.; Peiró, F.; et al. Learning from Nature

to Improve the Heat Generation of Iron-Oxide Nanoparticles for Magnetic Hyperthermia Applications. Sci. Rep. 2013, 3, 1652. [CrossRef] [PubMed]

18. Reyes-Ortega, F.; Delgado, Á.V.; Iglesias, G.R. Modulation of the Magnetic Hyperthermia Response Using Different Super-paramagnetic Iron Oxide Nanoparticle Morphologies. Nanomaterials 2021, 11, 627. [CrossRef] [PubMed]
19. Augusto, P.A.; Castelo-Grande, T.; Merchan, L.; Estevez, A.M.; Quintero, X.; Barbosa, D. Landfill leachate treatment by sorption in magnetic particles: Preliminary study. Sci. Total Environ. 2019, 648, 636–668. [CrossRef] [PubMed]
20. Castelo-Grande, T.; Augusto, P.A.; Rico, J.; Marcos, J.; Iglesias, R.; Hernández, L.; Barbosa, D. Magnetic water treatment in a wastewater treatment plant: Part I—Sorption and magnetic particles. J. Environ. Manag. 2021, 281, 111872, ISSN 0301-4797. [CrossRef]
21. Castelo-Grande, T.; Augusto, P.A.; Rico, J.; Marcos, J.; Iglesias, R.; Hernández, L.; Barbosa, D. Magnetic water treatment in a wastewater treatment plant: Part II—Processing Waters and Kinetic Study. J. Environ. Manag. 2021, acepted.
22. Hernández, L.; Augusto, P.A.; Castelo-Grande, T.; Barbosa, D. Regeneration and Reuse of Magnetic Particles for Contaminant Degradation in Water. J. Environ. Manag. 2021, acepted. [CrossRef] [PubMed]
23. Estevez, A.M.; Rodriguez, J.M.; Alvaro, A.; Augusto, P.A.; Jimenez, O.; Castelo-Grande, T.; Barbosa, D. Preparation, Characterization, and Testing of Magnetic Carriers for Arsenic Removal From Water. IEEE Trans. Magn. 2008, 44, 4436–4439. [CrossRef]
24. Kim, D.; Lee, N.; Park, M.; Hyo, B.; Kwangjin, K.; Hyeon, T. Synthesis of Uniform Ferrimagnetic Magnetite Nanocubes. J. Am. Chem. Soc. 2009, 131, 454–455. [CrossRef] [PubMed]
25. Gómez Roca, A. Preparación de Nanopartículas Magnéticas Uniformes y de Alta Cristalinidad para Biomedicina; Instituto de Ciencia de Materiales de Madrid, Departamento de Química Física I, Universidad Complutense de Madrid: Madrid, Spain, 2009.

03 DECIPHERING COLORECTAL CENCER'S MOLECULAR PATHOGENESIS: ADVANCES IN BIOMARKERS AND NANOTECHNOLOGY-ENHANCED DIAGNOSTICS & THERAPEUTICS

1. INTRODUCTION: AN OVERVIEW

Colorectal cancer stands as a significant global health concern, necessitating a comprehensive understanding of its molecular pathogenesis for effective diagnosis and treatment. In recent years, substantial progress has been made in unraveling the intricate mechanisms underlying the development and progression of colorectal cancer. This exploration has not only identified novel biomarkers that enhance our ability to detect the disease at early stages but has also witnessed remarkable strides in the realm of nanotechnology, leading to innovative diagnostic and therapeutic approaches. This chapter delves into the molecular intricacies of colorectal cancer, placing particular emphasis on the latest advancements in biomarker research and the transformative potential offered by nanotechnology in refining diagnostic precision and therapeutic interventions. Colorectal cancer (CRC) is a serious health issue in developed and developing countries alike, with the third highest incidence rate of all tumour-causing diseases [1]. This kind of cancer is a leading cause of morbidity and mortality in Western nations. Only a small percentage of CRC instances are caused by basic genetic abnormalities, with most cases occurring due to specific regular factors and growing age [2]. If left untreated, it can spread to the intestinal wall and the muscles beneath the surface of the skin. Environmental and genetic variables can also combine in many ways to accelerate cancer development and make it more difficult to treat.

In most cases, patients appear beyond the age of 60, with precursor starting adenomas that develop into cancer over for 1 to 2 decades [3]. Generalised screening of this population at age 50 years and older has been shown to reduce death from the disease, detect cancer at an earlier stage, and reduce the incidence of the disease. There have been four major hypotheses put out on the aetiology of colorectal cancer. To begin, the same genetic and epigenetic changes that cause colon cancer also increase the risk of developing colorectal cancer. Second, cancer develops in a staged manner, beginning with a precancerous condition and progressing to malignancy. Third, cancer development necessitates a lack of genetic stability. Fourth, in hereditary cancer syndromes, the germ line forms of significant genetic abnormalities are generally associated with the development of rare colon cancers, which are caused by somatic occurrences of hereditary diseases [4].

Epigenetics, defined as heritable changes in gene expression without permanent changes in DNA sequence, plays an important role in the aetiology of different malignancies, including colorectal cancer. Recent epigenetic research has found a missing connection between certain gene expression patterns associated with CRC and the lack of genetic abnormalities. This missing link has now been discovered during the last two decades [5]. As an example, microsatellite in-stability (MSI), which is a hallmark of a molecular subgroup of CRC, results from an inability to repair DNA mismatches (MMRs) [6] properly. A genetic mutation can cause this defect in one of the MMR genes, the hypermethylation of the MLH1 gene's promoter, or both. Chromosome instability (CIN) in CRC has also been linked to global hypomethylation [7]. There is further evidence that cancer-related pathways at the post-transcriptional level are influenced by microRNAs (miRNAs), which play a role in nearly all phases of CRC, from initiation through progression and metastasis. Cell development is inhibited by miR-143, which down regulates in CRC because it tar-gets the KRAS mRNA transcript [8]. Aside from improving our knowledge of CRC pathogenesis, these discoveries have also provided new opportunities to find disease biomarkers and treatment targets. The various CRC biomarkers such as BRAF mutations, NRAS, and KRAS, microsatellite instability (MSI), DNA mismatch repair (MMR) status, and CpG island methylation have been studied. A KRAS or NRAS mutation is associated with a worse prognosis, and anti-epidermal growth factor receptor (anti-EGFR) antibody treatment will be ineffective [9]. Mutation in BRAF V600E is seen in 8–10% of individuals with CRC. These patients have a more aggressive illness and a worse prognosis in the adjuvant and metastatic context. People with BRAF mutation and microsatellite instability-high (MSI-H) tumours had a better overall prognosis than people with BRAF mutation and microsatellite stable (MSS) illness, and it turns out, using MSI to screen for Lynch syndrome and identify patients who could benefit from immunotherapy is becoming increasingly important in CRC [10].

A variety of drug-loaded nanoparticles in the 20–400-nm size range have made significant contributions to chemotherapeutic drug delivery in recent years (e.g., liposomes, dendrimers, polymeric nanoparticles, and micelles) [11]. Many potential nanomedicine advancements are based on these systems, which have evolved from basic drug-loaded nanoparticles to multifunctional nanoparticles that target particular cancer cells by attaching them to specific cell-surface proteins. Antigens such as integrin and folic acid receptors, which are differentially expressed on the surface of cancer cells, can be targeted using targeted nanoparticles [12]. Several of these nanoparticles are now in clinical trials. A great deal has been learned about using nanoparticles as therapeutic platforms for treating malignancies ranging from the prostate to the ovary. The clinical development of nanoparticles for cancer therapy is restricted, despite the high death and morbidity rates associated with CRC. In this review, we emphasise in detail the pathogenesis of colorectal cancer and their suitable candidate as biomarkers and nanotechnology-based diagnostic methods and therapeutic approaches for malignant colorectal cancer. Additionally, it provided an outlook on therapeutic nanoparticles for the management of CRC.

2. PATHOGENESIS OF COLORECTAL CANCER

Genetic alterations in oncogenes and tumour suppressor genes occur sequentially from normal to the dysplastic epithelium in the adenoma–carcinoma process. This results in the development of CRC. Genomic changes, both genetic and epigenetic, convert normal glandular epithelium into adenocarcinoma when they accumulate [13]. While molecular mechanisms behind the cancer's unabated development and spread remain a mystery, several genetic pathways have been proposed to explain CRC aetiology (Figure 1). CRC involves numerous molecular pathways, including the three major pathways: CIN, MSI, and CIMP, all of which play a crucial role in the development of CRC due to genetic and epigenetic changes (Table 1). Three molecular pathways that indicate progression from cancer have been identified: microsatellite instability (associated with hypermutated group), chromosomal instability (completely within the non-hypermutated category), and the CpG Island Methylator Phenotype (CIMP; associated with both the hypermutated and non-hypermutated groups) [14]. These molecular pathways may dictate tumour development and metastatic time, as the epidemiology, mutational trials, and immune responsiveness information vary by route. The methods used to treat an illness might also differ. A biomarker known as "EMAST" refers to elevated microsatellite alterations at selected tetranucleotide repeats recently demonstrated to predict patient survival as a modulator of the three biological pathways. Individuals with sporadic CRC cannot predict whether they develop a metachronous adenoma or cancer based on the previously identified pathways

[15,16]. Clinical monitoring is critical in detecting and eliminating metachronous lesions in these patients.

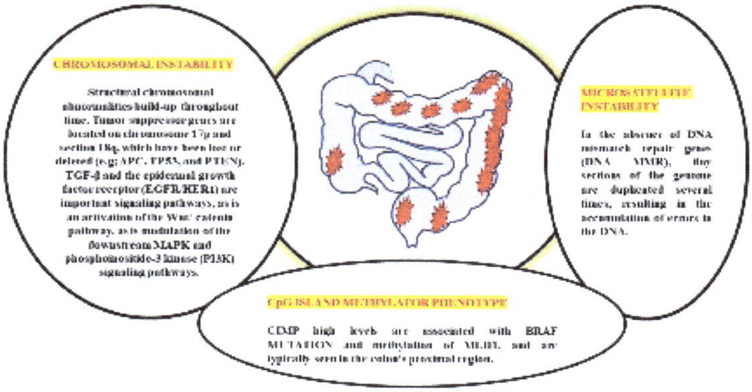

Figure 1. Multiple genetic pathways of CRC pathogenesis.

The biological behaviour, prognosis, and therapeutic response of CRC are all very variable. CRC has historically been divided into three subgroups based on the three pathophysiological routes of carcinogenesis: CIN, MSI, and CIMP from a molecular perspective [17]. CIN is the distinguishing feature for most CRCs, accounting for ~80–85% percent of the cases. CIN is characterised by the activation of growth-promoting path- ways while simultaneously decreasing the activity of apoptotic pathways, with the latter being more common [18]. These tumours begin as adenomatous polyps due to the APC gene's deactivation mutation [encodes the WNT pathway effector adenomatous polyposis coli (APC)].

Table 1. An assessment of the MSI-H, EMAST, CIMP, and CIN CRC indices.

CIN	Causes to the mutation and copy number variation; MSS; aneuploid	Tumour margin, lamina propria, and intraepithelial sites all have varying degrees of differentiation	Referent	Genetic changes that lead to heterozygosity loss
CIMP	Leading hypermethylation at DNA loci	Without hMLH1 hypermethylation: varied	Poor survival without hMLH1 hypermethylation	Without hMLH1 hypermethylation: unknown
MSI-H	Microsatellite instability (MSI) and diploid is appeared	Crohn’s-like around tumour (tumour margin)	Better survival; early stage	Target gene frameshift mutation; BRAFV600E
EMAST	Instability found in mostly at MSS and MSI-L, includes MSI-H	Tumour nests surrounding epithelial components have been linked to this condition.	Poor survival; later stage	Chromosome instability in combination with a frameshift mutation on a target gene

Further, the progress to adenocarcinomas by activating mutations in the KRAS (responsible for the receptor tyrosine kinase signalling) and deactivating mutations in SMAD4 (accounting for the TGFβ, it played an important role in the cell cycle control). Due to the initial event, this pathway is sometimes referred to as the APC pathway [19]. There are no mutations of MSI, MLH1, methylation, or BRAF in the CIN pathway. Moreover, Lynch syndrome (hereditary CRC's most frequent type) is caused by the germline mutations in the MMR genes, including MLH1, MSH2, MSH6, and PMS2. It is believed that MSI is initiated by a lack in the MMR DNA, causing mutations in the MLH1, MSH2, and MSH6 MMR genes. It is possible to have MMR deficiency due to MLH1 deactivation induced by biallelic hypermethylation of the MLH1 promoter or by dual somatic mutations in the MMR genes [20]. It is noticed that the individual with MMR-deficient and MSI-high CRC (dMMR–MSI-H) demonstrated a good prognosis but did not acquire therapeutics benefits from 5-fluorouracil (5-FU) and has already been characterised based on MSI status [21]. A subset of colorectal cancers known as CIMPs, which are defined by a substantial percentage of CpG island hypermethylation around the promoter of multiple tumour-suppressor genes, is identified by epigenetics rather than by CIN or MSI. Due to its close association with colorectal carcinogenesis's serrated route, hypermethylation of the MLH1 promoter, as well as its association with female gender, advanced age, and poor histology, CIMP has a significant risk of being found in patients with the disease [22]. However, the lack of a defined definition of CIMP-high makes it difficult to translate this pathophysiological feature into treatment. Although CIMP testing is most commonly done on MLH1, MINT1, CACNA1G, and CDKN2A (could potentially up to 16 distinct genes identified), there is currently does not agree about which cut-off values will be applied to distinguish between CIMP high and CIMP+ on certain scientific procedures must be employed to conduct assessments for CIMP tests [23].

As early as lesions in the polyp cancer development sequence, gene alterations have a role in cancer creation and progression. The abnormal crypt focus is the initial histological lesion linked to CRC development [24]. Tumour suppressor gene APC is often altered in colorectal cancers, and dysplastic aberrant crypt foci can spread mutations. The Wingless/Wnt pathway is activated by inhibiting APC, a systematic approach to start the polyp cancer development sequence. The Wnt signalling pathway harbours the clonal development of polyp cells to cancer due to subsequent mutations in genes like KRAS or TP53 (Figure 2). The transforming growth factor b (TGFB1)-mediated cell signalling pathway may also be modulated by KRAS and TP53 mutations, further accelerating CRC development [25].

Oncogene activation and tumour suppressor gene inactivation are attributed to genetic and epigenetic alterations that occur during colon cancer development. B-Raf sends EGFR signals to KRAS, a proto-oncogene that activates the MAPK pathway [26]. Over half of all CRCs have mutations in KRAS or B-Raf, which trigger the MAPK signalling pathway and promote proliferation while also stopping cell death. The PI3 K, WNT-APC-CTNNB1, and TGFB1-SMAD signalling pathways are also part of CRC. It also includes the RAS–RAF– MAPK pathway [27].

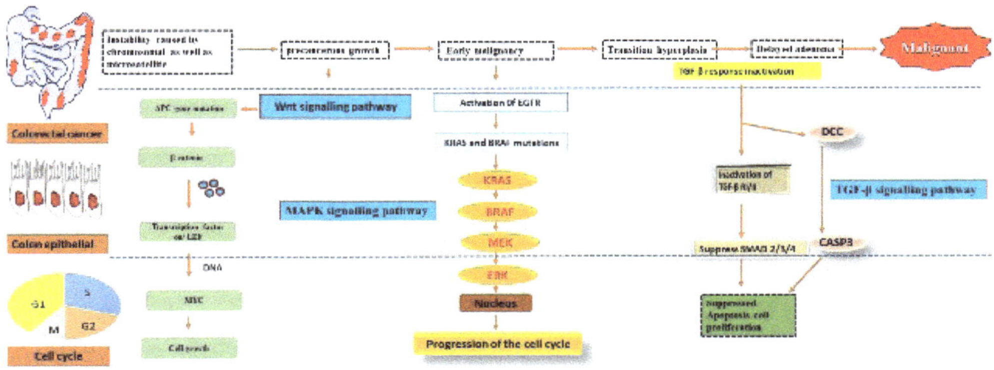

Figure 2. Pathways of signalling that have been genetically changed by CRC.

The study of epigenetic changes in cancer has led to the identification of genes commonly altered in these specific pathways, leading to in-depth investigations into the biology of CRCs. Novel CRC diagnostic and prognostic tests and new treatments are being developed based on these findings.

3. MSI, CPG AND CIMP INVOLVEMENT IN THE PROGRESSION OF CRC

CRC is a group of several illnesses, some purely molecular, while others include morphological manifestations. Colorectal carcinogenesis has been linked to three distinct molecular pathways, including chromosomal instability (CIN), microsatellite instability (MSI), and chromosomal mismatch repair (CIMP) (Figure 3). For example, chromosomal size and structure are altered in the CIN route. In contrast, changes in the number of mono and dinucleotide repeats in the coding and regulatory regions are seen in the MSI pathway. CRCs grow in two distinct routes, including the conventional adenoma–carcinoma sequence and the serrated neoplasia pathway, morphologically multistep processes [28]. An adenocarcinoma sequence that dates back to antiquity describes the progression of benign adenomas, including tubular adenomas, tubulovillous adenomas, or villous adenomas, progress and transform into invasive adenocarcinomas. The phrase "serrated neoplasia pathway" refers to the concept that sessile or conventional serrated adenomas can develop into invasive adenocarcinomas [29].

Even though Lynch syndrome is associated with inherited MSI-H tumours, the CIN pathway is the underlying biological mechanism for the typical adenoma–carcinoma sequence. Tubular adenomas with variable villosity are Lynch syndrome premalignant lesions. The great majority of spontaneous MSI-H CRCs, in contrast to Lynch syndrome tumours, are considered to develop in CIMP-H sessile serrated adenomas with BRAF mutation through methylation-associated inactivation MLH1. Hereditary and spontaneous MSI-H CRCs take the conventional adenoma–carcinoma sequence and the serrated neoplasia pathway, respectively, whereas MSI-H tumours follow a distinct morphological multistep approach. CIMP-H/ non-MSI-H CRCs, in contrast to CIMP-H/MSI-H CRCs, originate from pre-existing conventional serrated adenomas [30].

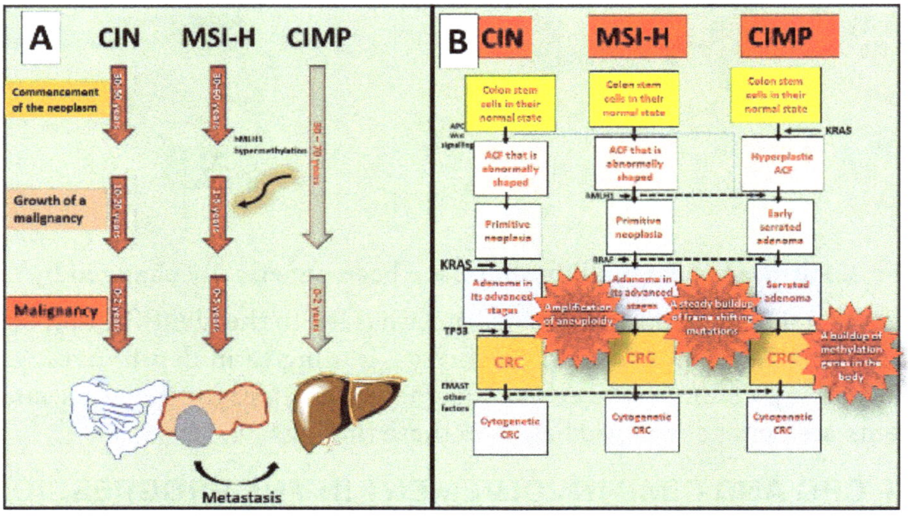

Figure 3. Involvement of MSI, CpG, and CIMP in the progression of CRC. (A) Every age groups are susceptible to malignancy, and progression of colorectal cancer shows after the occurrence of metastasis; (B) In the case of CIN, various molecular pathways play crucial roles such as Wnt, KRAS, TP53, and EMAST and forms CRC progression. While MSI-H shows progression via hMLH1 and BRAF. Whereas, CIMP is interconnected with KRAS mutation and causes cytogenetic to normal cells.

On the whole, 65–75 percent of CRCs are caused by the CIN route, whereas 20–30 percent are caused by the CIMP pathway and 5 percent by the hereditary MSI pathway. Females have a larger percentage of the CIMP route than men. Older persons have a higher percentage of the CIMP pathway than younger people. The CIMP pathway is more prevalent in the proximal intestine than the distal colon or the rectum [31].

4. EPIDEMIOLOGY

The frequency of colorectal cancer (CRC) has risen alarmingly in recent years around the globe (Figure 4). In 2020, an estimated 1.93 million new CRC cases were identified, and 0.94 million people died as a result of CRC-related causes globally, accounting for 10% of global cancer incidence (total 19.29 million new cases) and 9.4% of deaths due to cancer (total 9.96 million deaths). It is projected that CRC will be the third-greatest cause of cancer-related deaths in men and women globally in 2020, with 515,637 deaths in men and 419,536 in women, respectively [32]. Over 5.25 million individuals globally (5-year prevalence) have CRC, making it the second-most common disease after breast cancer (7.79 million new cases per year). Approximately 1.15 million new instances of colon cancer, 0.7 million cases of rectal cancer, and 50,000 cases of anal cancer in 2020 were found, according to the database from GLOBOCAN 2020. These figures are expected to rise to 1.92 million, 1.16 million, and 78,000 by 2040. Countries have different rates of colorectal cancer (CRC) [33]. Among countries, the greatest age-standardised incidence rates were observed in Hungary (45.3 cases per 100,000 people in 2020), Slovakia (43.9%), Norway (41.9 cases per 100,000), the Netherlands (41.0 cases per 100,000), and Denmark (40.9 cases per 100,000). Guinea, Gambia, Bangladesh, Bhutan, and Burkina Faso will have the lowest age-standardised incidence rates in 2020 at 3.3, 3.7, 3.8, 3.8, and 3.8 cases per 100,000 people, respectively.

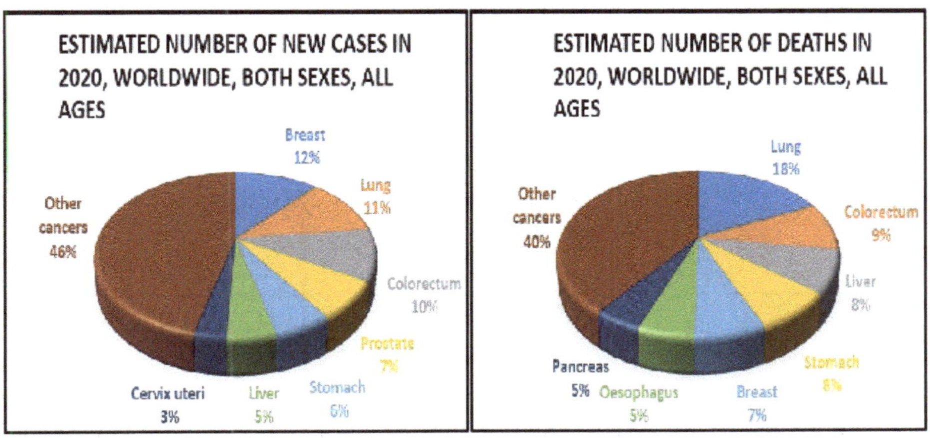

Figure 4. Epidemiology of colorectal cancer.

5. BIOMARKERS BASED ON EPIGENETIC CHANGES FOR CRC

When CRC progresses from an adenomas or serrated lesions stage, genetic and epigenetic changes accumulate in the lesions, giving progressively dysplastic characteristics until they become an adenocarcinoma [34]. To avoid this illness, it is critical to find precancerous lesions and early-onset colorectal cancer in people at average risk who are otherwise healthy during routine screening. As far as CRC screening is concerned, colonoscopy is the gold standard, since it can detect and eradicate precursor lesions [35]. For these reasons, many people choose not to get colonoscopies. They are painful, costly with low compliance rates, and can cause complications, including haemorrhage and perforation. For precursor lesions like adenomas, non-invasive screening procedures like the faecal occult blood test (FOBT) and faecal immunochemical test (FIT), widely used in Europe and other Western nations, have lower sensitivity and specificity than colonoscopy. As a result of these drawbacks, new non-invasive detection methods for precursor lesions and early-stage CRCs are urgently required [36].

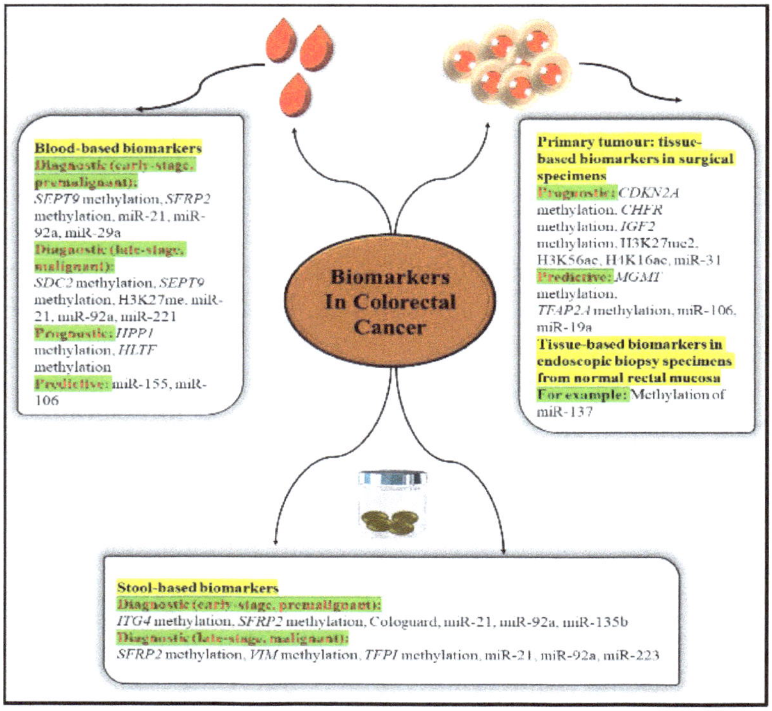

Figure 5. Epigenetic biomarkers in CRC.

At intermediate stages of CRC staging, the currently used tumour-node-metastasis (TNM) classification approach is inadequate for prognostication and clinical decision-making [37]. Individuals with a high risk of illness recurrence or mortality (prognostic biomarkers) and those who would benefit from chemotherapy, immunotherapy, or targeted therapy are urgently in need of biomarkers that aid identify these patients (which may be unnecessary). Many epigenetic biomarkers, including DNA methylation, histone alterations, miRNAs, and long noncoding RNA, have shown promise as clinically significant biomarkers for CRC diagnosis, prognosis, and therapy response prediction (lncRNA) [38]. Blood and other body fluids and tissues can include quantifiable indicators of pathological or physiological processes or conditions or diseases, such as molecular markers (biomarkers). Molecular markers can be used to identify conditions or diseases. Biomarkers are important diagnostic, prognostic, and therapeutic tools for cancer detection, diagnosis, and therapy selection (Figure 5). It can also be used to pinpoint the exact location of a tumour and identify any areas that are at an advanced stage of growth or are receptive to a particular therapy [39]. These molecular markers represent the mechanisms of neoplastic cell exfoliation and mucus secretion of aberrant glycoproteins in CRC by reflecting the mechanisms of cell exfoliation and mucus secretion. Shortly, researchers hope to find novel, non-invasive DNA, RNA, or protein-based molecular markers that may be used to detect colorectal cancer in the blood, faeces, and other body fluid. The prognostic, predictive, and diagnostic markers have been developed for the determination of the CRC. Diagnostic biomarkers indicate the likelihood of the illness progressing. Patient treatment measures can be predicted with the use of these molecular markers. Predictive biomarkers are utilised for forecasting the efficacy of treatment in cancer patients [40]. Colorectal polyps can be detected early with the use of the following biomarkers.

5.1 Molecular Markers For Diagnosis

5.1.1 HNPCC (Hereditary Nonpolyposis CRC)

HNPCC is the most well-known autosomal dominant hereditary colon cancer condition. It is a screening test for the early identification of colorectal cancer (CRC). HNPCC is present when a germline mutation is identified in one of the four MMR genes, such as MSH6, MLH1, MSH2, and PMS2, all found in the human body. Many young patients with a new genetic condition come from families where colon cancer runs in the blood.

5.1.2 Telomerase

The telomerase is key enzyme that preserves the telomere (repeating DNA (TTAGGG/ AATCCC in the human telomeres) that protects from chromosomal breakdown, elimination of operational genes, and cell death. Telomerase is necessary for the maintenance of the telomere. The activation of telomerase is an independent diagnostic and prognostic marker for malignant tumours that can be used independently.

5.1.3 Insulin Like Growth Factor Binding Protein 2 (IGFBP2)

According to research, compared to healthy individuals, CRC patients had higher serum and plasma levels of IGFBP2. It has emerged as a diagnostic device for the primary identification and development of CRC, since it is implicated in cancer cell proliferation, migration, and invasion.

5.1.4 Pyruvate Kinase M2 (PKM2)

PKM2 refers to a cytosolic enzyme expressed in both normal and malignant cells and plays an important function in energy metabolism. PKM2 expression is upregulated in CRC and other gastrointestinal malignancies. It may also be relevant for CRC inquiry and/or diagnosis based on stool or blood samples as a dismal prognostic sign.

5.2 Molecular Markers for Prognostic

5.2.1. p53

The existence of the tumour inhibitor p53 is also considered an independent prognostic factor for colon cancer. The alteration of the tumour-suppressor gene p53 is an initial and critical step in ulcerative colitis-associated carcinogenesis (sometimes referred to as TP53). The development of carcinoma in an adenoma coincides with a p53 mutation and a loss of heterozygosity (LOH) in the wild-type allele, further demonstrating its function in the regulation of malignancy. Patients with CRC who are missing this gene have a bad prognosis.

5.2.2 18 q Loss of Heterozygosity

Patients with colorectal cancer (CRC) who have chromosomal 18q LOH have a reduced chance of survival. Patients with stage II or stage III colon cancer who lose their somatic heterozygosity have a worse prognosis than those whose tumours retain both of their parents' alleles on chromosome 18 q. Patients with stage II or stage III colon cancer with this 18qLOH have a poorer prognosis than those with normal alleles on chromosome 18q.

5.2.3.MLH1 Methylation

DNA microsatellite unstable recognition or the absence of the MLH1 protein production on immuno-histochemical investigation confirm somatic MLH1 inactivation in primary colorectal malignancies, which is more common in early-stage colorectal cancers than in late illness. A more indolent illness or a better prognosis might result from this inactivation in the absence of adjuvant treatment.

5.2.4 VEGF

One of the angiogenic agents in CRC is vascular endothelial growth factor (VEGF), which is expressed in roughly 50% of CRCs but only in extremely low amounts in normal colonic mucosa and adenomas. As a result, VEGF-1 expression appears to deliver useful predictive information in CRC patients.

5.3 Predictive Molecular Markers

5.3.1 KRAS

The discovery of KRAS mutations is employed as the foremost potential prognostic marker in anti-EGFR (epidermal growth factor receptor) antibody-based therapy in CRC, such as panitumumab and cetuximab.

2.3.2 B-Raf V600E

The activation of this kinase by somatic mutation leads to an unconstrained MAPK signalling pathway in humans. The BRAF V600E activating mutation does not respond to EGFR inhibitor treatment in individuals with stage IV CRC. Although the significance of BRAF mutant level as a predicting marker gene is uncertain, BRAF mutant as a biomarker of anti-EGFR antibody resistance is perhaps the extensively researched predicting potential of BRAF alteration.

5.3.3 PIK3CA Status

The PIK3CA mutation causes an abnormal stimulation of the AKT pathway, leading to cancer cell growth. If the PI3K/AKT pathway is persistently activated, it is likely to affect CRC progression significantly. Colon cancer cells often with an induced PIK3CA mutant were more sensitive to cetuximab in preclinical studies even than PIK3CA wild-type cells.

5.3.4 ERCC—1

Repair protein for DNA excision ERCC-1 is a nucleotide excision repair gene that is involved in DNA repair. A new analysis of metastasis and stage II or III CRC and evaluating ERCC-1 as a possible predictive biomarker in direct response to oxaliplatin by studying dysregulation of ERCC-1 expression in between oxaliplatin treatment has been presented [41–44].

6. DIAGNOSTIC/MATERIALISTIC TOOL FOR THE TREATMENT OF COLORECTAL CANCER

A diet rich in vegetables, fruit, and whole-grain cereal goods, as well as regular exercise, lowers the chance of developing colorectal cancer. CRC risk has been linked to environmental variables such as obesity, smoking, and high alcohol intake. Calcium and vitamin D3 supplementation may have a disease-fighting effect. It has also been shown that long-term usage of acetylsalicylic acid (NSAIDS) decreases the risk of cancer. Colorectal cancer is just one of many cancers that can be prevented by quitting smoking or abstaining from smoking. Adenomas (primary prevention) and malignancies (early detection) can be detected with screening tests (secondary prevention).

This is why regular screenings in high-risk people can help detect CRC even if symptoms are not present. After the diagnosis of colon cancer, the diagnostic tools are also useful for the TNM staging, which is used to determine the appropriate treatment. These include colonoscopy, sigmoidoscopy, magnetic resonance imaging (MRI), CTCG, transrectal ultra-sonography, stool testing, double-contrast barium enema, and many others. However, all techniques have some of the other related limitations and remain inaccurate for effective diagnosis in the early phases of the disease process. There is a pressing need to develop more advanced diagnostic methods in order to increase survival rates, reduce disease-related problems due to diagnosis at later stages, and properly stage the disease so that appropriate therapy may be prescribed. Here are some of the most commonly used diagnostic tools, along with their advantages and disadvantages (Figure 6) [45,46].

6.1 CT Colonography

When referring to a computerised X-ray imaging procedure, "virtual colonoscopy" or "CT colonography", both terms refer to the same thing: a method that produces two- and three-dimensional images of the colon. It is possible that CT colonography can be used as an alternative to colonoscopy in patients with obstructive or stenotic tumours. After the colon has been completely evacuated (bowel preparation), it is insufflated with air or carbon dioxide to increase its size for better imaging during the CT scanning procedure. After insufflation, the subject is scanned in both the supine and prone positions. CT colonography is comparable to colon endoscopy in terms of patient pain because of the need for colon evacuation and insufflation, but it has the advantage of not requiring anaesthesia. For polyps larger than 10 mm in diameter (big polyps), the detection sensitivity and specificity are excellent (82 percent), but the sensitivity is poor (63 percent) for polyps between 6 and 9 mm in diameter (small polyps). Furthermore, the method cannot distinguish between flat and serrated lesions. For diagnosing the local stage of a CRC tumour, MRI is preferred over CT colonography, which performs poorly in this regard. Due to its high level of accuracy in determining the tumour's invasion and spread, preoperative MRI is useful in planning surgery. The CTCG approach has many drawbacks, including radiation exposure, a lack of sensitivity to detect small polyps, the necessity for intestinal preparation, and the need for an extra procedure to remove polyps if they are discovered. As a result, the patient's concern rises and the expense rises as a result of unneeded monitoring, which is exacerbated by extracolonic lesion identification [45,47].

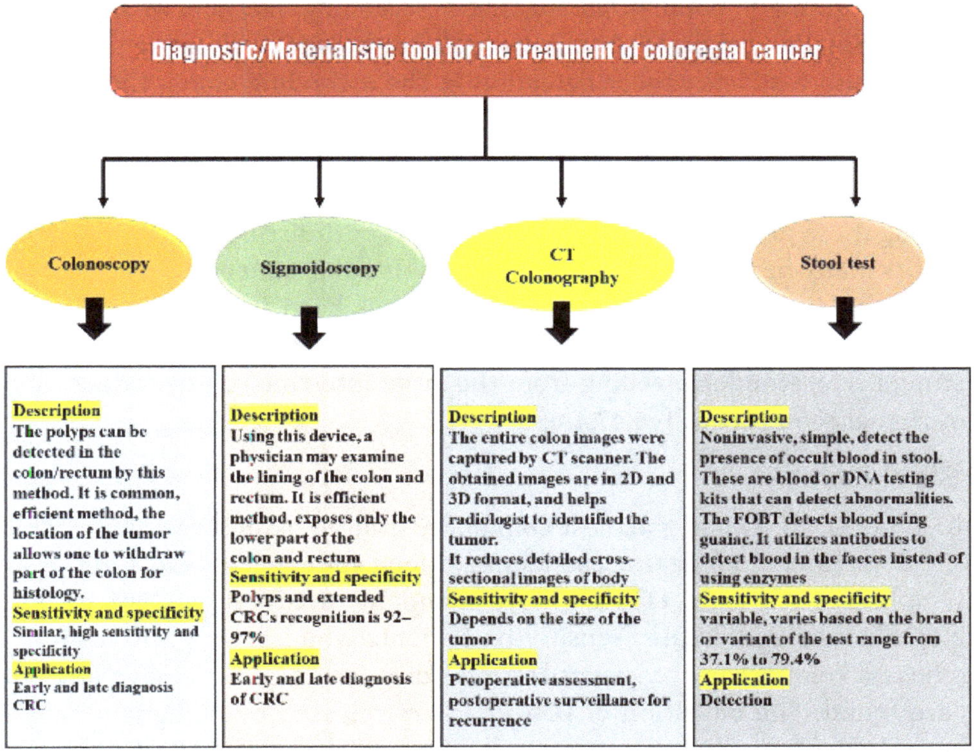

Figure 6. Diagnostic/materialistic tool for the treatment of colorectal cancer.

6.2 Colonoscopy and Sigmoidoscopy

Endoscopy is the most often used and most effective procedure for CRC diagnosis. Colonoscopy and sigmoidoscopy are included in this procedure. These tests enable the detection of a tumour and the removal of a portion of the large intestine for histological analysis to be performed. Sigmoidoscopy has a sensitivity and specificity of 92–97 percent for the detection of polyps and expanded CRCs. Sigmoidoscopy only reveals the lowest portion of the colon and the rectum when performed with a flexible endoscope. Colonoscopy provides the same level of sensitivity and specificity as a colonoscopy in terms of obtaining an image of the entire intestine. Colonoscopy has several advantages over other screening tests, including the fact that it may be completed in a shorter period and that it improves the patient's acceptance and tolerance of new sedative procedures. In comparison to the control group, the study that included individuals with an average risk of CRC following colonoscopy revealed a 67 percent of overall decrease in morbidity and a 65 percent of the overall decrease in mortality. Preliminary screening with sigmoidoscopy and colonoscopy was found in a randomised trial to reduce CRC incidence and mortality by 33% and 38–59%. As a result of this discovery, sigmoidoscopy has become increasingly popular as the main screening method in England and elsewhere. Smaller polyps (6–10 mm) are difficult to identify with this invasive method. It also restricts cancer detection in the ascending colon, the transverse colon, and the cecum. Endoscopy (sigmoidoscopy) can be used to diagnose and treat CRC in patients who are ineligible for surgery (according to the severity of the tumour or comorbidities). Methods for removing the cancer-related blockage are among these procedures. Patients typically report discomfort from endoscopy despite its many benefits. Due to the potential for consequences, people avoid or postpone the evaluation. Virtual colonoscopy has been increasingly popular in recent years. The use of computed tomography is it provides a 3D view of the large intestine. Perforation or bleeding from the large intestine can be reduced by the use of virtual colonoscopy [45,47,48].

6.3 Stool Test

GFOBT and FIT are two of the most commonly utilised main screening modalities for CRC detection. Stool testing can identify blood spilling from CRCs in the stools of asymptomatic patients. FITs are more commonly used than gFOBTs because of their ease of use and greater sensitivity. Patients who have a positive stool test must have a colonoscopy to get an accurate diagnosis and to remove any polyps that are found. The detection of DNA markers released by malignant cells in a person's stool is a promising technique for CRC screening. Stool DNA testing has a high level of sensitivity (71–91%) and specificity (93–100%) for the identification of colon cancer. False results are common because of the low sensitivity and specificity of the FOBT. Additionally, these tests fail to detect non-bleeding tumours. Faecal occult blood testing is a straightforward, inexpensive, and non-invasive diagnostic procedure. Haemoglobin in faeces implies gastrointestinal bleeding, according to the test. Since it can come from both malignant alterations and polyps larger than 1 to 2 cm, blood in faeces is an unspecific sign of colon cancer. The test's sensitivity can be increased by as much as 90% by repeating it. Heme and human globin are the two building blocks of haemoglobin, and the immuno-histochemistry faecal occult blood test (FIT) confirms its presence in CRC diagnoses. Aside from sDNA, which indicates DNA changes in colorectal adenocarcinomas, molecular diagnostics, such as PCR, may be used to separate and discriminate DNA from that of bacteria in faeces. The usefulness of CRC molecular diagnostics based on genetic and epigenetic testing is restricted. Costs are expensive, because it is not widely available. Molecular diagnoses of CRC have some drawbacks, which have prompted researchers to look for biomarkers that can be found in biological samples at a cheaper cost [45,47,49,50].

7. NANOTECHNOLOGY-BASED CRC DIAGNOSTIC TECHNIQUES

Medicines based on nanoparticles are being developed as therapeutic methods for cancer therapy, which has led to significant improvements in pharmaceutical research [51]. These advancements have reduced the adverse effects of cytotoxic drugs while simultaneously increasing their efficacy and solubility. The encapsulation of anticancer cargo, including as siRNA, antibiotics, and chemotherapeutics has proven highly effective in the last 50 years using a variety of distinct nanoparticles with various forms, sizes, and chemical natures [52]. Using the increased permeability and retention effect afforded by tumour vascular and lymphatic drainage, these first-generation anticancer nanoparticles passively penetrate tumour tissue, allowing nanoparticle extravasation and storage within cancer cells, improving therapeutic effectiveness. Early diagnosis and therapeutic efficacy monitoring can both benefit from the use of nanoparticles [53]. Incorporating various contrast agents (such as radioactivity, super-paramagnetic, or fluorescent), as well as targeting groups and biocompatible coatings, into a nanoparticle design is a feasible option [54]. Due to the disadvantages of low tissue specificity, rapid clearance, and nonspecific extracellular distribution associated with small molecular weight gadolinium and metal chelate-based contrast agents, nanotechnology may be used to modify these contrast agents to improve CRC diagnostic sensitivity and specificity [55]. Various nanostructures are being used as inputs of in vitro diagnostic assays for protein indicators or nucleic acid targets. In vivo imaging with nanostructures, similar to in vitro diagnostics, seems to be a potentially intriguing area in diagnosing colon cancer [56]. There are, however, just a few examples of how nanotechnology has been used in CRC imaging. Novel MRI contrast agents are among the most readily useful pieces of colorectal cancer technology. The introduction of nanostructures that modify standard contrast agents like gadolinium or imaging agents like iron oxide could increase the detection capability of medical image [57]. These nanostructures enhance the capabilities of traditional MRI imaging, but they also provide new possibilities for the early diagnosis and healing of the CRC. Tumour early diagnosis and healing techniques not possible with current conventional technologies have been found as beneficial nano-technological applications in cancer biology [58]. When it comes to diagnosing, producing, and healing some malignancies, nanometre-sized particles of various structures and conformations have shown to be highly effective and promising new approaches. A patient's long-term survival can be affected by the early identification of CRC, which is critical for prevention [59]. Here are a few examples of recent Nps accomplishments that have opened up new possibilities for the early identification and successful treatment of CRC.

7.1 Quantum Dots (QDs)

QDs are semiconductor nanocrystals that generate fluorescence when excited by light and have remarkable optical properties, such as high brightness, photobleach resistance, and the ability to emit fluorescence at many wavelengths [60]. QD-based nanotechnology, with its optical and chemical benefits, is a developing platform for cancer research, particularly colorectal cancer [61]. Its bandgap energy, the degree of energy required to transport electrons from one electronic zone to the next, this time at a greater frequency, is an important characteristic of a semiconductor nanocrystal. An exciton is a pair of electrons and holes formed as a result of this excitation strategy. The unstable exciton returns its ground state and releases fluorescence photons in the form of energy [62]. When the semiconducting nanocrystal size and shape approach the bulk Bohr exciton radius, which is still generally 2–10 nm in length, electrical and optical properties are added to the particle. Scientists know that as a nanocrystal's size decreases, its bandgap increases, and its stimulation and emissions wavelength decrease because of this inverse relationship [63].

Park et al. used an enzyme-sensitive fluorescent dye in conjunction with antibody-QD conjugates to produce a multiplexed detection method for the fast and precise diagnosis of CRC. The instrument incorporated a Cresyl violet–glutamic acid (CV–Glu) derivative as the fluorescent molecular probe, as it easily transits between two fluorescent colours in reply to k-gluta myltranspeptidase enzymatic activity (GGT). A biocompatible, 690-nm emitting AgInS2/ZnS (core/shell) QD probe was created by bioconjugating an anti-matrix metalloproteinase-14 antibody (MMP14). The probe provided significant findings in ex vivo and in vitro applications by employing human CRC cell lines, ex vivo patient tumour samples, and ex vivo experimental animal model tissues. The researchers utilised two-photon microscopy to show that the co-application of both probes penetrated the tissue at depths of 10–20 mm quickly. The study's findings indicated that using both probes simultaneously allowed for quick (within 5 min) and accurate imaging of tumour lesions that would otherwise be impossible to identify using traditional colonoscopy procedures. When CRC was in its beginning phases, the dual-probe approach helped find hyperplastic polyps and adenomas [64].

Carbary-Ganz et al., created a dual-modality imaging technique that uses optical coherence tomography and laser-induced fluorescence to provide non-destructive endoscopic viewing of CRC with just a minimal amount of invasiveness. While CRC develops and progresses, this method allows for the concurrent longitudinal surveillance of biochemical and morphological changes. With great sensitivity and specificity, QDot655-VEGFR2 localised to the colon in carcinogen-treated animals and provided a significant difference between sick and healthy tissue ex vivo [65].

7. 2 Iron Oxide Nanocrystals

Iron oxide NPs are incorporated in MRI as a contrast agent for human use because of their dual magnetic and photo-thermal properties. Apart from that, they are highly biodegradable in vivo, and then, the iron salts delivered after disintegration can be ingested into the body system through a well-coordinated biochemical mechanism [66]. There is a magnetic core to the iron oxide nanocrystals and a polymer covering, which will contain the various medicinal molecules. These nanoparticles have various biological properties due to their relatively small dimension, efficient surface region, minimal deposition rate, and ease of cellular transport [67]. The severe toxicity and oxidation sensitivity of strong magnetic metals like cobalt and nickel seem to offer limited biomedical utility. Using nanoparticles, a cancer cell is transformed into a tiny magnetic magnet that may be attached to the biopsy needle's tip. These nanoparticles are unstable in aqueous environments and easily collected and deposited [68]. It cannot detect the target tumour using nanoparticles of the iron oxide in vivo due to the diverse over expression of the highly targeted receptor in cancer cells. Using magnetic nanoparticles, it is possible to discriminate between malignant and normal tissue. It is possible to create functional groups on these nanoparticles that crosslink with tumour-targeting molecules such as peptides or monoclonal antibodies or small molecules for the diagnostic testing or remedial agent administration due to their great surface area [69].

Kim et al. used a 3D multi-echo gradient echo to compare the particular performance of three distinct parametric techniques (normalised signal intensity, R2* (R-lymph node Region of interest), and susceptibility) to measure the number of ultra-small super-paramagnetic iron oxide (USPIO) particles in lymph nodes. Before and after USPIO injection, nine rabbits with VX2 tumour implants were scanned. All of this was accomplished by combining 3D GRE amplitude and phase data to produce multi-echo combined T2* (weighted multi-echo combined magnitude image of 3D mGRE 24 h after USPIO injection)-weighted pictures and an R2* map. USPIO build-up in 18 lymph nodes (nine metastatic and nine reactive) was found to have a significant impact on signal intensity. R2* differences before and after USPIO injection were statistically significant between lymph nodes that were reactive and metastatic ($p < 0.05$), although signal intensity and vulnerability were not significantly different between the nodes. USPIO-enhanced MRI employing R2* mapping from 3D multi- echo GRE may be used to identify lymph node metastases and to perform a parametric analysis of the lymph node status in a rabbit model, as demonstrated in their work [70].

Wu *et al.* conducted a meta-analysis to examine how well USPIO-enhanced MRI and non-enhanced MRI, as well as USPIO-enhanced MRI in various body areas and postcontrast MRI performed in the diagnosis of lymph node metastases. As a result of this meta-analysis, USPIO-enhanced MRI is more accurate than conventional

MRI in the diagnosis of lymph node metastases. For lymph node characterisation, postcontrast pictures alone are sufficient to provide the same diagnostic performance as pre-contrast MRI. There is still a lot of research to be done on the use of USPIO enhanced MRI in clinical practice [71].

Kuo et al. developed an antibody-targeting peptide AP-1(MPVA-AP1) and anticancer medicines by incorporating the magnetic nano-vehicles. The magnetic nano-vehicles disseminated in the dilute solution demonstrated great hemocompatibility and toxic-free to L929 fibroblasts, indicating that they might be used in treatments. The researchers found significant encapsulated hydrophobic and hydrophilic low-molecular-weight medicines and protein-like pharmaceuticals using a straightforward manufacturing approach. Magnetic nano-vehicles were also used to immobilise the antibody-targeting peptide AP-1, which was verified using electron spectroscopy. The results of a CRC cell (CT26-IL4R□) assay demonstrated that the AP-1-bound nano-vehicles (MPVA-AP1) show exceptional targeting and selectivity. In the absence of a magnetic stimulation, a stable storage test revealed virtually little leakage of the encapsulated medicines. Instead of rupturing in the presence of a high-frequency magnetic field, doxorubicin-loaded nano-vehicles released the drug with pinpoint accuracy. There was further evidence from in vivo experiments that the magnetic nano-vehicles showed significant chemotherapeutic and thermotherapeutic effects. Due to this, magnetic nano-vehicles with smart properties, such as MPVA-AP1, hold great promise for anticancer applications that need precise dosing and regulated release [72].

It was found that the coprecipitation technique proposed by Syu et al., (2019) was simple and ligand-assisted to manufacture biocompatible iron oxide (IO) nanocrystals with NIR absorbance for cancer theranostics. Nanocrystals had little cytotoxicity in HT-29 colorectal cancer cells; however, the vitality of cells treated with nanocrystals dropped substantially following laser irradiation at a 808-nm wavelength. Cell death is thought to be caused by changes in protein secondary structure and membrane permeability, among other things. Nanocrystals with magnetic field (MF) application significantly enhanced tumour aggregation by fourfold in the in vivo studies, leading to a thrice-larger T2-weighted MR signal than that provided by a conventional T2-weighted MR contrast agent (Resovist®) and improved photothermal potency for the therapy of cancer. The novel iron oxide nanocrystals were shown to be highly biocompatible and show significant promise as a cancer theranostic agent [73].

8. SYNTHETIC/HERBAL NANOCRYSTAL USED IN DRUG DELIVERY SYSTEM

Biomedicine's use of nanotechnology is expanding quickly. The field of nanotechnology serves as a link between various NPs and nanophases in terms of both technology and science [74]. Particles with a size range of 1–100 nm with great specific surface area and specialised surface properties are referred to as nanoparticles (NPs) [75]. It is possible to obtain controlled drug release, site-specific drug delivery, and reduced toxicity by using nanoparticles (NPs). The NP's size significantly impacts how well it is taken in by cells and excreted from them in vivo. It has been demonstrated that greater particulates have much reduced concentrations of intracellular delivery than small nanoparticles at a similar quantity and that NPs with such a dimension of 100 nm had 2.5-fold more absorption than 1-m size nanoparticles [76]. Since the glomerular capillary sieve coefficient is 10 nm, NPs with a smaller diameter exhibit better renal clearance. Larger NPs, on the other hand, are unable to diffuse into intracellular gaps because of the increased permeability and retention (EPR) effect, lacking in a approach cancer cells [77]. The toxicology of NPs is influenced by physicochemical characteristics, including size, shape, and concentration, which affect absorption, circulation time, and total toxicity. The phospholipid bilayer structure gives the cell membrane a negative surface charge, making it even more susceptible to positively charged particles [78].

In other words, because they immediately enter the cell membrane and cause membrane rupture or damage, highly positive charged NPs have higher cytotoxicity. NP surfaces have been modified to hold a negatively charged initially to allow them to the exits specific macrophage organelles that is later converted to a positive charge by environmental cues to enhance drug administration [79]. To counteract the cationic charge, scientists used serum albumin in preincubated NPs, including shape and size, toxic effects, quantity, and charge density, which should be considered while designing nanomedicine. For example, MRI imaging, cancer diagnostics, precise cell and subcellular treatment, and the detection of genetic mutations NPs have enormous promise, with suitable surface changes in biomedical and therapeutic applications such as nanoparticle surface modification [80].

9. PHYTOMEDICINE DRUG DELIVERY USING NANOPARTICLES

Developing resistance to chemotherapeutic drugs and radiation therapy is a key challenge in cancer treatment. Phytochemicals with low side effects, namely lycopene, genistein, epigallocatechin gallate, anthocyanin, curcumin, and resveratrol, are being investigated by researchers as an alternative therapy to help overcome this problem [81,82]. On the other hand, phytochemicals have several disadvantages, including poor absorption, limited water solubility, and rapid metabolism. Consequently, using NPs to formulate these phytochemicals can improve phytochemical medication delivery while also minimising adverse effects [82].

9. 1 Curcumin

As a chemoprevention and anticancer agent, curcumin obtained from Indian spices turmeric is highly effective. Inhibiting different signalling pathways such as NF-κB, STAT3 and EGFR, which contribute to tumour development and metastasis, slows the advancement of many malignancies, including CRC [83]. Due to the nanoformulation, curcumin has improved water solubility and distribution while causing no harm to healthy cells. To create a nanocurcumin formulation for use in CRC treatment, NPs made of liposomes, polymers, cyclodextrin, nanogels, gold, lipids, and micelles can be used [84].

In addition to its ability to act as a free radical scavenger, curcumin also changes the expression of several stress proteins and angiogenesis-related genes. It inhibits the activity of numerous key transcription factors, namely activator protein 1 (AP-1) and nuclear factor κ-light chain enhancer of activated B cells (NF-κB) [85]. Additionally, the size of 10 μm showed significant antioxidant activity. It triggered cell death when they reached a concentration of around 50% at 50 μm, which may be due to the production of superoxide radicals [86]. Human plasma levels can be as low as 0.41–1.75 μmol/L after taking 4–8 g of turmeric orally each day. The curcumin concentration is critical for evaluating the biological consequences of frequent oral intake. Bio-transformed moieties, such as tetra hexahydro–curcumin, have drawn attention because of their poor systemic bioavailability [87].

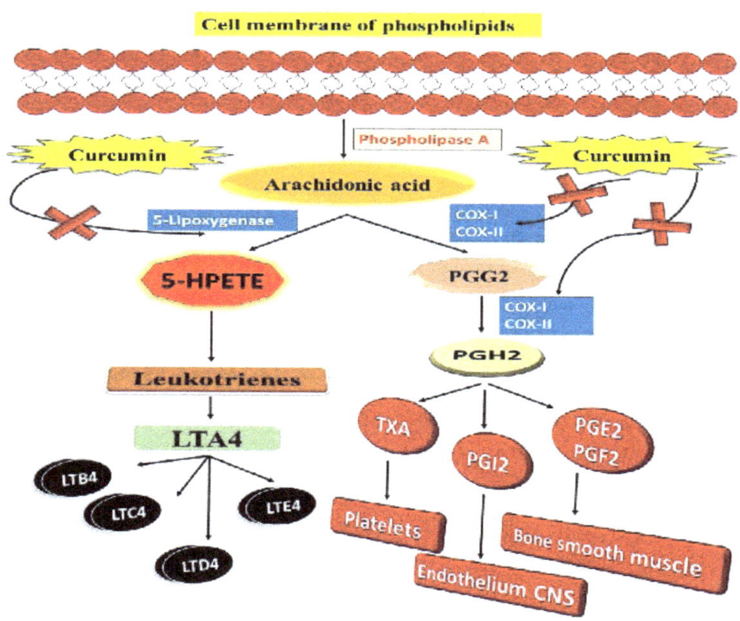

Figure 7. Curcumin and the arachidonic acid metabolism: A potential inhibitory effect.

Additionally, research shows that curcumin is an efficient chemopreventive drug in addition to its antitumour properties (Figure 7). It works by inhibiting cyclooxygenase, phospholipase A2, and phospholipase-Cr1 in tumours of the colon, stomach, and skin. It is clear that antagonists of COX-2 are effective in suppressing polyp development and cancer growth in FAP, and were it not for the negative consequences, COX-2 inhibitors would be strongly suggested for cancer polyp chemoprevention [88]. In order to avoid carcinogenesis, we require drugs that block cellular pathways that cause or promote cancer. With these characteristics, curcumin is a promising chemo-preventive agent. It has recently been shown that 5-LOX plays a crucial function in controlling cell proliferation and apoptosis [89]. Some research shows 5-LOX over expression in cancer cells and human tumours, such as those of the gastrointestinal tract and liver and those of the breast and prostate. 5-LOX has also been shown to be over expressed in the brain and the mesothelium. 5-LOX over expression has been connected to enhanced proliferation and tumour growth in investigations done in the lab and on mice in the wild [90].

Ohd et al. indicate that 5-LOX and cysteinyl leukotriene receptors are over expressed in human colon adenocarcinomas and that individuals with these mutations have a poor prognosis. It indicates that increased 5-LOX expression is related to malignant transformation, although the exact processes that link 5-LOX gene expression to colon cancer development are unclear. The use of COX-2 knockout mouse models has demonstrated that COX-2 plays a function in colon carcinogenesis. However, in colon carcinogenesis, similar data from 5-LOX knockout animals are missing [91]. According to Raveendran et al., curcumin- encapsulated PCL exhibits hydrophobicity and effective NP stability due to its amphiphilic block copolymer nature. It was an effective technique, because it targeted the tumour and made tumour cells more susceptible to treatment resistance. Additionally, this formulation enhanced water solubility, resulting in increased in vitro curcumin absorption by Caco-2 CRC cells. There was significantly less inhibition in the micelles of bigger amphiphilic block copolymers than smaller ones [92].

9.2 Resveratrol

Natural phenols such as resveratrol are found in fruits such as red grapes, blue- berries, and raspberries [93]. Resveratrol has been discovered to have anticancer, anti- cardiovascular, anti-inflammatory, and anti-aging biological and pharmacological effects [94]. Further studies have shown that resveratrol has anti-proliferative properties in vitro against

CRC via modulating intracellular signalling pathways such as PTEN/PI3K/Akt, Wnt/β- catenin, and others [95]. It is possible to discover therapeutic medication's biological targets and molecular processes by using network pharmacology as an approach in clinical illnesses (Figure 8). As a result, identifying biomarkers and treatment molecular pathways is crucial even before diagnosing a patient with a relapsed CRC [96]. Resveratrol belongs to the phytoalexin stilbene compound and is isolated from the various plants extracts. Chemo-sensitising, anti-oxidative, and anti-inflammatory characteristics are some of the things it does. Due to the poor physicochemical profile of resveratrol, it is not generally recommended therapeutically despite its positive features [97]. The iNOS, COX-2, and other inflammatory signalling pathways are all reduced by resveratrol, inhibiting proinflammatory mediators, including IL-1β and TNF-α [98]. Soo et al. created and improved an innovative drug carrier by co-encapsulating resveratrol and cyclodextrin-resveratrol inclusion complexes in liposomes' lipophilic and hydrophilic compartments. The final formulation's particle size, polydispersity index, and zeta potential were measured at 131 ± 1.30 nm, 0.089 ± 0.005 and -2.64 ± 0.51 mV, respectively. The drug release profiles of 40–60% were found to be acceptable for the standard liposomal formulations and free resveratrol, while the innovative nanoformulations demonstrated total (100%) release in 24 h. It remained steady at 4 °C for 14 days. In HT-29 colon cancer cell lines, resveratrol-encapsulated liposomes showed in vitro cytotoxicity. Liposomes were shown to have a dose-dependent cytotoxicity profile that was also boosted compared to free resveratrol (in DMSO). Using liposomal formulations to deliver hydrophobic chemotherapy agents is possible because of the study's findings, which show that the co-encapsulation of resveratrol in its purest form with its cyclodextrin complex is feasible [99]. Summerlin et al. enclosed resveratrol in colloidal MSN, and they discovered that the encapsulation efficiency and loading capacity of MCM-48–resveratrol were superior. In CRC cell lines, the formula promoted tumour cell killing via the PARP and cIAP1 pathways. Furthermore, it slowed down the production of NF-κB. As a result, this new approach to medication development may be useful in the future [100].

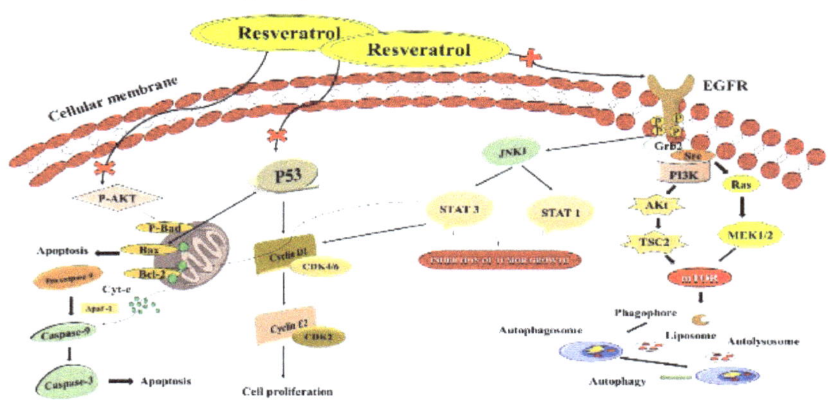

Figure 8. Schematic illustration of resveratrol as a novel healing method for treating CRC by targeting distinct signalling pathways.

10. CHEMOTHERAPEUTIC DRUG DELIVERY USING NANOPARTICLES

Most cancers are still treated with radiotherapy and chemotherapy. It is a serious issue that a combination of high dosages and combinational therapy produces severe toxic effects to healthy colon cells. Moreover, it is also associated with a variety of cardiac and digestive side effects. Hence, it is required to reach the target location because of the short half-life of these therapies and the tendency towards resistance [101]. Since the drug's circulation time and pharmacokinetic characteristics may be improved via nanocarrier-based delivery, it may overcome these adverse effects [102].

10.1 5-Fluorouracil (Capecitabine)

5-Fluorouracil (5-FU) was applied to heal CRC as a single agent and in combination with other drugs for more than 40 years [103]. Researchers had to develop an oral version of 5-FU because of its short half-life, the necessity of using a central line, and the demand for continual infusions [104]. When compared to conventional 5-FU, Capecitabine provides many benefits. After passing through the digestive tract, it undergoes three enzymatic processes before transforming to 5-FU (Figure 9). Thymidine phosphorylase (TP), the pathway's last enzyme, is thought to be present in tumour tissue at disproportionately high levels, which increases the agent's effectiveness and tolerance through tailored delivery [105]. Most CRC chemotherapy regimens start with 5-FU as the main active ingredient. Deoxythymidine monophosphate (dTMP) plays key role in DNA replication and repair and is inhibited by 5-FU by altering into fluorodeoxyuridine monophosphate (FdUMP) in cells [106]. FdUMP creates a stable complex mostly with the enzyme thymidylate synthase, inhibiting the production of dTMP. It often interacts mostly with the formation and activity of nucleic acids by substituting uracil or thymine in DNA and RNA as a secondary method of action. The 5-FU-related gene TYMP (formerly known as ECGF1), genes for the thymi- dine phosphorylase (TP) protein, varied amongst the groups. After being converted to fluorodeoxyuridine (FUDR) by the action of TP, the active metabolite FdUMP is produced. In vitro, the over expression of TP increases the 5-FU sensitivity, which is thought to be due to the increased production of FdUMP as a result [107].

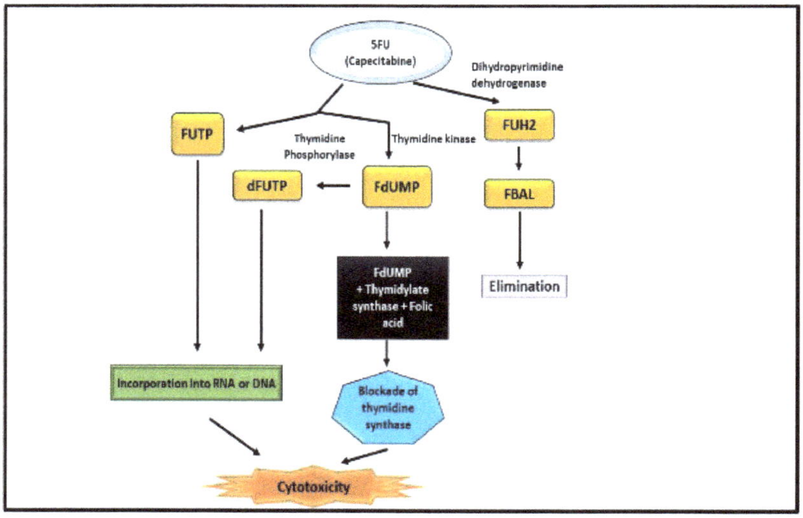

Figure 9. Mechanism of 5-FU for the treatment of CRC.

Capecitabine was used in a phase II study by Mizumoto et al. The quality of life and disease-free survival of individuals treated with Capecitabine may improve if the adverse effects are reduced throughout the drug. Capecitabine (2500 mg/m2/day) was given to CRC patients for five days, followed two days off (5 days on/2 days off regimen). It took three weeks to complete one course, and it took 24 weeks to complete all eight courses. Retrospectively, included patients on the traditional regimen received substantially more treatment courses than those on the 5 days on/2 days off regimen, with a p-value of 0.0438. There was a substantially greater rate of patients in the 5 days on/2 days off regimen completing their planned therapy (p = 0.0389). However, according to the results of this phase II research, toxic effects related to the 5 days on/2 days off regimen are less severe than those associated with the more traditional regimen, and adverse events are more common. However, there are fewer reports of high-grade toxicities. The novel regimen had a shorter time to treatment failure and showed high practicality. There was no significant difference in overall quality of life or side effects (mainly mild); therefore, the 5 days on/2 days off regimen must be a plan to be studied in future randomised controlled trials to confirm its viability [108].

10.2 Oxaliplatin

Oxaliplatin is a third-generation diamino cyclohexane (DACH) platinum compound. Oxaliplatin creates intrastrand connections between two adjacent guanine residues or between guanine and adenine, inhibiting DNA replication and transcription [109]. Oxaliplatin is the third generation of the DACH platinum compound. Oxaliplatin is successful in the treatment of colorectal cancer, but platinum compounds cisplatin and carboplatin have been shown to be ineffective [110]. It is frequently used to treat patients who have not responded to 5-fluorouracil (5FU)-based therapy. Oxaliplatin's cytotoxic effects, on the other hand, are still poorly studied in detail. Many chemotherapeutic drugs, including oxaliplatin, have been demonstrated to cause apoptosis in tumour cells [111]. Since a medication's cellular target is dependent on the accumulation in cancer cells, the clinical efficacy of oxaliplatin may be influenced by variables that regulate its accumulation due to its expression by tumour cells or by biological barriers that impact drug disposition. (e.g., enterocytes and hepatocytes) (Figure 10).

Oxaliplatin can enter cells passively or by transporters that carry it in from the outside. Its physicochemical properties are important to determine if passive diffusion is preferable to other routes [112]. Due to these characteristics, it may accumulate better in tumour cells than cisplatin, even in drug-resistant cells with a lower permeability toward the more hydrophilic drug. Efflux and influx transporters work together to determine how much medication is stored in cells. Oxaliplatin's therapeutic efficacy and side effects may be influenced by the tissue-specific expression of influx transporters or decreased efflux transporter activity when these variables are present [113].

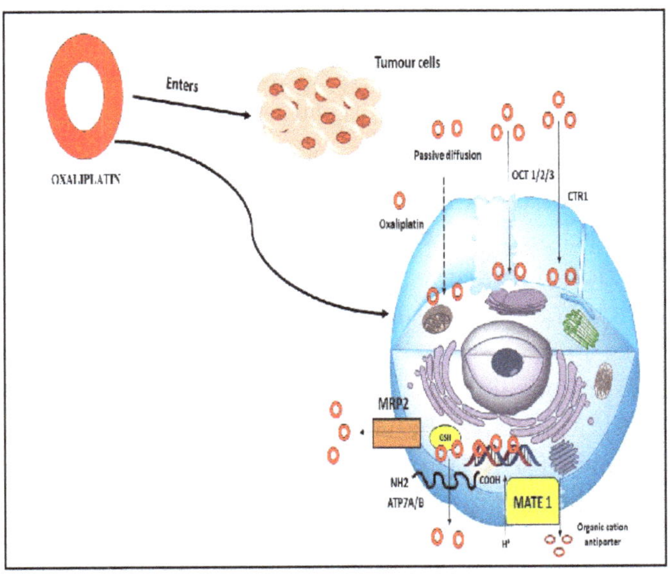

Figure 10. Numerous influx and efflux transporters that allow oxaliplatin to enter the cell passively (dashed arrow). There are several kinds of organic cation transporters, such as the OCT 1/2/3 trans- porters, copper transporters, MRP2 multidrug resistance-associated proteins, glutathione transporters, MATE 1 multidrug and toxin extrusion transporter, and ATP7A/B ATP-binding cassette transporters.

Trifluridine/tipiracil FTD/TPI and oxaliplatin were used synergistically by Limagne et al. to enhance the immunogenic cell death (ICD). Even though all therapy groups indicated T-cell dependence, only the combination effectively induced ICD in vivo. Furthermore, neither FTD/TPI or oxaliplatin affected regulatory T cells or myeloid-derived suppressor cells. However, they did remove type-2 tumour-associated macrophages (TAM2), resulting in increased cytotoxic CD8+ T-cell infiltration and activation. Secondary T-cell exhaustion was related to both tumour cell PD-L1 expression and CD8+ T-cell PD-1 induction.

Last, but not least, even when used alone, anti–PD-1 did not boost the antitumor efficacy of FTD/TPI or oxaliplatin monotherapy, but when combined with the drugs, it did. FTD/TPI and oxaliplatin depleted TAM2 in a new immunomodulatory manner. A rationale for using oxaliplatin and FTD/TPI to remove immunosuppressive cells and boost checkpoint efficacy in patients with metastatic colorectal cancer has been provided by the induction of ICD in vivo by combining these medicines [114

11. CURRENT COLORECTAL CANCER CLINICAL TRIALS AND STATE OF NANOTECHNOLOGY

Clinical endpoints and biomarkers are two issues that were particularly interested in when it comes to contemporary clinical studies. The alterations in OS amid two treatment groups in research evaluating a single line of therapy are becoming increasingly challenging to accomplish in advanced cancer patients with all types of clinical end goals [115]. There are a rising variety of viable therapeutic choices available in second and subsequent chemotherapy lines in addition to the regularly utilised and often ethically obligatory salvage therapies. However, the initial treatment effect may be weakened or confounded, eventually precluding the proof of a statistical significance overall survival improvement [116]. A growing number of patients will have to be included in clinical studies over an ex- tended period to obtain significant statistical and clinically acceptable OS impacts from long-term patient survival and the ensuing mild OS risk ratio variations within the treated group [117]. Induction chemotherapy combined with a targeted drug is the standard treatment for metastatic colorectal cancer [118]. Clinical trials have evaluated various techniques for maintaining cytotoxic treatment until progression against periods of observation or using various maintenance medicines in several randomised trials. On the other hand, clinical trials have shown wildly divergent effectiveness findings, making it difficult to draw any firm conclusions about which method is best. Table 2 shows the summary of various current and finished clinical studies of CRC.

Table 2. Illustration of completed and ongoing clinical trials evidence for CRC

Metastatic Colorectal Cancer Database	Methodology, Biostatistics and Data Management Dijon, France	NCT04031625	Not applicable, Recruiting Maintenance
Maintenance Therapy for Metastatic Colorectal Cancer After First-Line Treatment with Fruquintinib Plus Capecitabine Versus Bevacizumab Plus Capecitabine	Medical College of Zhejiang University Hangzhou, China	NCT04733963	Phase 2, Recruiting
CRC Patients having liver metastases are being treated with TAS-102 and radiation therapy	Massachusetts General Hospital, United States	NCT03223779	Phase 1, Recruiting
Anlotinib Combined With mXELIRI as Second-line Treatment of Advanced Colorectal Cancer	Guangdong Provincial Hospital of Chinese Medicine Guangzhou, China	NCT05035914	Phase 1, Recruiting
Do Colorectal Cancer Risk Estimates Affect Screening Behavior?	Stanford University School of Medicine Palo Alto, US	NCT03819920	Not Applicable, Completed
Study of Fruquintinib Efficacy and Safety in Patients with 3+ Line Colorectal Cancer (FRESCO) in Phase III	Hutchison Medi Pharma Investigational Site Hefei, Anhui, China	NCT02314819	Phase 3, Completed
Gut Microbiome Dynamics in Metastasized or Irresectable Colorectal Cancer	Wilhelmina Ziekenhuis Assen, Netherlands	NCT03941080	Not Applicable, Recruiting
Dabrafenib + Trametinib + PDR001 In Colorectal Cancer	Massachusetts General Hospital Cancer Center	NCT03668431	Phase 2, Recruiting
Colorectal Cancer Research Consortium (NCRCC) Study: National Colorectal Cancer Research Consortium	Zhejiang University College of Medicine Hangzhou, China	NCT04074538	Not Applicable, Recruiting

A Translational Study Examines the Impact of the MET Oncogene in Human Colorectal Cancer	Fondazione del Piemonte per l'Oncologia Candiolo, Italy	NCT02238821	Not Applicable, Completed
The Use of a Patient Navigator to Increase Colorectal Cancer Screening Uptake	Group Health Centre Sault Ste. Marie, Canada	NCT01506687	Phase 3, Completed
Regorafenib in the Treatment of Patients with Metastatic Colorectal Cancer: Real-World Effectiveness	Henan Cancer Hospital Zhengzhou, China	NCT05023720	Not Applicable, Recruiting
Investigating the effects on colorectal cancer patients of a walking programme	National Taiwan University Hospital Taipei, Taiwan	NCT01595256	Not Applicable, Completed
Metabolomics-Based Detection of Colorectal Cancer	Indiana University Cancer Center Indianapolis, US	NCT00507598	Not Applicable, Completed
Exploration into how often patients with colorectal cancer who are receiving irinotecan-based therapy experience nausea and vomiting	Caritas St. Elizabeth Medical Center Brighton, US	NCT00713128	Not Applicable, Completed
A Colorectal Cancer Screening Decision Aid for American Indians That Is Culturally Adapted	Robeson Health Care Corporation Lumberton, US	NCT03569761	Not Applicable, Completed
Vaccination Against MSI Colorectal Cancer	Krankenhaus Nordwest Frankfurt/Main, Germany	NCT01461148	Phase 2, Completed
Fecal Occult Blood Tests for Colorectal Cancer Screening: A Comparison of the Different Methods	Qilu Hospital Jinan, China	NCT04454099	Not Applicable, Completed
Concern for Positron Emission Tomography (PET) in Colorectal Cancer Stage II and III Follow-Up	Cancérologie et Hépato-Gastro-Entérologie Bordeaux, France	NCT00199654	Phase 3, Completed
Relapsed/refractory Colorectal Cancer Patients Receiving IMMU-130 Trial	Memorial Sloan-Kettering Cancer Center New York, US	NCT01270698	Phase 1, Completed

12. FUTURE PROSPECTS AND RECOMMENDATION

Colorectal cancer is still the leading cause of cancer-related death in the globe. To enhance the overall survival, minimise the disease-free progress, and reduce the recurrence risk, early detection is critical. Oncogenes and tumour suppressor genes undergo successive mutations in the adenoma-carcinoma pathway, progressing from normal to dysplastic epithelium. Normal glandular epithelium becomes adenocarcinoma when genetic and epigenetic alterations to the genome accumulate. Even though the molecular processes responsible for the cancer's growth and spread remain a mystery, numerous genetic pathways have been postulated to explain CRC aetiology. A biomarker is a useful tool for detecting diseases early on and predicting how well a patient will respond to therapy. Although biomarkers employed at various stages of the illness make it easier to control CRC, each one has its limits. To make advances in biomarker creation and validation, understanding molecular interactions that aid pathogenesis is necessary. The use of nanotechnology in the treatment of colorectal cancer is critical. When it comes to the elimination or, at the very least, the chronic treatment of cancer, nanotechnology has shown to be a vital and optimistic technique. The use of nanotechnology in cancer diagnosis and therapy has been game-changing. It has been a priority in biomedicine to create drug delivery techniques that can affect bio-distribution, tissue absorption, and pharmacokinetics, among other things.

Contrary to popular belief, nanomedicine can have a significant impact on human CRC diagnosis and treatment and even improve normal human physiology in some cases. We hope that this article will help scientists, research scholars, industrialists, etc. with their research work and provide ease in searching through the findings of the current research work in CRC. It also helps to identify the diagnostic, prognostic, and predictive biomarkers. Thus, it provides a conceptual knowledge of nano-based medicine for the management of CRC.

13. CONCLUSIONS

Many people around the world are still affected by CRC. mCRC management remains a challenge despite screening recommendations and advancements in treatment options. A variety of genetic and epigenetic factors contribute to the heterogeneity of CRC. There is also much variability in the tumour microenvironment in CRC, making it difficult to find therapeutic targets for people with metastatic CRC. The heterogeneity of tumours within and across primary and metastatic sites can be extremely diverse. This could be a contributing factor to a treatment failure when a patient develops a resistance to targeted medicines. Since the effectiveness of more recent focused treatments depends on a good understanding of the patient's previous response, a tumour specimen analysis is also required for understanding the underlying illness mechanism and/or tumour resistance traits. As our understanding of the colorectal carcinogenesis molecular landscape expands, new molecular biomarkers with prognostic and predictive information are being identified. For the real-time examination of tumour clonal development, medication response, the existence of minimum residual illness, and acquired resistance, liquid biopsies are a potential approach. CRC patients may benefit from real-time molecular categorisation and customised treatment using non-invasive biomarkers. With a better understanding of the epigenetic regulatory mechanisms, including cancer-specific epigenetic changes, prospective clinical implications as diagnostics or therapeutic targets in colorectal cancer can now be explored more thoroughly.

The diagnostic efficacy of these biomarkers in early-stage CRC screening, prognostication, or treatment response prediction has yet to be improved, and most of these biomarkers have not yet been validated in large independent patient cohorts. A significant step towards individualised medicine should, in our opinion, include the consideration of non-cancer-related aspects such as nutrition and lifestyle. The existing methods in the cancer management sector have been rapidly replaced by new ones that offer greater diagnostic capabilities, increased sensitivity for tumour staging, and better therapeutic approaches.

There are various advantages to using nanotechnology in colon cancer treatment. We can see early indications of nanotechnology's value in evaluating colorectal cancer imaging, medicines, and ablation methods. The first measures have already been taken to improve these areas. There should be no doubt that nanotechnology-based techniques have the potential to treat cancers other than CRC, even if this opinion has concentrated on CRC as a particularly attractive target for early application.

14. CONCISE SUMMARY

Colorectal cancer (CRC) is a serious disease that affects millions of people throughout the world, despite considerable advances in therapy. The formation of colorectal adenomas and invasive adenocarcinomas is the consequence of a succession of genetic and epigenetic changes in the normal colonic epithelium. Genetic and epigenetic processes associated with the onset, development, and metastasis of sporadic CRC have been studied in depth, resulting in identifying biomarkers that might be used to predict behaviour and prognosis beyond staging and influence therapeutic options. A novel biomarker, or a group of biomarkers, must be discovered in order to build an accurate and clinically useful test that may be used as an alternative to conventional methods for the early detection of CRC and to identify prospective new therapeutic intervention targets. To minimise the mortality burden of colorectal cancer, new screening methods with higher accuracy and nano-based diagnostic precision are needed. Cytotoxic medication has negative side effects and is restricted by medication resistance. One of the most promising cancer treatment techniques is the use of nano-based carrier system as a medication delivery mechanism. To deliver cytotoxic medicines, targeted nanoparticles might take advantage of differently expressed molecules on the surface of cancer cells. The use of different compounds as ligands on the surface of nanoparticles to interact with cancer cells, enabling the efficient delivery of antitumor medicines. Formulations based on nanoparticles might aid in early cancer diagnosis and help to overcome the limitations of traditional treatments, including low water solubility, nonspecific bio-distribution, and restricted bioavailability. This article addresses about the molecular pathogenesis of CRC and highlights about biomarkers. It also provides conceptual knowledge of nanotechnology-based diagnostic techniques and therapeutic approaches for malignant colorectal cancer.

REFERENCES

1. Issa, I.A.; Noureddine, M. Colorectal cancer screening: An updated review of the available options. World J. Gastroenterol. 2017, 23, 5086–5096. [CrossRef]
2. Jeon, J.; Du, M.; Schoen, R.E.; Hoffmeister, M.; Newcomb, P.A.; Berndt, S.I.; Caan, B.; Campbell, P.T.; Chan, A.T.; Chang-Claude, J.; et al. Determining risk of colorectal cancer and starting age of screening based on lifestyle, environmental, and genetic factors. Gastroenterology 2018, 154, 2152–2164. [CrossRef]
3. Carbone, M.; Arron, S.T.; Beutler, B.; Bononi, A.; Cavenee, W.; Cleaver, J.E.; Croce, C.M.; D'Andrea, A.; Foulkes, W.D.; Gaudino, G.; et al. Tumour predisposition and cancer syndromes as models to study gene–environment interactions. Nat. Rev. Cancer 2020, 20, 533–549. [CrossRef]
4. Ishaque, N.; Abba, M.L.; Hauser, C.; Patil, N.; Paramasivam, N.; Huebschmann, D.; Leupold, J.H.; Balasubramanian, G.P.; Kleinheinz, K.; Toprak, U.H.; et al. Whole genome sequencing puts forward hypotheses on metastasis evolution and therapy in colorectal cancer. Nat. Commun. 2018, 9, 1–4. [CrossRef]
5. Guo, M.; Peng, Y.; Gao, A.; Du, C.; Herman, J.G. Epigenetic heterogeneity in cancer. Biomark. Res. 2019, 7, 23. [CrossRef]
6. Ganesh, K.; Stadler, Z.K.; Cercek, A.; Mendelsohn, R.B.; Shia, J.; Segal, N.H.; Diaz, L.A. Immunotherapy in colorectal cancer: Rationale, challenges and potential. Nat. Rev. Gastroenterol. Hepatol. 2019, 16, 361–375. [CrossRef]
7. Mármol, I.; Sánchez-de-Diego, C.; Pradilla Dieste, A.; Cerrada, E.; Rodriguez Yoldi, M.J. Colorectal carcinoma: A general overview and future perspectives in colorectal cancer. Int. J. Mol. Sci. 2017, 18, 197. [CrossRef]

8. Huang, F.T.; Chen, W.Y.; Gu, Z.Q.; Zhuang, Y.Y.; Li, C.Q.; Wang, L.Y.; Peng, J.F.; Zhu, Z.; Luo, X.; Li, Y.H.; et al. The novel long intergenic noncoding RNA UCC promotes colorectal cancer progression by sponging miR-143. Cell Death Dis. 2017, 8, e2778. [CrossRef]

9. Bupathi, M.; Wu, C. Biomarkers for immune therapy in colorectal cancer: Mismatch-repair deficiency and others. J. Gastrointest. Oncol. 2016, 7, 713. [CrossRef]

10. Mauri, G.; Bonazzina, E.; Amatu, A.; Tosi, F.; Bencardino, K.; Gori, V.; Massihnia, D.; Cipani, T.; Spina, F.; Ghezzi, S.; et al. The evolutionary landscape of treatment for BRAFV600E mutant metastatic colorectal cancer. Cancers 2021, 13, 137. [CrossRef]

11. Chenthamara, D.; Subramaniam, S.; Ramakrishnan, S.G.; Krishnaswamy, S.; Essa, M.M.; Lin, F.H.; Qoronfleh, M.W. Therapeutic efficacy of nanoparticles and routes of administration. Biomater. Res. 2019, 23, 1–29. [CrossRef]

12. Kalaydina, R.V.; Bajwa, K.; Qorri, B.; Decarlo, A.; Szewczuk, M.R. Recent advances in "smart" delivery systems for extended drug release in cancer therapy. Int. J. Nanomed. 2018, 13, 4727. [CrossRef]

13. Müller, M.F.; Ibrahim, A.E.; Arends, M.J. Molecular pathological classification of colorectal cancer. Virchows Arch. 2016, 469, 125–134. [CrossRef]

14. Aghagolzadeh, P.; Radpour, R. New trends in molecular and cellular biomarker discovery for colorectal cancer. World J. Gastroenterol. 2016, 22, 5678. [CrossRef]

15. Carethers, J.M. Microsatellite instability pathway and EMAST in colorectal cancer. Curr. Colorectal Cancer Rep. 2017, 13, 73–80. [CrossRef]

16. Langner, C. Serrated and non-serrated precursor lesions of colorectal cancer. Dig. Dis. 2015, 33, 28–37. [CrossRef]

17. Fischer, J.; Walker, L.C.; Robinson, B.A.; Frizelle, F.A.; Church, J.M.; Eglinton, T.W. Clinical implications of the genetics of sporadic colorectal cancer. ANZ J. Surg. 2019, 89, 1224–1229. [CrossRef]

18. Raskov, H.; Søby, J.H.; Troelsen, J.; Bojesen, R.D.; Gögenur, I. Driver gene mutations and epigenetics in colorectal cancer. Ann. Surg. 2020, 271, 75–85. [CrossRef]

19. Koveitypour, Z.; Panahi, F.; Vakilian, M.; Peymani, M.; Forootan, F.S.; Esfahani, M.H.; Ghaedi, K. Signaling pathways involved in colorectal cancer progression. Cell Biosci. 2019, 9, 1–14. [CrossRef]

20. Boland, P.M.; Yurgelun, M.B.; Boland, C.R. Recent progress in Lynch syndrome and other familial colorectal cancer syndromes. CA A Cancer J. Clin. 2018, 68, 217–231. [CrossRef]

21. Watson, M.M. Microsatellite Instability at Tetranucleotides (EMAST) in Colorectal Cancer: Clinical Relevance, Mechanisms and Immune Markers. Ph.D. Thesis, University of Bergen, Bergen, Norway, 2020.

22. Mehta, R.S.; Song, M.; Nishihara, R.; Drew, D.A.; Wu, K.; Qian, Z.R.; Fung, T.T.; Hamada, T.; Masugi, Y.; da Silva, A.; et al. Dietary patterns and risk of colorectal cancer: Analysis by tumor location and molecular subtypes. Gastroenterology 2017, 152, 1944–1953. [CrossRef]

23. Mitsuhashi, K.; Yamamoto, I.; Kurihara, H.; Kanno, S.; Ito, M.; Igarashi, H.; Ishigami, K.; Sukawa, Y.; Tachibana, M.; Takahashi, H.; et al. Analysis of the molecular features of rectal carcinoid tumors to identify new biomarkers that predict biological malignancy. Oncotarget 2015, 6, 22114. [CrossRef]

24. Szylberg, Ł.; Janiczek, M.; Popiel, A.; Marszałek, A. Serrated polyps and their alternative pathway to the colorectal cancer: A systematic review. Gastroenterol. Res. Pract. 2015, 2015, 573814. [CrossRef]

25. Novellasdemunt, L.; Antas, P.; Li, V.S. Targeting Wnt signaling in colorectal cancer. A review in the theme: Cell signaling: Proteins, pathways and mechanisms. Am. J. Physiol.-Cell Physiol. 2015, 309, C511–C521. [CrossRef]

26. Amodio, V.; Yaeger, R.; Arcella, P.; Cancelliere, C.; Lamba, S.; Lorenzato, A.; Arena, S.; Montone, M.; Mussolin, B.; Bian, Y.; et al. EGFR blockade reverts resistance to KRASG12C inhibition in colorectal cancer. Cancer Discov. 2020, 10, 1129–1139. [CrossRef]

27. Ahronian, L.G.; Sennott, E.M.; Van Allen, E.M.; Wagle, N.; Kwak, E.L.; Faris, J.E.; Godfrey, J.T.; Nishimura, K.; Lynch, K.D.; Mermel, C.H.; et al. Clinical acquired resistance to RAF inhibitor combinations in BRAF-mutant colorectal cancer through MAPK pathway alterations. Cancer Discov. 2015, 5, 358–367. [CrossRef]

28. Ito, M.; Kanno, S.; Nosho, K.; Sukawa, Y.; Mitsuhashi, K.; Kurihara, H.; Igarashi, H.; Takahashi, T.; Tachibana, M.; Takahashi, H.; et al. Association of Fusobacterium nucleatum with clinical and molecular features in colorectal serrated pathway. Int. J. Cancer 2015, 137, 1258–1268. [CrossRef]

29. Rhee, Y.Y.; Kim, K.J.; Kang, G.H. CpG island methylator phenotype-high colorectal cancers and their prognostic implications and relationships with the serrated neoplasia pathway. Gut Liver 2017, 11, 38. [CrossRef]

30. Valo, S.; Kaur, S.; Ristimäki, A.; Renkonen-Sinisalo, L.; Järvinen, H.; Mecklin, J.P.; Nyström, M.; Peltomäki, P. DNA hyperme- thylation appears early and shows increased frequency with dysplasia in Lynch syndrome-associated colorectal adenomas and carcinomas. Clin. Epigenet. 2015, 7, 71. [CrossRef]

31. Suraweera, N.; Mouradov, D.; Li, S.; Jorissen, R.N.; Hampson, D.; Ghosh, A.; Sengupta, N.; Thaha, M.; Ahmed, S.; Kirwan, M.; et al. Relative telomere lengths in tumor and normal mucosa are related to disease progression and chromosome instability profiles in colorectal cancer. Oncotarget 2016, 7, 36474. [CrossRef]

32. Yu, H.; Hemminki, K. Genetic epidemiology of colorectal cancer and associated cancers. Mutagenesis 2020, 35, 207–219. [CrossRef]

33. Siegel, R.L.; Jakubowski, C.D.; Fedewa, S.A.; Davis, A.; Azad, N.S. Colorectal cancer in the young: Epidemiology, prevention, management. Am. Soc. Clin. Oncol. Educ. Book 2020, 40, e75–e88. [CrossRef]

34. De Palma, F.D.; D'argenio, V.; Pol, J.; Kroemer, G.; Maiuri, M.C.; Salvatore, F. The molecular hallmarks of the serrated pathway in colorectal cancer. Cancers 2019, 11, 1017. [CrossRef]

35. Bibbins-Domingo, K.; Grossman, D.C.; Curry, S.J.; Davidson, K.W.; Epling, J.W.; García, F.A.; Gillman, M.W.; Harper, D.M.; Kemper, A.R.; Krist, A.H.; et al. Screening for colorectal cancer: US Preventive Services Task Force recommendation statement. Jama 2016, 315, 2564–2575.

36. Mousavinezhad, M.; Majdzadeh, R.; Sari, A.A.; Delavari, A.; Mohtasham, F. The effectiveness of FOBT vs. FIT: A meta-analysis on colorectal cancer screening test. Med. J. Islam. Repub. Iran 2016, 30, 366.

37. Li, M.; Zhang, J.; Dan, Y.; Yao, Y.; Dai, W.; Cai, G.; Yang, G.; Tong, T. A clinical-radiomics nomogram for the preoperative prediction of lymph node metastasis in colorectal cancer. J. Transl. Med. 2020, 18, 1–10. [CrossRef]

38. Nikolouzakis, T.K.; Vassilopoulou, L.; Fragkiadaki, P.; Mariolis Sapsakos, T.; Papadakis, G.Z.; Spandidos, D.A.; Tsatsakis, A.M.; Tsiaoussis, J. Improving diagnosis, prognosis and prediction by using biomarkers in CRC patients. Oncol. Rep. 2018, 39, 2455–2472. [CrossRef]

39. Ortiz-Quintero, B. Cell-free microRNAs in blood and other body fluids, as cancer biomarkers. Cell Prolif. 2016, 49, 281–303. [CrossRef]

40. Erstad, D.J.; Tumusiime, G.; Cusack, J.C. Prognostic and predictive biomarkers in colorectal cancer: Implications for the clinical surgeon. Ann. Surg. Oncol. 2015, 22, 3433–3450. [CrossRef]

41. Moosavi, F.; Giovannetti, E.; Saso, L.; Firuzi, O. HGF/MET pathway aberrations as diagnostic, prognostic, and predictive biomarkers in human cancers. Crit. Rev. Clin. Lab. Sci. 2019, 56, 533–566. [CrossRef]

42. Yao, Q.; Wang, W.; Jin, J.; Min, K.; Yang, J.; Zhong, Y.; Xu, C.; Deng, J.; Zhou, Y. Synergistic role of Caspase-8 and Caspase-3 expressions: Prognostic and

predictive biomarkers in colorectal cancer. Cancer Biomark. 2018, 21, 899–908. [CrossRef]

43. Masuda, T.; Hayashi, N.; Kuroda, Y.; Ito, S.; Eguchi, H.; Mimori, K. MicroRNAs as biomarkers in colorectal cancer. Cancers 2017, 9, 124. [CrossRef]

44. Zarkavelis, G.; Boussios, S.; Papadaki, A.; Katsanos, K.H.; Christodoulou, D.K.; Pentheroudakis, G. Current and future biomarkers in colorectal cancer. Ann. Gastroenterol. 2017, 30, 613. [CrossRef]

45. Vinchhi, P.; Patel, M.M. Triumph against cancer: Invading colorectal cancer with nanotechnology. Expert Opin. Drug Deliv. 2021, 18, 1169–1192. [CrossRef]

46. Ladabaum, U.; Dominitz, J.A.; Kahi, C.; Schoen, R.E. Strategies for colorectal cancer screening. Gastroenterology 2020, 158, 418–432. [CrossRef]

47. Fattepur, S.; Setia, A.; Prashar, D.; Sahu, R.K.; Nilugal, K.C.; Abdullah, I.; Sreeharsah, N. Understanding diverse signaling path- ways involved in colorectal carcinoma and potentials of nanotechnology in their diagnosis and therapeutics. PharmacologyOnLine 2021, 2, 1226–1243.

48. Wu, Z.; Li, Y.; Zhang, Y.; Hu, H.; Wu, T.; Liu, S.; Chen, W.; Xie, S.; Lu, Z. Colorectal Cancer Screening Methods and Molecular Markers for Early Detection. Technol. Cancer Res. Treat. 2020, 19, 1–9. [CrossRef]

49. Guimarães, D.P.; Fregnani, J.H.; Reis, R.M.; Taveira, L.N.; Scapulatempo-Neto, C.; Matsushita, M.; Silva, S.R.; Oliveira, C.Z.; Longatto-Filho, A.; Eklund, C.; et al. Comparison of a new-generation fecal immunochemical test (FIT) with guaiac fecal occult blood test (gFOBT) in detecting colorectal neoplasia among colonoscopy-referral patients. Anticancer Res. 2019, 39, 261–269. [CrossRef]

50. Segev, L.; Kalady, M.F.; Church, J.M. Left-sided dominance of early-onset colorectal cancers: A rationale for screening flexible sigmoidoscopy in the young. Dis. Colon Rectum 2018, 61, 897–902. [CrossRef]

51. Bobo, D.; Robinson, K.J.; Islam, J.; Thurecht, K.J.; Corrie, S.R. Nanoparticle-based medicines: A review of FDA-approved materials and clinical trials to date. Pharm. Res. 2016, 33, 2373–2387. [CrossRef]

52. Feng, W.; Jin, M.; Yang, K.; Pei, Y.; Pei, Z. Supramolecular delivery systems based on pillararenes. Chem. Commun. 2018, 54, 13626–13640. [CrossRef]

53. Elsabahy, M.; Heo, G.S.; Lim, S.M.; Sun, G.; Wooley, K.L. Polymeric nanostructures for imaging and therapy. Chem. Rev. 2015, 115, 10967–11011. [CrossRef]

54. Naseri, N.; Ajorlou, E.; Asghari, F.; Pilehvar-Soltanahmadi, Y. An update on nanoparticle-based contrast agents in medical imaging. Artif. Cells Nanomed. Biotechnol. 2018, 46, 1111–1121. [CrossRef]

55. Cisterna, B.A.; Kamaly, N.; Choi, W.I.; Tavakkoli, A.; Farokhzad, O.C.; Vilos, C. Targeted nanoparticles for colorectal cancer. Nanomedicine 2016, 11, 2443–2456. [CrossRef]

56. Xiao, M.; Lai, W.; Man, T.; Chang, B.; Li, L.; Chandrasekaran, A.R.; Pei, H. Rationally engineered nucleic acid architectures for biosensing applications. Chem. Rev. 2019, 119, 11631–11717. [CrossRef]

57. Siddique, S.; Chow, J.C. Application of nanomaterials in biomedical imaging and cancer therapy. Nanomaterials 2020, 10, 1700. [CrossRef]

58. Banerjee, A.; Pathak, S.; Subramanium, V.D.; Dharanivasan, G.; Murugesan, R.; Verma, R.S. Strategies for targeted drug delivery in treatment of colon cancer: Current trends and future perspectives. Drug Discov. Today 2017, 22, 1224–1232. [CrossRef]

59. Chaturvedi, V.K.; Singh, A.; Singh, V.K.; Singh, M.P. Cancer nanotechnology: A new revolution for cancer diagnosis and therapy. Curr. Drug Metab. 2019, 20, 416–429. [CrossRef]

60. Wagner, A.M.; Knipe, J.M.; Orive, G.; Peppas, N.A. Quantum dots in biomedical applications. Acta Biomater. 2019, 94, 44–63. [CrossRef]

61. Wu, X.; Xiao, T.; Luo, Z.; He, R.; Cao, Y.; Guo, Z.; Zhang, W.; Chen, Y. A micro-/nano-chip and quantum dots-based 3D cytosensor for quantitative analysis of circulating tumor cells. J. Nanobiotechnol. 2018, 16, 1–9. [CrossRef]

62. Ai, X.; Evans, E.W.; Dong, S.; Gillett, A.J.; Guo, H.; Chen, Y.; Hele, T.J.; Friend, R.H.; Li, F. Efficient radical-based light-emitting diodes with doublet emission. Nature 2018, 563, 536–540. [CrossRef]

63. Tamang, S.; Lincheneau, C.; Hermans, Y.; Jeong, S.; Reiss, P. Chemistry of InP nanocrystal syntheses. Chem. Mater. 2016, 28, 2491–2506. [CrossRef]

64. Park, Y.; Ryu, Y.M.; Wang, T.; Jung, Y.; Kim, S.; Hwang, S.; Park, J.; Bae, D.J.; Kim, J.; Moon, H.; et al. Colorectal Cancer Diagnosis Using Enzyme-Sensitive Ratiometric Fluorescence Dye and Antibody–Quantum Dot Conjugates for Multiplexed Detection. Adv. Funct. Mater. 2018, 28, 1703450. [CrossRef]

65. Carbary-Ganz, J.L.; Welge, W.A.; Barton, J.K.; Utzinger, U. In vivo molecular imaging of colorectal cancer using quantum dots targeted to vascular endothelial growth factor receptor 2 and optical coherence tomography/laser-induced fluorescence dual-modality imaging. J. Biomed. Opt. 2015, 20, 096015. [CrossRef]

66. Nosrati, H.; Salehiabar, M.; Fridoni, M.; Abdollahifar, M.A.; Manjili, H.K.; Davaran, S.; Danafar, H. New insight about biocompati- bility and biodegradability of iron oxide magnetic nanoparticles: Stereological and in vivo MRI monitor. Sci. Rep. 2019, 9, 7173. [CrossRef]

67. Prasad, M.; Lambe, U.P.; Brar, B.; Shah, I.; Manimegalai, J.; Ranjan, K.; Rao, R.; Kumar, S.; Mahant, S.; Khurana, S.K.; et al. Nanotherapeutics: An insight into healthcare and multi-dimensional applications in medical sector of the modern world. Biomed. Pharmacother. 2018, 97, 1521–1537. [CrossRef]

68. Ebrahimi, N.; Bi, C.; Cappelleri, D.J.; Ciuti, G.; Conn, A.T.; Faivre, D.; Habibi, N.; Hošovský, A.; Iacovacci, V.; Khalil, I.S.; et al. Magnetic actuation methods in bio/soft robotics. Adv. Funct. Mater. 2021, 31, 2005137. [CrossRef]

69. Huang, J.; Li, Y.; Orza, A.; Lu, Q.; Guo, P.; Wang, L.; Yang, L.; Mao, H. Magnetic nanoparticle facilitated drug delivery for cancer therapy with targeted and image-guided approaches. Adv. Funct. Mater. 2016, 26, 3818–3836. [CrossRef]

70. Kim, S.H.; Oh, S.N.; Choi, H.S.; Lee, H.S.; Jun, J.; Nam, Y.; Lee, S.H.; Lee, J.K.; Lee, H.G. USPIO enhanced lymph node MRI using 3D multi-echo GRE in a rabbit model. Contrast Media Mol. Imaging 2016, 11, 544–549. [CrossRef]

71. Wu, L.; Cao, Y.; Liao, C.; Huang, J.; Gao, F. Diagnostic performance of USPIO-enhanced MRI for lymph-node metastases in different body regions: A meta-analysis. Eur. J. Radiol. 2011, 80, 582–589. [CrossRef]

72. Kuo, C.Y.; Liu, T.Y.; Chan, T.Y.; Tsai, S.C.; Hardiansyah, A.; Huang, L.Y.; Yang, M.C.; Lu, R.H.; Jiang, J.K.; Yang, C.Y.; et al. Magnetically triggered nanovehicles for controlled drug release as a colorectal cancer therapy. Colloids Surf. B Biointerfaces 2016, 140, 567–573. [CrossRef]

73. Syu, W.J.; Huang, C.C.; Hsiao, J.K.; Lee, Y.C.; Huang, Y.T.; Venkatesan, P.; Lai, P.S. Co-precipitation synthesis of near-infrared iron oxide nanocrystals on magnetically targeted imaging and photothermal cancer therapy via photoablative protein denature. Nanotheranostics 2019, 3, 236. [CrossRef]

74. Patra, J.K.; Das, G.; Fraceto, L.F.; Campos, E.V.; del Pilar Rodriguez-Torres, M.; Acosta-Torres, L.S.; Diaz-Torres, L.A.; Grillo, R.; Swamy, M.K.; Sharma, S.; et al. Nano based drug delivery systems: Recent developments and future prospects. J. Nanobiotechnol. 2018, 16, 71. [CrossRef]

75. Muhammad, Q.; Jang, Y.; Kang, S.H.; Moon, J.; Kim, W.J.; Park, H. Modulation of immune responses with nanoparticles and reduction of their immunotoxicity. Biomater. Sci. 2020, 8, 1490–1501. [CrossRef]

76. Behzadi, S.; Serpooshan, V.; Tao, W.; Hamaly, M.A.; Alkawareek, M.Y.; Dreaden, E.C.; Brown, D.; Alkilany, A.M.; Farokhzad, O.C.; Mahmoudi, M. Cellular uptake of nanoparticles: Journey inside the cell. Chem. Soc. Rev. 2017, 46, 4218–4244. [CrossRef]

77. Tee, J.K.; Yip, L.X.; Tan, E.S.; Santitewagun, S.; Prasath, A.; Ke, P.C.; Ho, H.K.; Leong, D.T. Nanoparticles' interactions with vasculature in diseases. Chem. Soc. Rev. 2019, 48, 5381–5407. [CrossRef]

78. Gupta, R.; Rai, B. Effect of size and surface charge of gold nanoparticles on their skin permeability: A molecular dynamics study. Sci. Rep. 2017, 7, 1–13. [CrossRef]

79. Fang, R.H.; Kroll, A.V.; Gao, W.; Zhang, L. Cell membrane coating nanotechnology. Adv. Mater. 2018, 30, 1706759. [CrossRef]

80. Tang, C.; He, Z.; Liu, H.; Xu, Y.; Huang, H.; Yang, G.; Xiao, Z.; Li, S.; Liu, H.; Deng, Y.; et al. Application of magnetic nanoparticles in nucleic acid detection. J. Nanobiotechnol. 2020, 18, 62. [CrossRef]

81. Singh, V.K.; Arora, D.; Ansari, M.I.; Sharma, P.K. Phytochemicals based chemopreventive and chemotherapeutic strategies and modern technologies to overcome limitations for better clinical applications. Phytother. Res. 2019, 33, 3064–3089. [CrossRef]

82. Ahmad, R.; Srivastava, S.; Ghosh, S.; Khare, S.K. Phytochemical delivery through nanocarriers: A review. Colloids Surf. B Biointerfaces 2021, 197, 111389. [CrossRef]

83. Allen, C.T.; Ricker, J.L.; Chen, Z.; Van Waes, C. Role of activated nuclear factor-κB in the pathogenesis and therapy of squamous cell carcinoma of the head and neck. Head Neck J. Sci. Spec. Head Neck 2007, 29, 959–971. [CrossRef]

84. Wong, K.E.; Ngai, S.C.; Chan, K.G.; Lee, L.H.; Goh, B.H.; Chuah, L.H. Curcumin nanoformulations for colorectal cancer: A review. Front. Pharmacol. 2019, 10, 152. [CrossRef]

85. Pistollato, F.; Giampieri, F.; Battino, M. The use of plant-derived bioactive compounds to target cancer stem cells and modulate tumor microenvironment. Food Chem. Toxicol. 2015, 75, 58–70. [CrossRef]

86. Liczbin´ski, P.; Michałowicz, J.; Bukowska, B. Molecular mechanism of curcumin action in signaling pathways: Review of the latest research. Phytother. Res. 2020, 34, 1992–2005. [CrossRef]

87. Canistro, D.; Chiavaroli, A.; Cicia, D.; Cimino, F.; Curro, D.; Dell'Agli, M.; Ferrante, C.; Giovannelli, L.; Leone, S.; Martinelli, G.; et al. The pharmacological basis of the curcumin nutraceutical uses: An update. PharmAdvances 2021, 3, 421–466. [CrossRef]

88. Zhang, I.; Cui, Y.; Amiri, A.; Ding, Y.; Campbell, R.E.; Maysinger, D. Pharmacological inhibition of lipid droplet formation enhances the effectiveness of curcumin in glioblastoma. Eur. J. Pharm. Biopharm. 2016, 100, 66–76. [CrossRef]

89. Che, X.H.; Chen, C.L.; Ye, X.L.; Weng, G.B.; Guo, X.Z.; Yu, W.Y.; Tao, J.; Chen, Y.C.; Chen, X. Dual inhibition of COX-2/5-LOX blocks colon cancer proliferation, migration and invasion in vitro. Oncol. Rep. 2016, 35, 1680–1688. [CrossRef]

90. Tuncer, S.; Banerjee, S. Eicosanoid pathway in colorectal cancer: Recent updates. World J. Gastroenterol. 2015, 21, 11748. [CrossRef]

91. Öhd, J.F.; Nielsen, C.K.; Campbell, J.; Landberg, G.; Löfberg, H.; Sjölander, A. Expression of the leukotriene D4 receptor CysLT1, COX-2, and other cell survival factors in colorectal adenocarcinomas. Gastroenterology 2003, 124, 57–70. [CrossRef]

92. Raveendran, R.; Bhuvaneshwar, G.S.; Sharma, C.P. In vitro cytotoxicity and cellular uptake of curcumin-loaded Pluronic/ Polycaprolactone micelles in colorectal adenocarcinoma cells. J. Biomater. Appl. 2013, 27, 811–827. [CrossRef]

93. Basli, A.; Belkacem, N.; Amrani, I. Health Benefits of Phenolic Compounds Against Cancers. In Phenolic Compounds—Biological Activity; IntechOpen: London, UK, 2017; pp. 193–210.

94. Yang, T.; Wang, L.; Zhu, M.; Zhang, L.; Yan, L. Properties and molecular mechanisms of resveratrol: A review. Pharm.-Int. J. Pharm. Sci. 2015, 70, 501–506.

95. Pashirzad, M.; Johnston, T.P.; Sahebkar, A. Therapeutic Effects of Polyphenols on the Treatment of Colorectal Cancer by Regulating Wnt β-Catenin Signaling Pathway. J. Oncol. 2021, 2021, 3619510. [CrossRef]

96. Azer, S.A. Overview of molecular pathways in inflammatory bowel disease associated with colorectal cancer development. Eur. J. Gastroenterol. Hepatol. 2013, 25, 271–281. [CrossRef]

97. Raghav, N.; Sharma, M.R. Ginger (Zingiber officinale Roscoe): A mini-review of constituents and biological activities. Chem. Biol. Interface 2021, 11, 81–93.

98. Bigagli, E.; Cinci, L.; Paccosi, S.; Parenti, A.; D'Ambrosio, M.; Luceri, C. Nutritionally relevant concentrations of resveratrol and hydroxytyrosol mitigate oxidative burst of human granulocytes and monocytes and the production of pro-inflammatory mediators in LPS-stimulated RAW 264.7 macrophages. Int. Immunopharmacol. 2017, 43, 147–155. [CrossRef]

99. Soo, E.; Thakur, S.; Qu, Z.; Jambhrunkar, S.; Parekh, H.S.; Popat, A. Enhancing delivery and cytotoxicity of resveratrol through a dual nanoencapsulation approach. J. Colloid Interface Sci. 2016, 462, 368–374. [CrossRef]

100. Summerlin, N.; Qu, Z.; Pujara, N.; Sheng, Y.; Jambhrunkar, S.; McGuckin, M.; Popat, A. Colloidal mesoporous silica nanoparticles enhance the biological activity of resveratrol. Colloids Surf. B Biointerfaces 2016, 144, 1–7. [CrossRef]

101. Schirrmacher, V. From chemotherapy to biological therapy: A review of novel concepts to reduce the side effects of systemic cancer treatment. Int. J. Oncol. 2019, 54, 407–419.

102. Zhang, M.; Zang, X.; Wang, M.; Li, Z.; Qiao, M.; Hu, H.; Chen, D. Exosome-based nanocarriers as bio-inspired and versatile vehicles for drug delivery: Recent advances and challenges. J. Mater. Chem. B 2019, 7, 2421–2433. [CrossRef]

103. Vodenkova, S.; Buchler, T.; Cervena, K.; Veskrnova, V.; Vodicka, P.; Vymetalkova, V. 5-fluorouracil and other fluoropyrimidines in colorectal cancer: Past, present and future. Pharmacol. Ther. 2020, 206, 107447. [CrossRef]

104. Sara, J.D.; Kaur, J.; Khodadadi, R.; Rehman, M.; Lobo, R.; Chakrabarti, S.; Herrmann, J.; Lerman, A.; Grothey, A. 5-fluorouracil and cardiotoxicity: A review. Ther. Adv. Med. Oncol. 2018, 10, 1–18. [CrossRef]

105. Che, J.; Pan, L.; Yang, X.; Liu, Z.; Huang, L.; Wen, C.; Lin, A.; Liu, H. Thymidine phosphorylase expression and prognosis in colorectal cancer treated with 5-fluorouracil-based chemotherapy: A meta-analysis. Mol. Clin. Oncol. 2017, 7, 943–952. [CrossRef]

106. Huehls, A.M.; Huntoon, C.J.; Joshi, P.M.; Baehr, C.A.; Wagner, J.M.; Wang, X.; Lee, M.Y.; Karnitz, L.M. Genomically incorporated 5-fluorouracil that escapes UNG-initiated base excision repair blocks DNA replication and activates homologous recombination. Mol. Pharmacol. 2016, 89, 53–62. [CrossRef]

107. Reynolds, I.S.; O'Connell, E.; Fichtner, M.; McNamara, D.A.; Kay, E.W.; Prehn, J.H.; Furney, S.J.; Burke, J.P. Mucinous adenocarci- noma is a pharmacogenomically distinct subtype of colorectal cancer. Pharm. J. 2020, 20, 524–532. [CrossRef]

108. Mizumoto, Y.; Yokoyama, S.; Matsuda, K.; Iwamoto, H.; Mitani, Y.; Tamura, K.; Nakamura, Y.; Murakami, D.; Oka, M.; Kobayashi, Y.; et al. Modulation of capecitabine administration to improve continuity of adjuvant chemotherapy for patients with colorectal cancer: A phase II study. Mol. Clin. Oncol. 2020, 12, 126–133. [CrossRef]

109. Tolan, D.; Almotairy, A.R.; Howe, O.; Devereux, M.; Montagner, D.; Erxleben, A. Cytotoxicity and ROS production of novel Pt (IV) oxaliplatin derivatives with indole propionic acid. Inorg. Chim. Acta 2019, 492, 262–267. [CrossRef]

110. Cirri, D.; Bartoli, F.; Pratesi, A.; Baglini, E.; Barresi, E.; Marzo, T. Strategies for the Improvement of Metal-Based Chemotherapeutic Treatments. Biomedicines 2021, 9, 504. [CrossRef]

111. Ma, L.; Li, X.; Zhao, X.; Sun, H.; Kong, F.; Li, Y.; Sui, Y.; Xu, F. Oxaliplatin promotes siMAD2L2 induced apoptosis in colon cancer cells. Mol. Med. Rep. 2021, 24, 1–9. [CrossRef]

112. Ulbrich, K.; Hola, K.; Subr, V.; Bakandritsos, A.; Tucek, J.; Zboril, R. Targeted drug delivery with polymers and magnetic nanoparticles: Covalent and noncovalent approaches, release control, and clinical studies. Chem. Rev. 2016, 116, 5338–5431. [CrossRef]

113. Kandasamy, P.; Gyimesi, G.; Kanai, Y.; Hediger, M.A. Amino acid transporters revisited: New views in health and disease. Trends Biochem. Sci. 2018, 43, 752–789. [CrossRef]

114. Limagne, E.; Thibaudin, M.; Nuttin, L.; Spill, A.; Derangère, V.; Fumet, J.D.; Amellal, N.; Peranzoni, E.; Cattan, V.; Ghiringhelli, F. Trifluridine/tipiracil plus oxaliplatin improves PD-1 blockade in colorectal cancer by inducing immunogenic cell death and depleting macrophages. Cancer Immunol. Res. 2019, 7, 1958–1969. [CrossRef]

115. Wilson, T.A. Second Cures Draft Retains Proposals To Modernize Clinical Trials. Inside Health Policy Com's FDA Week 2015, 21, 1–9.

116. Hall, G.L. Patient-Centered Clinical Care for African Americans: A Concise, Evidence-Based Guide to Important Differences and Better Outcomes; *Springer Nature: Cham, Switzerland*, 2019.

117. Singer, P.; Blaser, A.R.; Berger, M.M.; Alhazzani, W.; Calder, P.C.; Casaer, M.P.; Hiesmayr, M.; Mayer, K.; Montejo, J.C.; Pichard, C.; et al. ESPEN guideline on clinical nutrition in the intensive care unit. Clin. *Nutr.* 2019, 38, 48–79. [CrossRef]

118. Sanz-Garcia, E.; Grasselli, J.; Argiles, G.; Elez, M.E.; Tabernero, J. Current and advancing treatments for metastatic colorectal cancer. Expert Opin. Biol. Ther. 2016, 16, 93–110. [CrossRef]

04 GOLD NANOPARTICLES LOADED WITH CIPROFLOXACIN FOR COMBATING ANTIMICROBIAL RESISTANCE: AN IN VIVO ASSESSMENT

1. INTRODUCTION: AN OVERVIEW

In the relentless pursuit of innovative strategies to address the escalating global challenge of antimicrobial resistance (AMR), the integration of nanotechnology into pharmaceutical advancements has emerged as a promising frontier. Among these advancements, the utilization of gold nanoparticles as carriers for potent antibiotics, such as ciprofloxacin, represents a novel approach in the battle against AMR. This synthesis of nanotechnology and antibiotic therapy not only enhances drug delivery efficiency but also holds the potential to overcome the resistance mechanisms that threaten the effectiveness of conventional treatments. In this context, the exploration of Ciprofloxacin-Loaded Gold Nanoparticles stands as a beacon of hope, offering a unique and targeted solution to combat the rising tide of antimicrobial resistance. This research endeavours to unravel the multifaceted implications and potential breakthroughs associated with this cutting-edge amalgamation in the ongoing fight against infectious diseases.

E. faecalis has become one of the most daunting bacteria among the opportunistic pathogens [1]. This bacterium is responsible for a major percentage of nosocomial infections. It can cause infections in the gastrointestinal, respiratory, and urinary tracts; root canal; etc. Its characteristic features include extraordinary survival capacities in extreme conditions and development of resistance against antibiotics [2–4] such as β-lactam antibiotics and vancomycin [5]. Hence, innovative and effective antibiotics are required to tackle drug resistance mediated by E. faecalis infections.

The development of antimicrobial resistance (AMR) is one of the most pressing global issues, and great efforts have been committed to explore new and advanced antibiotics. However, extensive and impeded regulatory approval has shifted the focus to improve the efficacy of existing blockbuster drugs [6]. Advances in nanotechnology offer exciting methods to provide novel materials that can support penetration into biofilms, promote bacteria entry of the drugs, and present synergism with their own functional attributes [7–9]. Metallic nanoparticles (NPs), such as gold nanoparticles (AuNPs), are gaining recognition as improved drug delivery vehicles. They have shown effectiveness for various medical illnesses, including cancer, chronic diseases, and antibacterial, antiviral, and antifungal infections [10]. AuNPs do not exert (inherent) antibacterial activity; however, they enhance antibacterial properties of the loaded antibacterial drugs [11]. Drug-loaded AuNPs collapse bacteria membrane potential, hinder ATPase activity, and halt binding of ribosomes to tRNA leading to deteriorated bacterial cell metabolism [12]. Moreover, their site directed activity can target biofilms and quorum sensing [13,14]. The stability and the activity of such AuNPs under physiological conditions (such as pH, temperature, etc.) have also been evaluated in various studies, as such factors can modify their antibacterial activity [15,16]. Over the last years, various nanocarrier systems against *E. faecalis* have been reported such as ciprofloxacin and metronidazole encapsulated nanomatrix gels [17], bismuth nanoparticles [18], chitosan–propolis nanoparticles [19], doxycycline-functionalized polymeric nanoparticles [20], methylene blue and biogenic gold nanoparticles [21], silver nanoparticles and calcium hydroxide mixtures [22], biosynthesized silver nanoparticles [23], and graphene oxide nanoparticles [24].

Ciprofloxacin [1-cyclopropyl - 6 - fluoro - 1,4 - dihydro - 4 - oxo - 7 piperazinylquinolone - 3- carboxylic acid] (CIP) is a large spectrum fluoroquinolone antibiotic that is widely used for the treatment of numerous bacterial infections in joints, bones, skin, tooth, gastrointestinal, and urinary and respiratory tracts [25]. CIP-loaded gold nanoparticles (CIP-AuNPs) were reported by Tom et al. in 2004 [26]. Recently reported nanocarriers for CIP include CIP-loaded nanocomposites [27], PLGA–chitosan-based ciprofloxacin [28], Ciprofloxacin- containing polymer-based nanofibers (scaffolds) [29], and CIP-loaded single-walled carbon nanotubes [30]. This study aims to use ciprofloxacin protected gold nanoparticles to eradicate bacteria in the tissues of mice. The physical characterization of the nanoformulated version of this drug along with in vitro antibacterial activity, against E. faecalis, were carried out. For in vivo studies, an infectious mouse model was developed for E. faecalis and anti-colonization through the use of CIP-AuNPs was determined.

2. MATERIALS AND METHODS

2.1 Bacterial Strains

The Gram-positive bacteria strain used was *E. faecalis* JH2-2 (derived from the parental strain JH2). GM17 medium was used for their growth at 37 □C [31].

2.2 Preparation of AuNPs and CIP-AuNPs

The AuNPs were prepared by a previously reported method [26]. Trisodium citrate (0.5 mM) and chloroauric acid (0.5 mM) (MERK, Munich, Germany) were fluxed together. Citrate acted as a stabilizer as well as a reducing agent. The particle formation was confirmed upon achieving a red wine color. For the loading of CIP on AuNPs, 20 mL of the above solution was mixed with 5 mL of CIP (0.5, 1.0, 1.5, 2, and 2.5 mM) at pH 6.5. Then, the solution was stirred until the red colour turned to blue-purple.

2.3 Characterization of AuNPs and CIP-AuNPs

The CIP-AuNPs were characterized by UV–Vis spectro-photometry using a UV-2800 (BMS Biotechnology Medical Services, Madrid, Spain) spectrophotometer. Zeta potential was recorded using a Malvern Zeta sizer (Malvern, UK). Scanning electron microscopy (SEM) and Energy-dispersive X-ray spectroscopy (EDS) analyses were carried out using a SEM VEG 3 LMU (Tescan, Czech Republic), while Fourier-Transform Infrared Spectroscopy (FTIR) analyses were carried out using a Bruker FTIR Spectrometer ALPHA II (Westborough, MA, USA).

2.4. Drug Loading Capacity and Encapsulation Efficiency

CIP loading capacity and encapsulation efficiency by the AuNPs were evaluated using Equations (1) and (2), respectively.

$$\text{Loading capacity (\%)} = \frac{\text{Weight of CIP in CIP} - \text{AuNPs}}{\text{Weight of CIP} - \text{AuNPs}} \times 100 \quad (1)$$

$$\text{Encapsulation Efficiency (\%)} = \frac{\text{Total CIP added} - \text{Free CIP}}{\text{Total CIP added}} \times 100 \quad (2)$$

2.5. Drug Release Efficiency

The in vitro release of CIP from the CIP-AuNPs formulated with various concentrations of CIP was studied over time using a UV–Vis spectrophotometer at 280 nm from 0 to 24 h (every 2 h) by adding 20 mL of a 20-mM PBS buffer to 20 mL of the CIP-AuNPs solutions. The cumulative drug release was calculated using Equation (3).

$$\text{Cumulative drug release (\%)} = \frac{\text{CIP released from the CIP} - \text{AuNPs at t}}{\text{The total amount of CIP loaded onto the AuNPs.}} \quad (3)$$

The total amount of CIP loaded onto the AuNPs.

To determine the quantity of the drug present at the absorption site, a significant predictor T60%, was calculated as the time taken to release 60% of the drug. For instance, Stineman interpolation using Minitab 17 software was applied. The T60% drug release data were fit into the Korsmeyer-Peppas model for non-swellable matrices (Section S1.3).

2.6 Kinetic Analysis of the Drug Release

To determine the drug release from the CIP-AuNPs, numerous mathematical models (zero-order, first-order, and Higuchi's model) were employed (Section S1.3) [27,32].

2.7 In Vitro Stability of CIP-AuNPs

The effect of temperature on CIP-AuNPs was found by heating CIP-AuNPs at different temperatures (25 °C, 50 °C, 75 °C, and 100 °C) for 30 min. The effect of pH values (4, 7, and 10) and different salt concentrations (0.05, 0.1, 0.5, and 1 M) on CIP-AuNPs were also determined. For salt concentration, 10 mL of CIP-AuNPs were centrifuged at 10,000× g for 10 min. The resulting pellets were suspended in NaCl solutions (50 mM–1 M) at 37 °C for 24 h.

2.8 In Vitro Antibacterial Potential of CIP-AuNPs

The Minimum Inhibitory Concentration (MIC) of free CIP was calculated using differ- ent concentrations of CIP and CIP-AuNPs (0.1–10 mg/mL) against *E. faecalis*. The bacterial cultures at exponential phase of OD600 were harvested and counted for 106 CFU/mL using the standard dilution method. Using a standard inoculum of 106 CFUs, cultures were incubated at standard conditions for 24 h. The concentrations with a 50% reduction in bacterial count were observed as MIC. With an inoculum of approximately 106 CFU/mL, 50 µL of a 10-fold-diluted culture were placed on M-17 agar plates for measuring the viable cells. Colonies were counted after 24 h [33,34]. Moreover, a zone of inhibition test was also conducted by using the disc diffusion method to compare the antibacterial activity of CIP-AuNPs and free CIP. An *E. faecalis* culture was developed on the nutrient agar plate. A zone of inhibition (in mm) was then measured for *E. faecalis* using the Kirby–Bauer method. Sterile Whatman filter paper discs were impregnated with CIP, AuNPs, and CIP-AuNPs at a concentration of 10 µg/disc each.

2.9 Hemolytic Activity of CIP-AuNPs

A hemolysis test was carried out to compare the hemolytic activity of different concentrations (i.e., 10, 25, 50, and 100 µg/mL) of CIP-AuNPs (2 mM CIP), AuNPs, and CIP. Blood samples were taken from healthy female donors. Red blood cells were incubated for 4 h using the method described by Zarmina et al. [35]. As a positive control, Triton X-100 (0.5%) was used, while PBS was used as a negative control. The absorbance was measured at a wavelength of 550 nm.

2.10 Colonization of *E. faecalis* in BALB/c Mice

The in vivo investigation was conducted using female BALB/c mice (eight weeks old, weighing 25–30 g; n = 15) purchased from the National Institute of Health (NIH), Islamabad Pakistan. They were kept at 25 ± 2 °C and presented with a natural light–dark cycle (10–14 h). The mice were given autoclaved tap water and a normal diet ad libitum. For bacterial colonization in hepatic and renal tissues, a well-established intravenous (IV) infection model was used [36]. The GM17 broth at a temperature of 37 °C was used and the preculture was grown overnight. A total of 100 µL of preculture was added to brain heart infused (BHI) medium supplemented with 40% filter-sterilized serum. Phosphate buffer saline (PBS), pH 7.4, was then used to wash the subsequent pellets from the cultures, optimized by colony counting for the number of cells. The bacterial pallets were then suspended in sterile PBS. A total of 100 µL of the suspension adjusted for 1 × 109 cells/mL of bacterial suspensions were (tail vein) injected into each of the female mice (n = 15).

2.11 In Vivo Antibacterial Activity of CIP-AuNPs

To assess the in vivo antibacterial activity of CIP-AuNPs and CIP, they were suspended or dissolved in PBS buffer. The infected group (n = 15) was treated with CIP-AuNPs (500 µg/Kg, n = 5) and with free CIP (10 mg/Kg, n = 5); the remaining five mice remained untreated. CIP-AuNPs and free CIP were delivered by the tail vein once a day for eight days starting from the seventh day of infection until the day of the challenge. After a week of treatment, all mice were sacrificed, and their liver and kidneys were removed to measure the viable bacterial count through the colony-forming unit (CFU) method. For instance, organs were weighed and homogenized in 10 mL of a PBS solution. 10-fold dilutions of the homogenate were placed on the agar plate. CFUs were counted after 24 h.

2.12 Statistical Analysis

Statistical analysis (such as the mean, standard deviation, and multiple group comparison analysis by one-way ANOVA) was carried out using Origin 2021 (OriginLab Corporation, Northampton, MA, USA) and Graph Pad Prism 9.2.

3. RESULTS

3.1 Synthesis of AuNPS and CIP-AuNPs

The synthesis of AuNPs was accomplished by a chemical reduction method. The red colour confirmed the formation of AuNPs upon the addition of citrate to chloroauric acid. The UV–Vis graph shows the absorbance peak of gold nanoparticles at 521 nm (Figure 1). Further, the addition of CIP into the AuNP solution resulted in a purple to bluish solution, indicating CIP adsorption onto the AuNPs. UV–Vis spectroscopy verified the adsorption of various CIP concentrations on AuNPs, as presented in Figure 2. The transition of the solution colour took place upon adding different CIP concentrations, typically from red to purple to bluish, and finally, to dark blue (Figure 1).

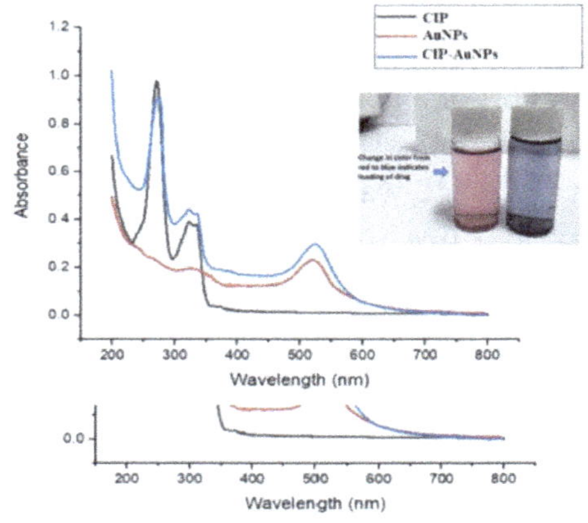

Figure 1. UV–Vis spectra of CIP, AuNPs, and CIP-AuNPs.

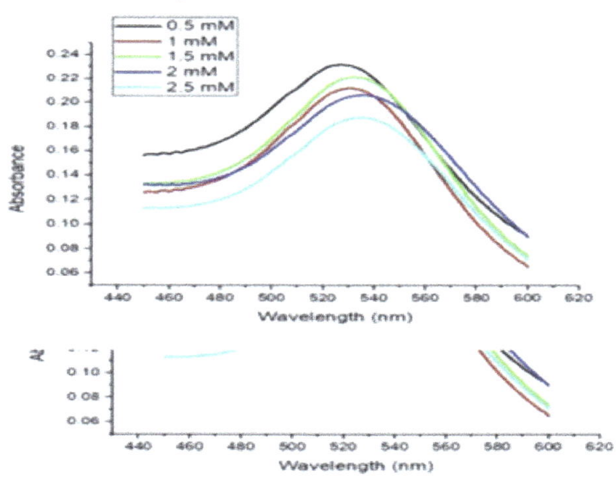

Figure 2. UV–Vis spectra of CIP-AuNPs loaded with various CIP concentrations.

3.2. CIP Encapsulation Efficiency and CIP Loading Capacity

Encapsulation Efficiency and CIP Loading Capacity CIP encapsulation efficiency was the highest (60.83%) at the highest CIP concentration used (2.5 mM); conversely, CIP encapsulation efficiency into AuNPs was the lowest (24.43%) at the lowest CIP concentration used (0.5 mM) (Table 1). Consistently, the drug loading capacity was also drug-concentration-dependent. Indeed, CIP loading capacity into AuNPs was the highest (34.54%) at the highest CIP concentration used (2.5 mM); conversely, CIP loading.

Table 1. Encapsulation efficiency (%) and loading capacity (%) of CIP onto the AuNPs at varying CIP.

0.5 mM	24.43	8.5
1.0 mM	29.30	15.60
1.5 mM	30.65	28.85
2.0 mM	48.92	33.81
2.5 mM	60.83	34.54

3.3. Particle Size and Zeta Potential of AuNPs and CIP-AuNPs

The average particle size (PS) of AuNPs was found to be 23 nm. The PS for the different CIP-AuNPs are mentioned in Table 2. Upon addition of CIP (0.5 mM and 1 mM), the size of the NPs remained almost the same at 24 nm. The PS increased to 41, 88, and 128 nm using 1.5, 2.0, and 2.5 mM CIP, respectively.

Table 2. Zeta potential values for CIP-AuNPs, AuNPs, and CIP.

0.5 Mm	24.43	0.26	6.21	-32.1
1.0 mM	24.09	0.301	6.044	-33.3
1.5 mM	41	068	10.21	-19.7
2.0 mM	88.2	1.000	57.4	-13.4
2.5 mM	128.2	0.48	79.18	-2.12

The zeta size and charge values of the CIP-AuNPs and AuNPs are mentioned in Table 2. The AuNP had a negative charge of -32.1 mV, which remained unchanged upon the addition of 0.5 mM CIP. However, the zeta potential (ZP) and PDI values of the CIPAuNPs at a 1.5 mM CIP concentration were -19.7 ± 6.65 mV and 0.680, respectively.

3.4. Surface Morphology and Elemental Chemical Composition of AuNPs by SEM–EDS

The surface morphology of the chemically prepared NPs (AuNPs and CIP-AuNPs) was analyzed by SEM at 10 kV. SEM images revealed spherically-shaped AuNPs (Figure 3a), 2 mM CIP-AuNPs (Figure 3b), and 2.5 mM CIP- uNPs (Figure 3c). The SEM analysis further presented a high polydispersity of 2.5 mM CIP-AuNPs.

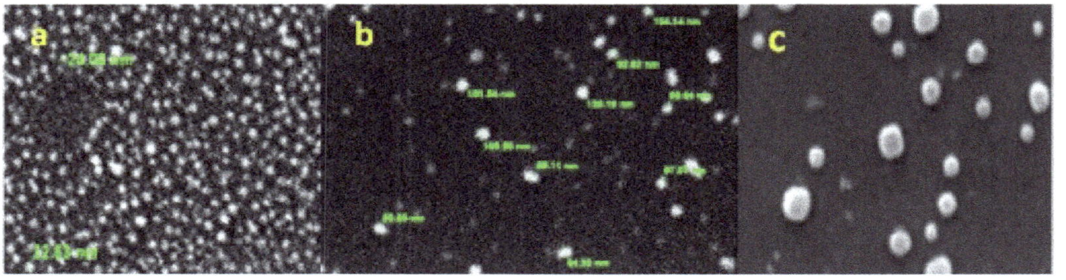

Figure 3. SEM micrographs of (a) AuNPs, (b) AuNPs (2 mM), and (c) CIP-AuNPs (2.5 mM).

3.5. Structural Analysis of CIP-AuNPs by FTIR Spectroscopy

The FTIR of CIP-AuNPs, AuNPs, and free CIP are presented in Figure 4. FTIR results showed the successful adsorption of CIP in AuNPs. The stretching of the N–H bond of the imino moiety on the piperazine group of the CIP was shown by the band at 3410 cm^{-1}. Absorption bands at 1655 cm^{-1} and 1070 cm^{-1} represent a primary amine (N–H) bend of the pyridone moiety and the C–functional group of the CIP, respectively [30]. The peak at 1634.84 cm^{-1} corresponded to C = O symmetric stretching, and the band at 1381.02 cm^{-1} was assigned to the C–H stretching vibration in the AuNPs. The CIP-AuNPs exhibited bands at 3341 cm^{-1}, which corresponded to the strong stretching vibration of the O–H group of alcohols and phenols, while the band at 2062 cm^{-1} referred to strong C–H stretching [33]. The stretching bands at 3410 cm^{-1} of the N–H bond of the imino moiety on the piperazine group of the CIP moved cm^{-1} the CIP-AuNPs, showing adsorption of CIP on AuNPs. It has been reported that the nitrogen atom of the NH moiety of the piperazine group of CIP binds to the AuNP surface to form CIP-AuNPs. The binding of CIP to the surface of AuNPs was confirmed by the fact that the corresponding NH2 peaks were broadened and shifted to higher wavelengths (1655 cm^{-1} to 1667 cm^{-1}). The broad stretching band of O–H was due to water traces in the CIP-AuNP and AuNP samples. samples. The weak bands around 2900 cm-1 in CIP-AuNPs are explained by traces of citrate.

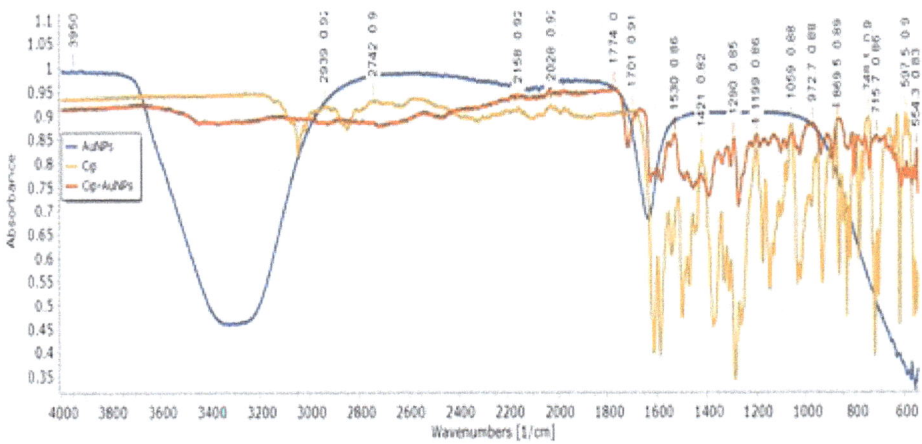

Figure 4. FTIR spectra of CIP-AuNPs, AuNPs, and free CIP.

3.6. Kinetics of CIP Drug Release from AuNPs

The drug release curve is represented by a normal biphasic drug release pattern, which can The drug release curve is represented by a normal biphasic drug release pattern, be split into two stages (Figure 5). In the initial stage (0–5 h), a rapid release of CIP which can be split into two stages (Figure 5). In the initial stage (0–5 h), a rapid release of occurred from all the formulations. This confirmed the adsorption of CIP on the surface CIP occurred from all the formulations. This confirmed the adsorption of CIP on the of AuNPs, in which the CIP-AuNPs displayed a high surface-to-volume ratio due to their surface of AuNPs, in which the CIP-AuNPs displayed a high surface-to-volume ratio due small size, which promoted burst release [34]. The second stage (5–24 h) was when the to their small size, which promoted burst release [34].

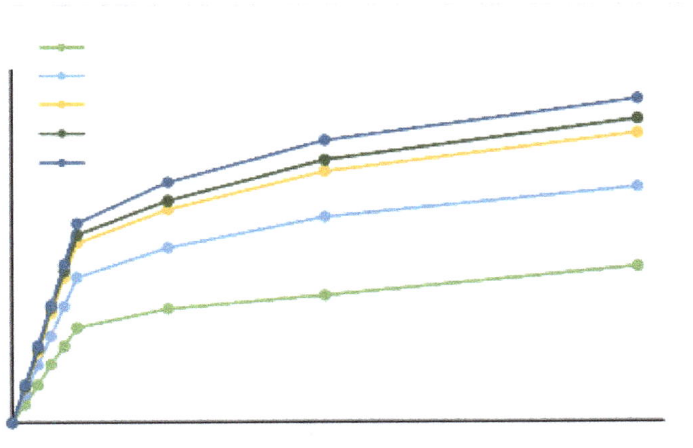

Figure 5. In vitro CIP release from AuNPs loaded with various CIP concentrations.

A first-order release was observed for CIP release from the CIP-AuNPs, which demon- strated that CIP release from CIP-AuNPs (2 mM) was concentration-dependent. Overall, the release studies demonstrated that CIP- AuNPs loaded with >1 mM CIP are suitable nano-formulations for CIP administration in a sustained manner. The initial burst release will act as a loading dose to suppress the disease spread, while the sustained release phase will contribute towards a better therapeutic effect.

3.7 Stability Tests on CIP-AuNPs

The CIP-AuNPs with targeted drug concentration (2 mM) showed stability at room temperature and in a basic medium. They were unstable at different salt concentrations. The colour change from blue to dark blue was observed with an increase in temperature for the CIP-AuNPs, and 2.0 and 2.5 mM of CIP, which is attributed to the increase in particle size (PS) due to aggregation. CIP-AuNPs (2 mM) were stable at room temperature (25 °C), and the Amax increases on the order of 25 °C < 50 °C < 75 °C < 100 °C. With different pH values and increasing salt concentrations, irregular trends in Amax (adsorption peak of CIP, 280 nm) of CIP-AuNPs in all formulations were observed. For further analysis, the CIP-AuNPs (2 mM) formulation was selected.

3.8 In Vitro Antibacterial Activity of CIP-AuNPs at the Optimized Dose

The MICs of free CIP and CIP-AuNPs were tested against *E. faecalis* JH2-2. For comparison of antibacterial activity of CIP and CIP-AuNPs (2 mM), the Minimum Inhibitory Concentration was obtained for our strain. The MIC of CIP and CIP-AuNPs were found to be 2 µg/mL and 1 µg/mL, respectively. The zone of inhibition using 10 µg/disc are presented in Figure 6. As such, a larger zone of inhibition of 23 mm was obtained for the Nanomaterials 2021, 11, x FOR PEER RCEVIPIE-AWuNPs compared with CIP (21 mm), whereas no zone of inhibition was obtain9edoffo1r5 the bare AuNPs.

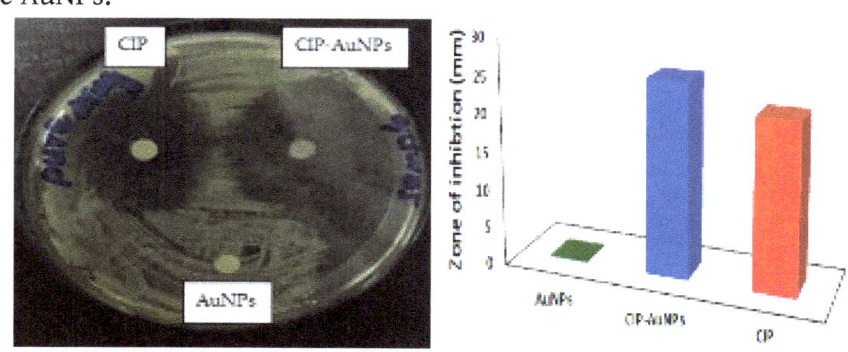

Figure 6. The zone of inhibition (ZI) displayed by CIP, AuNPs, and CIP-AuNPs for *E. faecalis*.

3.9. In Vivo Anticolonizing Potential of CIP-AuNPs in an Animal

The infected mouse liver and kidneys appeared to be inflamed and some lighter zones underneath the liver (Figure 7A) were observed. Since the strain was not biofilm-forming, no particular lesions were visualized in the infected liver Figure 7A(a). Mice treated with CIP and CIP-AuNPs had a normal liver appearance Figure 7A(b,c) respectively. The kidneys of infected, CIP treated and CIP-AuNPs treated mice presented no particular change in morphology or texture, Figure 7A(d–f) respectively. The average bacterial count (log_{10}CFU/gm organ) in the infected liver was 39.53± 1.93 (Figure 7B(a)). CIP and CIP-AuNPs treated mice harbored 29.049 ± 1.343 log_{10}CFU/mL and 28.4 ± 0.421 log_{10}CFU/mL ($p = 0.428$) respectively, in the liver. The bacterial load in the kidneys of infected mice was 13.98 ± 0.232 log_{10}CFU/mL (Figure 7B(b)) while kidneys of CIP and CIP-AuNPs treated mice presented significantly lower bacterial load, 7.91 ± 0.2132.76 ± 0.034 log_{10}CFU/mL (p < 0.0D0076), respectively. The in vivo data revealed that the mice treated with CIP-AuNPs significantly reduced bacterial colonization in the kidneys (Figure 7B,D) compared with those of CIP-treated mice. The CFU for the 10-fold serial dilutions in the infected, CIP treated and CIP AuNPs treated liver are shown in Figure 7C(a–c) respectively. Same dilutions for infected, CIP treated and CIP AuNPs treated kidneys are Nanomaterials 2021, 11, x FOR PEERs RhEVoIwEWn in Figure 7D(a–c) respectively

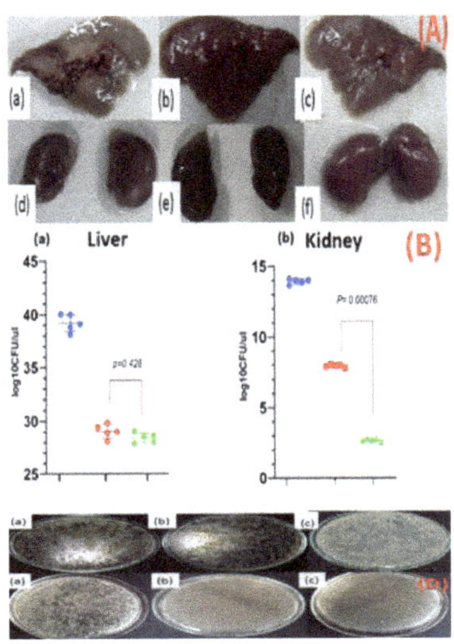

Figure 7. Effect of CIP (10 mg/Kg) and CIP-AuNPs (500 µg/Kg) on the colonization of *E. faecalis* in the (A) mouse organs—(a) infected liver, (b) CIP-treated liver, (c) CIP-AuNPs-treated liver, (d) infected kidney, (e) CIP-treated kidney, (f) CIP-AuNPs treated kidneys, infected kidneys, (B(a,b)) CFU *E. faecalis* in the infected liver and kidney; (C) CFU counting in the liver—(a) infected, (b) CIP-treated, and (c) CIP-AuNPs-treated; (D) CFU counting in the kidneys—(a) infected, (b) CIP-treated, and (c) CIP-

3.10. Hemolytic Activity of CIP-AuNPs

Hemolytic activity results of the CIP, AuNPs, and CIP-AuNPs are presented in Figure 8. According to ISO/TR 7406, the percentage hemolysis considered to be safe is <5%. These results present the greater hemolytic activity of CIP (6.4, 7.2 and 10%) compared with AuNPs and CIP-AuNPs (2 mM). Since the intravenous route gave an escape to the drug from first-pass metabolism, results indicate a hundred percent availability of the drug or nanoparticles in plasma. Considering the average mouse weight was 28 g, the average injected doses of CIP and CIP-AuNPs in the mice were 280 μg and 14 μg, respectively. The results indicated that CIP had hemolytic activity at concentrations of 20 μg/mL and above.

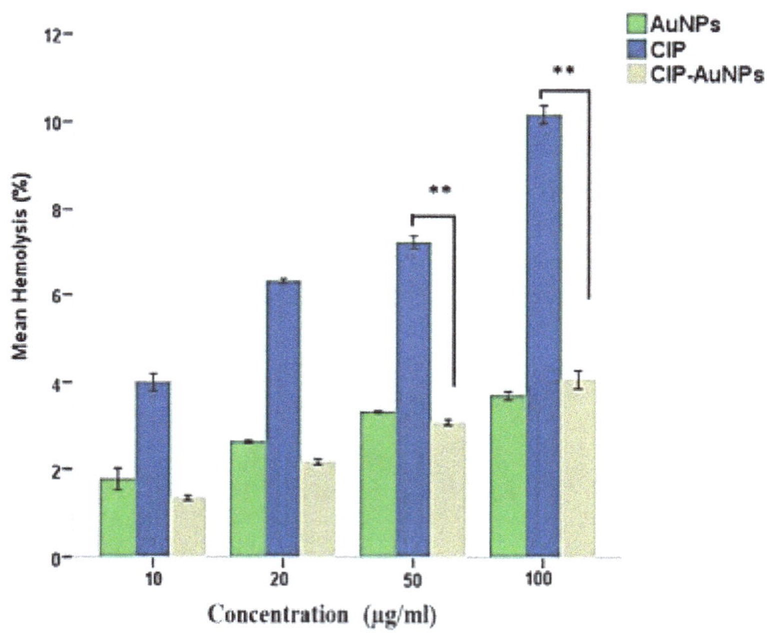

Figure 8. Hemolytic activity of CIP, AuNPs, and CIP-AuNPs, ** presents $p \leq 0.01$.

4. DISCUSSION

Ciprofloxacin is a second-generation fluoroquinolone that is active against many Gram-negative and Gram-positive bacteria. It acts through inhibition of bacterial DNA gyrase and topoisomerase IV. There has been no cross-resistance reported for CIP and other fluoroquinolones; therefore, it is of high clinical value. However, there are certain side effects associated with CIP including low blood sugar, headache, nerve damage causing numbness, tendon rupture, and joint pains. Gold nanoparticles are efficient drug carriers that have the ability to improve the antibacterial effects of loaded antibiotics as well as to reduce the amount of drug required to be effective because of their retention and penetration into bacterial biofilms and cell membranes at the infected sites.

This study reported the effectiveness of spherical CIP-AuNPs as an antibacterial platform against *E. faecalis* infection in the kidneys and liver of mice. The spherical CIP-AuNPs (Figures 1 and 3) were successfully prepared using a non-simple ionic interaction between CIP and negatively charged AuNPs, which maintained their therapeutic functions without chemical modification [28]. Drug adsorption or loading on the NP was determined by UV–Vis spectroscopy and FTIR In CIP-AuNPs, the zeta potential values changing from -32.1 mV to -19.7 mV and -13.4 mV indicated the adsorption of CIP onto AuNPs [37]. The negative charge on the AuNPs and the positive charge of the CIP [26] (at 6.5 pH) resulted in electrostatic interactions; hence, enhanced CIP encapsulation efficiency and loading capacity was observed. The addition of different concentrations of the drug to the nanoparticles changed the colour from red to purple to bluish purple, and finally, to blue. Increasing the concentration of CIP-AuNPs (>2 mM of CIP concentration) caused the zeta potential values to become less negative (Table 2). The zeta potential values ranging between -5 mV and 5 mV generally demonstrate fast aggregation [38]. The SPR spectra of CIP-AuNPs—2 mM (Figure 1) indicated an increase in the peak intensity of the AuNPs upon the addition of CIP. Moreover, there was a slight shift in the CIP peaks. These changes indicated loading of CIP onto the AuNPs. Considering the standard limits, the zeta potential values of CIP-AuNPs at a 2.5-mM CIP concentration showed fast aggregation. Optimization of the drug for loading is, thus, an important criterion for electro-statically loaded drugs. The encapsulation efficiency, which actually represents the amount of the drug that gets incorporated into a particle at a provided concentration, may help in this regard. The drug loading values present the percentage of drug in terms of weight of the nanoparticles. Considering this, we selected CIP-AuNPs (at a 2 mM CIP concentration) for the antibacterial and hemolysis studies.

The stability testing of the CIP-AuNPs illustrated a drastic shift in the stability of the CIP-AuNPs at 100 °C, which could be due to an increase in collision frequency between CIP-AuNPs. The activation energy produced from these collisions would be enough for CIP-AuNPs to react, which is fully depicted by the Arrhenius equation (Supplementary materials Figures S7–S12). From the observation of the effect of temperature on different CIP-AuNPs, the 2.5 mM ratio showed a change to dark blue. Thus broadening of the CIP-AuNPs peak at the 500 nm range shows particle aggregation. This is due to the fact that the particles destabilize causing a decrease in extinction peak intensity, resulting in a lower number of stable nanoparticles. In the case of pH, the CIP-AuNPs with 2 mM of CIP showed a peak shift from 537 nm to 521 nm at pH 7 and 10 (Figure S15). This is because of the fact that the CIP acquired a negative charge. This negative charge occurred due to deprotonation of the amine group, which then further led to drug unloading. The effect of salt showed the broadening of CIP-AuNPs peaks, this could be attributed to the fact of a very high molar concentrations of the salt used in this experiment. The submolar salt (100 mM and above) (Supplementary materials S17–S21) led to AuNP aggregation and drug unloading.

In the current study, the in vitro drug release was assessed by various kinetic models consisting of the zero-order, first-order, Higuchi, and Korsmeyer–Peppas models. A first- order release was seen for CIP release from the CIP-AuNPs. This shows that CIP release from CIP-AuNPs was directly dependent on the drug concentration. Overall, the release studies demonstrated the dependence of CIP-AuNPs on particle size. Smaller particles release drugs in a very slow fashion. The most suitable drug release kinetics was obtained by CIP-AuNPs (2 mM).

A higher level of antimicrobial activity is afforded by the CIP-AuNPs. This is due to the large surface-to-volume ratio of AuNPs, which facilitated the adsorption of CIP molecules onto the AuNP surfaces via electrostatic interaction between the amine groups of CIP and AuNPs. The MIC of the CIP-AuNPs was half to that of CIP. Considering that the overall particle seed was gold, the drug amount administered was much lower; thus, carrier driven therapies require less amount of the drug to achieve the same therapeutic outcome. It is well established that silver nanoparticles prepared via a chemical reduction procedure exhibited antibacterial activity. However, for the AuNPs in this study, no inherent antibacterial activity was obtained. This depicts that the surface-area-to-volume ratio of AuNPs served its purpose as a drug delivery system and supported the CIP approach to the target site, thus improving antibacterial activity. Studies revealed that both positively and negatively charged gold nanoparticles get localized in the bacterial membranes. For entry, the binding of negative charged particles to the bacteria's lipid bilayer results in local gelation. In contrast, the positively charged NPs get along the fluidity in the gelled regions of the bilayers [39–41]. Once CIP-AuNPs have reached the bacterial surface, they might have entered or fused with the cell membrane of E. faecalis and served as a reserve for the continuous release of CIP, which must have then diffused into the interior of the *E. faecalis*. It has also been reported that while acquiring resistance, *E. faecalis* has cationic membrane proteins on their surface, which alter the membrane fluidity and potential. At moderate levels of *E. faecalis* resistance, the cell surface becomes neutralized, while at further higher resistance levels, the cell surface acquires a positive charge. This positive charge on the membrane may further permit CIP-AuNPs to interact and enter the bacterial cell [39–41].

The in vivo antibacterial activity showed significant reduction in CFU by CIP-AuNPs compared with CIP. However, in the case of the liver, the difference was not significant but comparable to CIP; this could be attributed to the fact that the *E. faecalis* presented a very high hepatic load. Moreover, since the nanoparticles did not have a polymer coating such as polyethylene glycol, many of them can be uptake by the reticuloendothelial system and could have been taken for detoxification. In the case of the kidneys, the nanoparticles may have stayed or been entrapped at the glomerular filtration barrier (GFB), which is present within the glomerulus and sandwiched between the glomerular capillary and Bowman's space. Different layers of the GFB have pore sizes between 100–8 nm and, thus, act as a sieve for the nanoparticles. The nanoparticles, especially the anionic nanoparticles, have higher retention time in the kidneys and, in general, the nanoparticles disintegrate before passing through the GFB. The accumulation of nanoparticles with this size range could be the reason behind the efficiency of the significant decrease in bacterial count in the mice treated with CIP-AuNPs over mice treated with CIP.

Ciprofloxacin has also been associated with hemolytic anemia [42]. In addition to improved antibacterial activity, this study (Figure 7) also reported limited hemolytic activity of the CIP-AuNPs (2 mM) on human red blood cells (RBCs). These results presented the greater hemolytic activity of CIP (6.4, 7.2, and 10.6%) as compared to AuNPs and CIP-AuNPs. It has been reported that a positive charge on the particle's surface possesses a membrane-destabilizing and concomitantly destructive effect that results from the interaction of a positive charge and a negative charge of the cell membrane. Therefore, the limited hemolytic activity of the CIP-AuNPs (with a 2-mM CIP concentration) is attributable to the negative charge on its surface and the lack of free CIP in the medium. Considering the average mouse weight as 28 g, the average injected doses of CIP and CIP-AuNPs in the mice were 280 µg and 14 µg, respectively. The results indicated CIP to be hemolytic at a concentration of 20 µg/mL and above (Figure 8). Due to this, the mice experienced hyperactivity and irritation at the site of injection soon after the administration of CIP, and decreased eating after the second day of treatment.

Antibiotics, such as CIP, are an attraction because of their broad spectrum and lack of resistance development. *E. faecalis* has resistance acquisition capacity and this challenge can be sealed with CIP. Despite of the enormous work conducted on the development of CIP nanocarriers, there is a lack of studies investigating the effects of these nanoparticles in animal models. These studies are extremely important as bacteria residing in the tissue and forming biofilms might not be easily accessed by circulating nanoparticles and therefore they can obtain a colonization advantage. This in vivo study concluded showing efficiency of the CIP-loaded gold nanoparticles against intra-abdominal and urinary tract infections caused by *E. faecalis*. Further investigations for the biocompatibility and biofilm formation in the host tissues are necessary and will be vital for the further assessment of these novel nanoparticles.

5. CONCISE SUMMARY

Metallic nanoparticles, such as gold nanoparticles (AuNPs), have been extensively studied as drug delivery systems for various therapeutic applications. However, drug-loaded-AuNPs have been rarely explored in vivo for their effect on bacteria residing inside tissues. Ciprofloxacin (CIP) is a second-generation fluoroquinolone with a broad-spectrum of antibiotic properties devoid of developing bacteria resistance. This research is focused on the synthesis and physical characterization of Ciprofloxacin-loaded gold nanoparticles (CIP-AuNPs) and their effect on the colonization of *Enterococcus faecalis* in the liver and kidneys of mice. The successfully prepared CIP-AuNPs were stable and exerted enhanced in vitro antibacterial activity against *E. faecalis* compared with free CIP. The optimized CIP-AuNPs were administered (500 µg/Kg) once a day via tail vein to infected mice for eight days and were found to be effective in eradicating *E. faecalis* from the host tissues. Moreover, unlike CIP, CIP-AuNPs were non-hemolytic. In summary, this study demonstrated that CIP-AuNPs are promising and biocompatible alternative therapeutics for *E.-faecalis*-induced infections resistant to conventional drugs (e.g., beta-lactams and vancomycin) and should be further investigated.

REFERENCES

1. Wu, D.; Fan, W.; Kishen, A.; Gutmann, J.L.; Fan, B. Evaluation of the Antibacterial Efficacy of Silver Nanoparticles against Enterococcus faecalis Biofilm. J. Endod. 2014, 40, 285–290. [CrossRef]
2. Sedgley, C.; Buck, G.; Appelbe, O. Prevalence of Enterococcus faecalis at Multiple Oral Sites in Endodontic Patients Using Culture and PCR. J. Endod. 2006, 32, 104–109. [CrossRef] Saleh, I.M.; Ruyter, I.E.; Haapasalo, M.; Ørstavik, D. Survival of Enterococcus faecalis in infected dentinal tu-bules after root canal filling with different root canal sealers in vitro. Int. Endod. J. 2004, 37, 193–198. [CrossRef]
3. Liu, H.; Wei, X.; Ling, J.; Wang, W.; Huang, X. Biofilm Formation Capability of Enterococcus faecalis Cells in Starvation Phase and Its Susceptibility to Sodium Hypochlorite. J. Endod. 2010, 36, 630–635. [CrossRef] [PubMed]
4. Gilmore, M.S.; Clewell, D.B.; Ike, Y.; Shankar, N. Enterococci: From Commensals to Leading Causes of Drug Resistant Infection; Eye and Ear Infirmary: Boston, MA, USA, 2014.
5. Livermore, D.M.; Blaser, M.; Carrs, O.; Cassell, G.; Fishman, N.; Guidos, R.; Levy, S.; Powers, J.; Norrby, R.; Tillotson, G.; et al. Discovery research: The scientific challenge of finding new antibiotics. J. Antimicrob. Chemother. 2011, 66, 1941–1944. [CrossRef] [PubMed]

6. Pinto, R.; Lopes, D.; Martins, M.C.L.; Van Dijck, P.; Nunes, C.; Reis, S. Impact of nanosystems in Staphylococcus aureus biofilms treatment. FEMS Microbiol. Rev. 2019, 43, 622–641. [CrossRef] [PubMed]

7. Vassallo, A.; Silletti, M.F.; Faraone, I.; Milella, L. Nanoparticulate Antibiotic Systems as Antibacterial Agents and Antibiotic Delivery Platforms to Fight Infections. J. Nanomater. 2020, 2020. [CrossRef]

8. Pugazhendhi, A.; Kumar, S.S.; Manikandan, M.; Saravanan, M. Photocatalytic properties and antimicrobial ef-ficacy of Fe doped CuO nanoparticles against the pathogenic bacteria and fungi. Microb. Pathog. 2018, 122, 84–89. [CrossRef]

9. Uzair, B.; Liaqat, A.; Iqbal, H.; Menaa, B.; Razzaq, A.; Thiripuranathar, G.; Rana, N.F.; Menaa, F. Green and Cost-Effective Synthesis of Metallic Nanoparticles by Algae: Safe Methods for Translational Medicine. Bioengineering 2020, 7, 129. [CrossRef]

10. Wang, L.; Hu, C.; Shao, L. The antimicrobial activity of nanoparticles: Present situation and prospects for the future. Int. J. Nanomed. 2017, 12, 1227–1249. [CrossRef]

11. Dykman, L.; Khlebtsov, N. Gold nanoparticles in biomedical applications: Recent advances and perspectives. Chem. Soc. Rev. 2012, 41, 2256–2282. [CrossRef] [PubMed]

12. Abdel-Raouf, N.; Al-Enazi, N.M.; Ibraheem, I.B. Green biosynthesis of gold nanoparticles using Galaxaura elongata and characterization of their antibacterial activity. Arab. J. Chem. 2017, 10, S3029–S3039. [CrossRef]

13. Li, W.; Geng, X.; Liu, D.; Li, Z. Near-infrared light-enhanced protease-conjugated gold nanorods as a photo-thermal antimicrobial agent for elimination of exotoxin and biofilms. International journal of nanomedicine. Int. J. Nanomed. 2019, 14, 8047. [CrossRef] [PubMed]

14. Rhazouani, A.; Gamrani, H.; El Achaby, M.; Aziz, K.; Gebrati, L.; Uddin, S.; Aziz, F. Synthesis and Toxicity of Graphene Oxide Nanoparticles: A Literature Review of In vitro and In vivo Studies. BioMed Res. Int. 2021, 2021. [CrossRef] [PubMed]

15. Kaushik, S.N.; Scoffield, J.; Andukuri, A.; Alexander, G.C.; Walker, T.; Kim, S.; Choi, S.C.; Brott, B.C.; Eleazer, P.D.; Lee, J.-Y.; et al. Evaluation of ciprofloxacin and metronidazole encapsulated biomimetic nanomatrix gel on Enterococcus faecalis and Treponema denticola. Biomater. Res. 2015, 19, 9. [CrossRef]

16. Azad, A.; Rostamifar, S.; Modaresi, F.; Bazrafkan, A.; Rezaie, Z. Assessment of the Antibacterial Effects of Bismuth Nanoparticles against Enterococcus faecalis. BioMed Res. Int. 2020, 2020. [CrossRef]

17. Parolia, A.; Kumar, H.; Ramamurthy, S.; Davamani, F.; Pau, A. Effectiveness of chitosan-propolis nanoparticle against Enterococcus faecalis biofilms in the root canal. BMC Oral Heal. 2020, 20, 339. [CrossRef]

18. Arias-Moliz, M.T.; Baca, P.; Solana, C.; Toledano, M.; Medina-Castillo, A.L.; Toledano-Osorio, M.; Osorio, R. Doxycycline- functionalized polymeric nanoparticles inhibit Enterococcus faecalis biofilm formation on dentine. Int. Endod. J. 2021, 54, 413–426. [CrossRef]

19. Maliszewska, I.; Wróbel, J.; Wanarska, E.; Podhorodecki, A.; Matczyszyn, K. Synergistic effect of methylene blue and biogenic gold nanoparticles against Enterococcus faecalis. Photodiagnosis Photodyn. Ther. 2019, 27, 218–226. [CrossRef]

20. Balto, H.; Bukhary, S.; Al-Omran, O.; BaHammam, A.; Al-Mutairi, B. Combined Effect of a Mixture of Silver Nanoparticles and Calcium Hydroxide against Enterococcus faecalis Biofilm. J. Endod. 2020, 46, 1689–1694. [CrossRef]

21. Halkai, K.; Mudda, J.A.; Shivanna, V.; Rathod, V.; Halkai, R. Antibacterial efficacy of biosynthesized silver nanoparticles against Enterococcus faecalis Biofilm: An in vitro study. Contemp. Clin. Dent. 2018, 9, 237–241. [CrossRef] [PubMed]

22. Wu, S.; Liu, Y.; Zhang, H.; Lei, L. Nano-graphene oxide with antisense walR RNA inhibits the pathogenicity of Enterococcus faecalis in periapical periodontitis. J. Dent. Sci. 2019, 15, 65–74. [CrossRef] [PubMed]

23. Martini, C.; Longo, F.; Castagnola, R.; Marigo, L.; Grande, N.M.; Cordaro, M.; Cacaci, M.; Papi, M.; Palmieri, V.; Bugli, F.; et al. Antimicrobial and Antibiofilm Properties of Graphene Oxide on Enterococcus faecalis. Antibiotics 2020, 9, 692. [CrossRef] [PubMed]

24. Herizchi, R.; Abbasi, E.; Milani, M.; Akbarzadeh, A. Current methods for synthesis of gold nanoparticles. Artif. Cells Nanomed. Biotechnol. 2016, 44, 596–602. [CrossRef]

25. Tom, R.T.; Suryanarayanan, V.; Reddy, P.G.; Baskaran, S.; Pradeep, T. Ciprofloxacin-protected gold nanopar-ticles. Langmuir 2004, 20, 1909–1914. [CrossRef]

26. Topal, G.R.; Devrim, B.; Eryilmaz, M.; Bozkir, A. Design of ciprofloxacin-loaded nano-and micro-composite particles for dry powder inhaler formulations: Preparation, in vitro characterisation, and antimicro-bial efficacy. J. Microencapsul. 2018, 35, 533–547. [CrossRef]

27. Arafa, M.G.; Mousa, H.A.; Afifi, N.N. Preparation of PLGA-chitosan based nanocarriers for enhancing anti-bacterial effect of ciprofloxacin in root canal infection. Drug Deliv. 2020, 27, 26–39. [CrossRef]

28. Albuquerque, M.T.; Valera, M.C.; Moreira, C.S.; Bresciani, E.; de Melo, R.M.; Bottino, M.C. Effects of ciprof-loxacin-containing scaffolds on Enterococcus faecalis biofilms. J. Endod. 2015, 41, 710–714. [CrossRef]

29. Assali, M.; Zaid, A.N.; Abdallah, F.; Almasri, M.; Khayyat, R. Single-walled carbon nanotubes-ciprofloxacin nanoantibiotic: Strategy to improve ciprofloxacin antibacterial activity. Int. J. Nanomed. 2017, 12, 6647–6659. [CrossRef]

30. Rana, N.F.; Sauvageot, N.; Laplace, J.-M.; Bao, Y.; Nes, I.; Rincé, A.; Posteraro, B.; Sanguinetti, M.; Hartke, A. Redox Balance via Lactate Dehydrogenase Is Important for Multiple Stress Resistance and Virulence in Enterococcus faecalis. Infect. Immun. 2013, 81, 2662–2668. [CrossRef]

31. Dash, S.; Murthy, P.N.; Nath, L.; Chowdhury, P. Kinetic modeling on drug release from controlled drug delivery systems. Acta Pol. Pharm. Drug Res. 2010, 67, 217–223.

32. Bettini, R.; Catellani, P.L.; Santi, P.; Massimo, G.; Peppas, N.A.; Colombo, P. Translocation of drug particles in HPMC matrix gel layer: Effect of drug solubility and influence on release rate. J. Control. Release 2001, 70, 383–391. [CrossRef]=

33. Naz, S.S.; Islam, N.U.; Shah, M.R.; Alam, S.S.; Iqbal, Z.; Bertino, M.; Franzel, L.; Ahmed, A. Enhanced biocidal activity of Au nanoparticles synthesized in one pot using 2, 4-dihydroxybenzene carbodithioic acid as a reducing and stabilizing agent. J. Nanobiotechnol. 2013, 11, 13. [CrossRef] [PubMed]

34. Muhammad, Z.; Raza, A.; Ghafoor, S.; Naeem, A.; Naz, S.S.; Riaz, S.; Ahmed, W.; Rana, N.F. PEG capped methotrexate silver nanoparticles for efficient anticancer activity and biocompatibility. Eur. J. Pharm. Sci. 2016, 91, 251–255. [CrossRef]

35. Riaz, S.; Rana, N.F.; Hussain, I.; Tanweer, T.; Nawaz, A.; Menaa, F.; Janjua, H.A.; Alam, T.; Batool, A.; Naeem, A.; et al. Effect of Flavonoid-Coated Gold Nanoparticles on Bacterial Colonization in Mice Organs. Nanomaterials 2020, 10, 1769. [CrossRef]

36. Bhalerao, S.R.; Rote, A.R. Application of UV spectrophotometric methods for estimation of Ciprofloxacin and tinidazole in combined tablet dosage form. Int. J. Pharmacol. Pharm. Sci. 2012, 4, 4–7.

37. Bhattacharjee, S. DLS and zeta potential—What they are and what they are not? J Control. Release 2016, 235, 337–351. [CrossRef]

38. Canepa, E.; Salassi, S.; Simonelli, F.; Ferrando, R.; Rolandi, R.; Lambruschini, C.; Canepa, F.; Dante, S.; Relini, A.; Rossi, G. Non-disruptive uptake of anionic and cationic gold nanoparticles in neutral zwitterionic membranes. Sci. Rep. 2021, 11, 1–11. [CrossRef]

39. Okoampah, E.; Mao, Y.; Yang, S.; Sun, S.; Zhou, C. Gold nanoparticles-biomembrane interactions: From fundamental to simulation. Colloids Surf. B Biointerfaces 2020, 196, 111312. [CrossRef]

40. Dreaden, E.C.; Austin, L.; Mackey, M.A.; El-Sayed, M.A. Size matters: Gold nanoparticles in targeted cancer drug delivery. Ther. Deliv. 2012, 3, 457–478. [CrossRef]

41. Lim, S.; Alam, M.G. Ciprofloxacin-Induced Acute Interstitial Nephritis and Autoimmune Hemolytic Anemia. Ren. Fail. 2003, 25, 647–651. [CrossRef] [PubMed]

05 PHYSIOLOGICAL REACTIONS AT LOCAL AND SYSTEMIC LEVELS TO TITANIUM NANOTUBE SURFACES FACILITATING OSSEOINTEGRATION

1. INTRODUCTION: AN OVERVIEW

In the realm of biomaterials and implantology, the quest for optimal integration between artificial devices and living tissues has been a driving force in advancing medical technologies. Osseointegration, the direct structural and functional connection between bone and an implant, represents a paramount achievement in this pursuit. Titanium, renowned for its biocompatibility and mechanical properties, has become a cornerstone material for osseointegrative implants. Recent strides in nanotechnology have introduced a transformative element to this synergy, with the advent of titanium nanotube surfaces designed to enhance osseointegration. Titanium, both commercially pure and alloyed, has been used for decades in various biologic environments, including orthopedic implant designs [1–4]. With a combination of corrosion resistance, biocompatibility, mechanical properties approximating bone, and low cost, titanium continues to be a common material for fracture plates and screws as well as components requiring solid bone–implant fixation (e.g., knee arthroplasty tibial tray, hip arthroplasty femoral stem) [1–4]. To promote osseointegration, macroscale coatings have been applied to titanium implant surfaces via titanium plasma spray (TPS) as well as hydroxyapatite (HA) coating or sintering for powder- or bead-based coatings [5–8]. These coatings, however, are subjected to shear loads during surgical implantation, contact with surgical tools, and eventually micromotion at the bone–implant interface in vivo [5–8]. These coatings may then separate from the substrate, generating third-body wear debris that increases mechanical wear of bearing surfaces. Additionally, local phagocytic cells en- countering this debris may initiate a biologic cascade leading to periprosthetic osteolysis [9]. The body's immune response to wear debris is dependent on particulate composition, concentration, and morphology and may result in periprosthetic joint infection, bone fracture, catastrophic implant fracture, as well as osteoclastic bone resorption around the implant, component loosening, and, ultimately, revision surgery [5,10,11].

Titanium nanotube (TiNT) surfaces, which are electrochemically etched from the titanium implant substrate instead of applied as a coating, are an emerging technology that may enhance osseointegration of orthopedic implants [12–17], although literature is scarce regarding the in vivo immune response and toxicity to these nanostructured materials as implantable devices [18,19]. Due to improved interfacial strength, decreased thickness, and potential for more rapid osseointegration, TiNT surfaces may generate less third-body wear particulate compared to current coating technologies; however, thorough characterization of the immune and inflammatory response to new implantable materials is needed.

Two TiNT surface morphologies, termed aligned and trabecular, have been developed for orthopedic applications necessitating osseointegrative properties [17,20–22]. The aligned TiNT morphology comprises arrays of vertically oriented nanotubes, while the trabecular TiNT morphology exhibits a random surface resembling the three-dimensional porous network of trabecular bone. In this study, local and systemic responses to aligned and trabecular TiNT surfaces were assessed through in vitro cellular response on each material as well as in vivo performance in a clinically relevant, rodent model of long-term femoral intramedullary implant to simulate joint arthroplasty, including assessment of longitudinal animal weights, remote organ weights, metal ion levels in remote organs and whole blood, hematology, and nondecalcified histology [23,24]. Following on survivability testing and finite element modeling indicating that compression and shear loading to the TiNT surfaces were below yield failure strength, this study hypothesizes that TiNT surfaces will demonstrate equivalent local and systemic responses, equivalent to unmodified titanium alloy substrate surfaces [22]. Further, we hypothesize that aligned and trabecular TiNT surfaces yield comparable in vitro and in vivo results, as both morphologies provide additional surface area for cell attachment as well as resultant bony on growth and in growth.

2. MATERIALS AND METHODS

2.1 Implant Fabrication

Samples for in vitro and in vivo studies were fabricated with titanium alloy (Ti-6Al-4V ELI) sheet and wire (wire: Custom Wire Technologies, Inc., Port Washington, WI, USA), respectively, using a well-established electrochemical anodization process (Appendix A— Supplemental Methods, Figure A1) [17,21]. Following the initial etching process, aligned TiNT surfaces were then ultrasonicated in deionized (DI) water for 2 minutes; however, trabecular TiNT surfaces received no additional treatment (Figure 1). Control samples were rinsed with DI water and air-dried. On average, inside diameter (ID) of individual nanotubes was 60 nm and 1 μm length on aligned TiNT surfaces. Similarly, trabecular TiNT surfaces contained a 1 μm layer of 60 nm ID aligned nanotubes, on average, covered with an additional 1 μm of over-etched nanotubes. The over-etched layer mimics the porous structure and topography of trabecular bone.

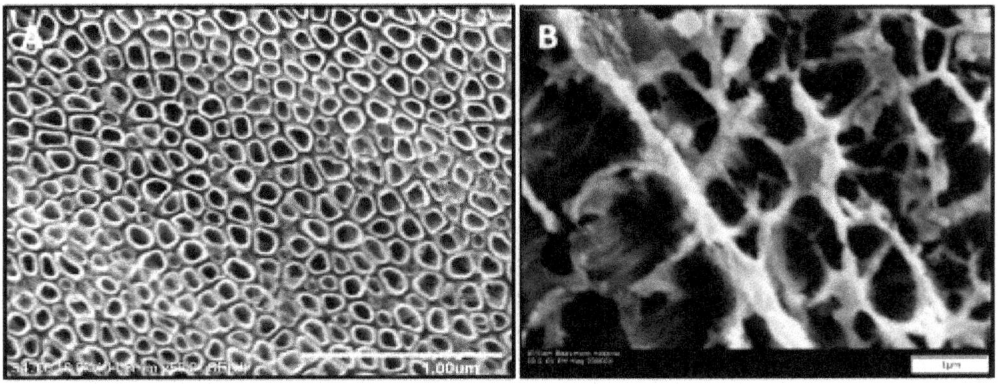

Figure 1. Aligned titanium nanotubes (TiNT) (A) and trabecular TiNT (B) surfaces. (A,B) Scale Bar: 1 μm.

2.2. In Vitro Experimentation to Assess Cell Attachment and Morphology

Cell for all in vitro experiments were isolated from the intramedullary cavities of the femora of 14 week-old female Sprague Dawley rats (SD; Charles River Laboratories, Wilmington, MA, USA). Following isolation and culture using previously described methods, the plastic-adherent fraction of bone marrow cells (BMC) was collected and used for all subsequent experiments (Appendix A—Supplemental Methods, Figure A2) [17]. This cell population was selected for experimentation due to the potential of mesenchymal stem cells (MSC) to differentiate toward numerous cell types, including osteoblasts.

Early cell attachment and morphology were compared between aligned TiNT, trabecular TiNT, and control surfaces, using three samples per group per time point (coupons sectioned from sheet: 10 mm x 10 mm; time points: 30 min, 2 h, 4 h) [15,25]. To increase attachment potential, all sample coupons were soaked in fetal bovine serum (FBS) for 30 minutes prior to drop-seeding (density: 40,000 BMCs/coupon), followed by a 6 h incubation before adding the remaining volume of media and incubating overnight. At each time point, cells were fixed and stained with Actin Green (Actin Green 488 ReadyProbes Reagent, Life Technologies, Carlsbad, CA, USA) for cytoskeleton visualization and morphology (1 drop stain; 40 min incubation) as well as 4', 6-Diamidino-2-Phenylindole, dihydrochloride (DAPI; Life Technologies, Carlsbad, CA, USA) for nucleus visualization (0.5 mL stain; 20 min incubation). Following cell fixation, fluorescence imaging (IX71, Olympus America,

Center Valley, PA, USA) was performed at thirteen standardized regions of interest to quantify the total number of adherent cells, cell equivalent diameter, and cell eccentricity. Because all non-adherent cells were removed prior to imaging, Actin Green and DAPI also demonstrated cell adhesion to the surfaces. Environmental scanning-electron microscopy (SEM; Vega3XMU, Tescan USA, Warrendale, PA, USA) was used to further document cell morphology; specifically, cells were fixed in 4% glutaraldehyde on coupons, and then coupon was sputter-coated with a thin layer of gold–palladium alloy. Samples were imaged at 20.0 kV and a magnification of 300×. Total cell number, cell equivalent diameter, and cell eccentricity (measured using onboard microscope measurement tool) were statistically compared using a one-way analysis of variance (ANOVA) model, with a Tukey post-hoc test and α= 0.05.

2.3. In Vivo Experimentation to Assess Biologic Response to Nanotube Surfaces

After preparing titanium alloy Kirschner wires (K-wire; CustomWire Technologies, Inc., PortWashington, WI; USA, material: Ti6Al4V ELI Hard, diameter: 1.25 mm, single trocar tip for insertion), aligned TiNT-etched, trabecular TiNT-etched, or unetched titanium (control) K-wire implants (n = 6 per group) were inserted retrograde into the femoral intramedullary canals of SD rats for a single, long-term endpoint of 12 weeks [24,26,27]. Three naïve/nonoperative animals were housed for the same duration, in order to establish baseline characteristics. Treatment groups were randomized just prior to surgery by nonoperative staff. Rats received daily veterinary care to identify complications and were weighed weekly; at endpoint, hematologic, metal ion, and histologic analyses were performed.

2.4. Surgical Procedure for Unilateral Intramedullary Implantation

Under an Institutional Animal Care and Use Committee (IACUC)-approved protocol, 14-week-old, female SD rats underwent unilateral femoral implants via retrograde insertion. After anesthetization, rats were placed supine on a sterile, heated operating table with the knee in maximum flexion. Following final sterile preparation, a lateral–patellar skin incision allowed access to the distal femur. After drilling a shallow, pilot hole in theintercondylar groove, a K-wire was inserted into the intramedullary canal to the greater trochanter (proximal femur). Placement of the implant was confirmed via fluoroscopy. The K-wire was then clipped and recessed at the insertion site, followed by closure of arthrotomy and skin incisions with appropriate suture materials. Animals were then recovered and allowed ad libitum activity under endpoint.

2.5 Postoperative Care and Health Assessments

All rats were evaluated daily for clinical signs of complications and overall health, including pain level, activity level, and food/water consumption. Rats were weighed preoperatively and weekly throughout the experiment. At endpoint, remote organs were collected and weighed, as another assessment of animal health. A one-way ANOVA model was used to compare longitudinal animal weights and endpoint organ weights between treatment groups, with $\alpha = 0.05$.

2.6 Hematologic Analysis

At endpoint, each anesthetized rat underwent antemortem cardiac puncture to collect blood for hematologic analyses (CBC; complete blood count with differential) to quantify the following thirteen parameters associated with either systemic inflammation/infection or anemia status: white blood cell count (WBC), lymphocyte concentration (Lymph), monocyte concentration (Mono), granulocyte concentration (Gran), hematocrit (HCT), mean cell volume of red blood cells (MCV), red blood cell distribution width (RDWa), hemoglobin concentration (Hgb), mean cell hemoglobin concentration (MCHC), mean cell hemoglobin (MCH), red blood cell count (RBC), total platelet count (PLT), and mean and platelet volume (MPV) (HemaTrue Hematology Analyzer, Heska, Loveland, CO, USA). Blood was collected in vacutainer tubes with Ethylenediaminetetraacetic acid (EDTA) and immediately processed. A one-way ANOVA model was used to compare hematologic parameters between treatment groups, with $\alpha = 0.05$.

2.7 Metal Ion Analysis of Whole Blood and Remote Organs

Inductively coupled plasma mass spectrometry (ICP-MS; HP 4500, Agilent Technologies, Santa Clara, CA, USA) was used to measure the titanium, aluminum, and vanadium concentrations in each remote organ and a whole blood sample from each animal (Appendix A—Supplemental Methods). ICP-MS calibration was performed using both ASTM standards of each target element as well as non-implanted samples of the rods from each testing condition. At endpoint, remote organs (i.e., spleen, liver, lungs, kidneys, brain) were harvested, weighed, and stored (4 °C in ultra-low leachable sample tubes) until analysis. Samples were analyzed in tandem with a laboratory reagent blank and duplicate laboratory fortified blanks to quantify background metal levels as well as process accuracy and precision, respectively. A one-way analysis of variance model was used to compare metal ion levels between treatment groups, with $\alpha = 0.05$.

2.7 Nondecalcified Histologic Analysis

Inductively coupled plasma mass spectrometry (ICP-MS; HP 4500, Agilent Technologies, Santa Clara, CA, USA) was used to measure the titanium, aluminum, and vanadium concentrations in each remote organ and a whole blood sample from each animal (Appendix A—Supplemental Methods). ICP-MS calibration was performed using both ASTM standards of each target element as well as non-implanted samples of the rods from each testing condition. At endpoint, remote organs (i.e., spleen, liver, lungs, kidneys, brain) were harvested, weighed, and stored (4 C in ultra-low leachable sample tubes) until analysis. Samples were analyzed in tandem with a laboratory reagent blank and duplicate laboratory fortified blanks to quantify background metal levels as well as process accuracy and precision, respectively. A one-way analysis of variance model was used to compare metal ion levels between treatment groups, with $\alpha = 0.05$.

2.8. Nondecalcified Histologic Analysis

At study endpoint, implanted femora were also harvested for nondecalcified histologic analysis. Femora from each animal were stripped of soft tissues and 10% zinc-buffered formalin for 96 h, followed by three rinses in phosphate buffered saline, and 70% ethanol storage to prepare for histologic processing. Femora were subsequently embedded in methyl methacrylate, then ground and polished longitudinally to the approximate center of the implant, followed by cutting and applying one 5 μm thick section to a slide. Each slide was stained with Stevenel's Blue and van Gieson picrofuchsin (SBVG) for visualization of presence of immune-related cellular activity. SBVG is a widely-used stain to assess bone formation, due to visualization of both mature bone and osteoid as well as fibrous tissue, which may signify inferior osseointegration and/or increased inflammatory response, especially at the bone–implant interface [28–30]. Stained sections were manually scanned at a magnification of 20× to facilitate both high-magnification and whole-mount analyses (90i, Nikon Instruments, Inc., Melville, NY, USA). Three 20× regions of interest (ROI) per location (i.e., distal, midshaft, proximal) per section were captured. Within each ROI, five cell types were identified: foreign body giant (FBG)/multinucleated cell, granulocyte (nonneutrophilic, including eosinophils and basophils), neutrophil, monocyte, and lymphocyte. Cell constituents within each ROI were then graded on a scale of 0 to 2, with 0 = cells comprising 0–25% of the field, 1 = cells comprising 25–50% of field, and 2 = cells comprising greater than 50% of field. For FBG/multinucleated or granulocyte, a Grade 2 was defined as three or more cells per field. Grades were statistically compared between TiNT groups and control, using a Mann–Whitney rank sum test, with $\alpha = 0.05$.

3. RESULTS

3.1. In Vitro Experimentation to Assess Cell Attachment and Morphology

API staining indicated that cells were viable and adhered on both the TiNT and control surfaces at the three early timepoints of 0.5, 2, and 4 h (Figure 2). The total number of adherent cells was significantly greater on the TiNT surfaces than on control surfaces, demonstrating more rapid cell attachment on both TiNT surfaces compared to control (Figure 3); specifically, total adherent cells on trabecular TiNT and aligned TiNT surfaces were significantly greater than control at 0.5 h ($p = 0.014$ and $p = 0.018$, respectively) and 4 h ($p = 0.008$ and $p = 0.044$). Over the 3.5-hour period, the number of cells increased on all surfaces, and the total number of adherent cells was equivalent or slightly greater on trabecular TiNT surfaces compared to aligned TiNT surfaces. Analysis of total cell count was unfeasible at later timepoints (3, 7, 14, 21 days) due to cell coalescence or superimposition. Actin Green staining showed active cell spreading on TiNT and control surfaces at all timepoints (Figure 2).

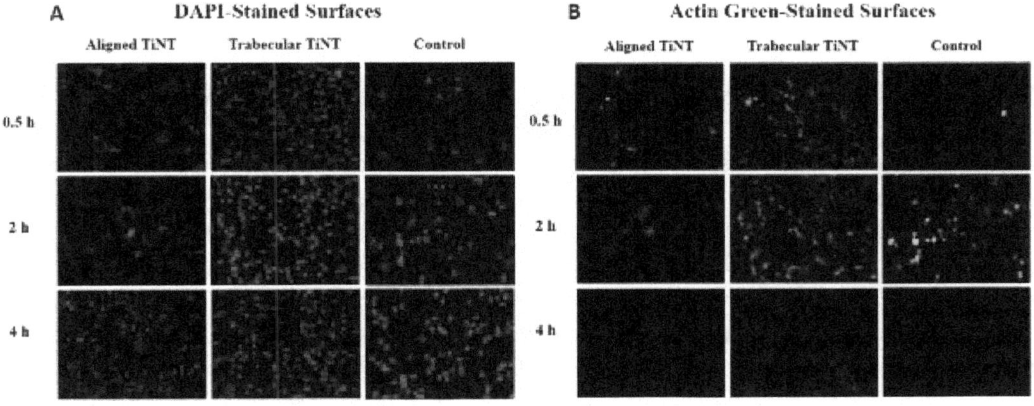

Figure 2. Representative fluorescent images of dihydrochloride (DAPI)-stained (A) and Actin Green-stained (B) aligned TiNT, trabecular TiNT, and control surfaces at three early timepoints, 0.5 h, 2 h, and 4 h.

At the 2- and 4-h timepoints, increased spreading was observed on TiNT surfaces compared to control surfaces. Additional imaging via SEM showed differential cell morphology patterns as a function of sample topography (Figure 4). On the TiNT surfacescells exhibited a globular shape, compared to the elongated, fibrillar cell morphology on control surfaces.

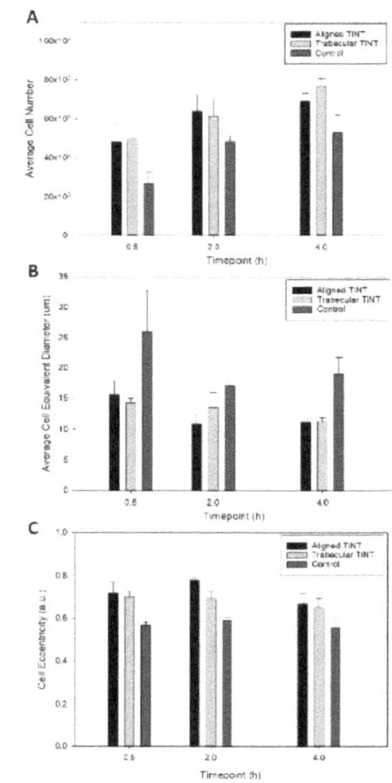

Figure 3. Average total cell number (A), cell equivalent diameter (B), and cell eccentricity (C) on TiNT and control surfaces at 0.5, 2, and 4 h. For Cell Number, significant comparisons were: aligned TiNT vs. control, p = 0.018; trabecular TiNT vs. control, p = 0.008; aligned TiNT vs. control, p = 0.044. For Cell Diameter, significant comparisons were: aligned TiNT vs. control, p = 0.048; trabecular TiNT vs. control, p = 0.031; aligned TiNT vs. control, p = 0.008; aligned TiNT vs. control, p = 0.003; trabecular TiNT vs. control, p = 0.003. For Cell Eccentricity, significant comparisons were: aligned TiNT vs. control, p = 0.004; trabecular TiNT vs. control, p = 0.007; aligned TiNT vs. control, $p < 0.001$; trabecular TiNT vs. control, p = 0.003.

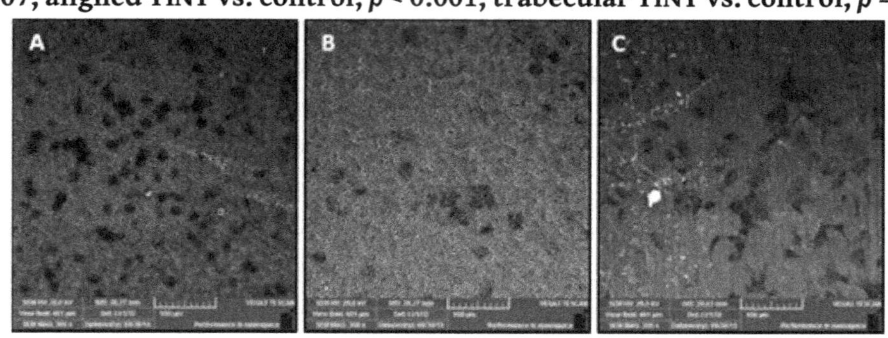

Figure 4. SEM micrographs demonstrating differential cell spreading patterns, via extracellular SEM micrographs demonstrating differential cell spreading patterns, via extracellular matrix aggregates, matrix aggregates, between aligned TiNT (A) and trabecular TiNT (B) versus control (C) surfaces iNT (A) and trabecular TiNT (B) versus control (C) surfaces at 3-day attachment timepoint.

3.2. In Vivo Experimentation to Assess Biologic Response to Nanotube Surfaces

Following the in vitro study, which demonstrated cell viability on the TiNT surfaces, the planned in vivo experiment was performed. Over the 12-week study, animals gained an average of 148 g, with control animals gaining the most weight on average (179 g) and rabecular TiNT animals gaining the least on average (118 g) (Figure 5). No animals lost more the 10% body weight, our institutional trigger for intervention. In week 9, the aligned TiNT group weighed significantly more than the trabecular TiNT group (p = 0.013), and in week 11, the control group weighed significant more than the trabecular TiNT group (p = 0.023). There were no significant differences at any other weekly time points. Mass of the spleen, brain, kidneys, lungs, and liver were obtained, and there were no significant differences between any groups for the specimens (Figure 5). differences between any groups for the specimens (Figure 5).

Hematologic analysis was performed, and no significant associations were found between the three groups for the thirteen parameters, and there were no significant differences between any groups for any parameters (Tables 1 and 2). Additionally, titanium, aluminum, and vanadium levels in each implant were analyzed to quantify the effect of the etching process on chemistry. Compared to control, aligned TiNT had 2.0 times the concentration of aluminum, approximately 1.8 times the titanium, and 2.9 times the vanadium; similarly, trabecular TiNT had 2.5 times the aluminum, 2.2 times the titanium, and 3.4 times the vanadium.

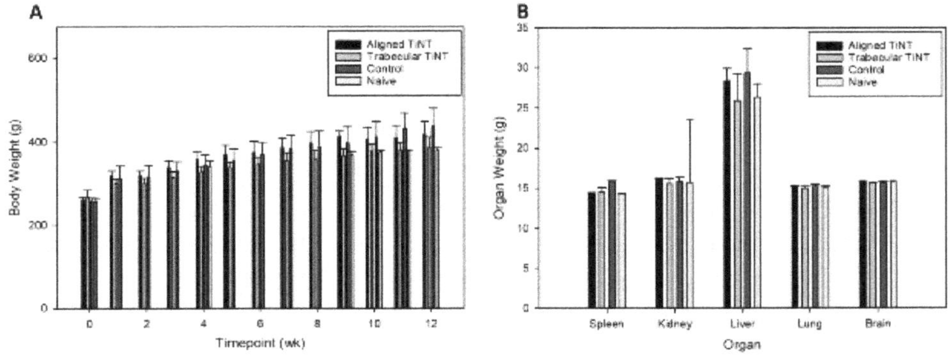

Figure 5. Average animal body weight over 12-week study (A) and organ weight at endpoint (B) per group. For Week 9,significant comparison: aligned TiNT vs. trabecular TiNT, p = 0.013; For Week 11, significant comparison: trabecular TiNT vs. control, *p* = 0.023.

Table 1. Hematologic analysis of white blood cell function of each treatment group at endpoint.*

Aligned TiNT	9.5 (2.7	7.3 (1.8)	0.4 (0.2)	1.7 (0.7)	77.9 (2.7)	3.9 (1.3)	18.2 (2.6)
Trabecular TiNT	8.02 (3.5)	6.2 (2.7)	0.3 (0.1)	1.5 (0.8)	77.9 (3.9)	3.3 (1.6)	18.8 (2.9)
Contro	8.9 (2.2)	7.0 (1.6)	0.3 (0.1)	1.5 (0.6)	79.7 (2.5)	79.7 (2.5)	17.3 (2.2)
Naïve	11.0 (6.1)	8.4 (3.9)	0.5 (0.4)	2.1 (1.8)	79.1 (6.1)	3.1 (0.8)	17.8 (5.3)

* Standard deviation listed in parentheses.

Table 2. Hematologic analysis of red blood cell function of each treatment group at endpoint

Aligned TiNT	T 40.1 (3.2)	49.7 (1.1)	32.8 (1.0)	16.4 (0.5)	15.3 (1.2)	38.3 (0.5)	19.0 (0.3)	8.1 (0.6)	318.7 (149.4)	6.4 (0.8)
Trabecular TiNT	38.5 (1.5)	50.0 (0.9)	32.9 (0.7)	16.4 (0.6)	14.6 (0.4)	38.3 (0.6)	19.1 (0.5)	7.7 (0.4)	275.8 (151.6)	6.7 (0.6)
Control	40.1 (3.5)	50.9 (1.2)	34.0 (1.6)	16.4 (0.4)	15.1 (1.2)	37.7 (0.5)	19.1 (0.4)	7.9 (0.6)	335.0 (89.7)	6.4 (0.5)
Naïve	41.4 (2.8)	50.6 (0.8)	33.7 (0.2)	16.5 (0.4)	15.6 (1.1)	37.7 (0.2)	19.1 (0.3)	8.2 (0.5)	305.7 (93.5)	6.1 (0.8)

* Standard deviation listed in parentheses.

After weighing each organ, metal concentrations were assessed (Figure 6). Aluminum levels in the lungs were significantly greater in the trabecular TiNT group compared to control ($p = 0.022$). No other organs exhibited significantly increased titanium, aluminum, or vanadium levels compared to control. Histologic analysis demonstrated a lack of inflammatory infiltrate in proximity to the nails within the intramedullary space, for all groups (Table 3; Appendix A—Supplemental Results, Table A1). There were significantly fewer granulocytes and neutrophils in the distal ROI of the femora implanted with trabecular TiNT-etched implants ($p = 0.040$ and $p = 0.019$, respectively). In the midshaft ROI, there were significantly fewer foreign body giant/multinucleated cells and neutrophils in the aligned TiNT group ($p = 0.039$ and $p = 0.019$, respectively). There were no observed necrotic or cytotoxic events observed within the marrow cavity of any of the animals (Figure 7).

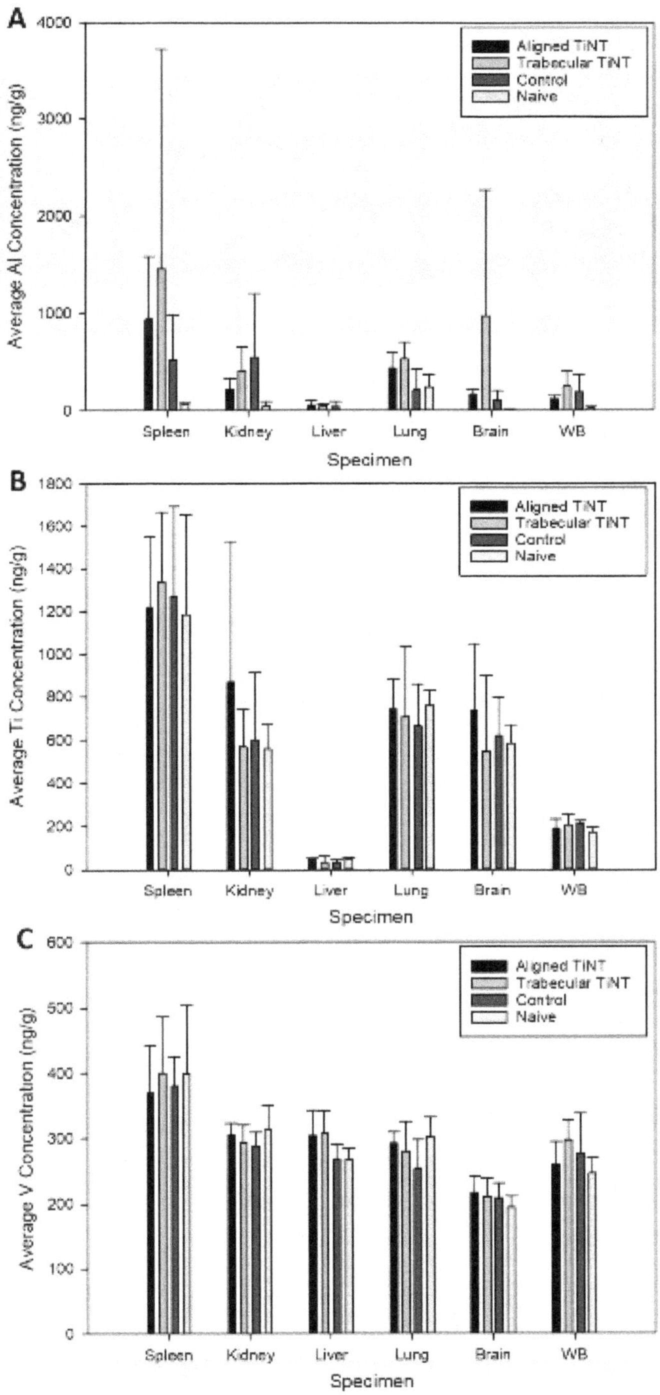

Figure 6. Concentration of aluminum (A), titanium (B), and vanadium (C) in remote organs and whole blood for each group. (WB = whole blood). For aluminum concentration, significant comparison: trabecular TiNT vs. control, p = 0.

Table 3. Statistical comparisons (p-values) between implant groups with control of average histologic grade for three regions of interest *.

[illegible]	[illegible]	[illegible]	[illegible]	[illegible]	[illegible]	[illegible]
Distal	Trabecular TiNT	0.126	0.040	0.019	0.720	0.391
	Aligned TiNT	0.287	0.476	0.345	0.345	0.260
Midshaft	Trabecular TiNT	0.142	0.603	1.000	0.314	0.982
	Aligned TiNT	0.039	0.682	0.019	0.736	0.260
Proximal	Trabecular TiNT	0.527	0.646	0.345	0.652	0.871
	Aligned TiNT	0.939	0.219	0.345	0.747	0.314

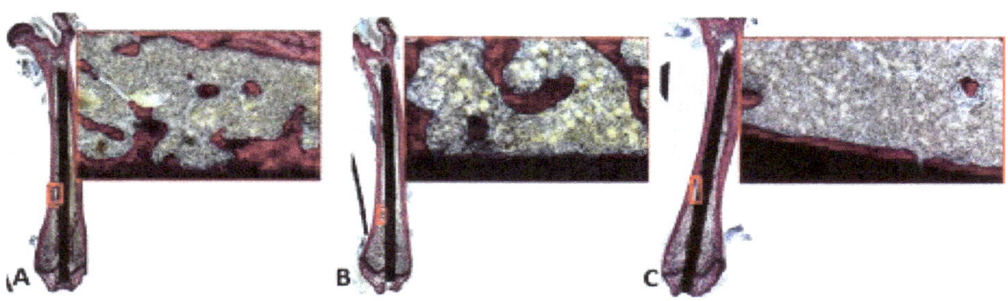

Figure 7. Longitudinal histologic sections and regions of interest of representative femora implanted with aligned TiNT (A), trabecular TiNT (B), and (C) control K-wires.

4. DISCUSSION

Local and systemic responses to aligned and trabecular TiNT surfaces were evaluated via an in vitro study focused on cell morphological and attachment behaviour, followed by a clinically relevant in vivo study of biologic response to implant materials. The in vivo study included multiple characterization methods, including a general health assessment (e.g., body weight) as well as hematologic, metal ion, and histologic analyses.

DAPI and Actin Green staining demonstrated cell attachment and spreading from 0.5 h through 21 days on all surfaces, an indication of the suitable environment of both TiNT surfaces and control. Total cell counts were greater on TiNT surfaces compared to unetched controls at the three early timepoints, which corresponds with current reports of more rapid cell attachment on nanotube surfaces [31–33]. Additionally, the diameter of BMC on TiNT surfaces was smaller compared to cells seeded onto controls, and cells on TiNT surfaces were more eccentric than cells on controls. These data correspond with SEM imaging of attached BMC, which showed a rounded cell morphology on TiNT surfaces and fibrillar-shaped cells on control surfaces. The diameter, eccentricity, and imaging findings may be indicative of cell development in different planes on the TiNT surfaces compared to controls. Shokuhfar et al. sectioned TiNT arrays seeded with osteoblasts (MC3T3-E1; mouse osteoblasts) and observed cell attachment on nanotube arrays as well as cell filopodia stretching downward into the hollow portion of individual nanotubes [20]; therefore, differences in cell morphologic behaviour between groups may be explained by the BMC interaction with TiNT surfaces, as the BMC extend into nanotubes and voids within the array, instead of across the top of the surface. Rangamani et al. performed modelling of cell shapes and determined that greater cell eccentricity was associated with enhanced signal modulation due to temporary collections of activated receptors in sections of greater curvature of cells, specifically in growth factor receptor pathways; therefore, greater cell eccentricity on nanotube surfaces may demonstrate amplified signals resulting in downstream effects, such as bone formation [34]. In the in vivo study, there were several significant differences in body weight between groups at two weeks during the experiment; however, the body weights recovered were not an established trend, and no interventions were required at any time for weight loss. At endpoint, remote organ masses were compared between groups, and these also showed no significant difference. Despite increased concentrations of aluminum, titanium, and vanadium in the aligned TiNT and trabecular TiNT wires versus control wires, only aluminum levels were significantly increased in the lungs of the rats implanted with trabecular TiNT-etched wires compared to controls. No other organs or whole blood samples showed significantly elevated metal levels, when the experimental groups were compared to con- trol. A study of intraarticular injection of TiO2 nanoparticles (45 nm average diameter; 0.2, 2, 20 mg/kg TiO2 in suspension) into rat knee joints showed nanoparticle migration to remote organs as well as pathologic changes in the heart, lung, liver, and knee at 7 days post-injection [35,36]. Although our study showed metal ion concentration in remote organs via ICP-MS and did not histologically assess remote organs, we theorize that the amount of TiO2 particulate debris shed in the rat femora was likely less than 0.2 mg/kg. At study endpoint, no hematologic markers of white and red blood cell function, which would signal systemic inflammation-, infection, or anemia-related complications, were significantly elevated in the experimental groups compared to control. Histologic analysis showed several significant differences in cell populations between the TiNT groups and control, with less foreign body giant/multinucleated, eosinophil/basophil, and neutrophil cell activity in TiNT-implanted femora than control. There were no significant differences between monocyte and lymphocyte activity.

In in vitro and in vivo studies of vascular toxicity, Bayat et al. established that titanium nanotubes with 30 nm diameter as well as ultra-small TiO2 nanoparticles (1-3 nm) were not cytotoxic, and nanoparticles did not possess oxidative potential [37]. In vivo studies in rat models have also shown that TiO2 nanoparticles are not genotoxic [38,39]. Neacsu et al. showed that after seeding murine macrophages onto TiNT and unetched titanium surfaces under pro-inflammatory and standard conditions, inflammatory activity related to cytokine and chemokine gene expression, foreign body giant cell products, and nitric oxide release all decreased on TiNT surfaces but not on unetched controls; therefore, the authors suggested that the TiNT surfaces may regulate macrophage response, thereby diminishing the overall inflammatory cascade [40]. Studies by Ostberg et al. and Latha et al., using models of leukocyte-seeded TiNT surface and mixed lymphocyte reaction (MLR), respectively, concurred with these findings [41,42]; specifically, Latha et al. showed a 30–35% suppression of splenocyte proliferation of titanium nanotubes (H2Ti3O7) [42]. Assessing hemocompatbility, Smith et al. showed increased adsorption of blood serum protein, platelet adhesion and activation, and clotting of whole blood as well as no evidence of monocyte activation and cytokine secretion on TiNT surfaces versus control [43]. Radizun et al. showed that aluminum nanoparticles, in concentrations up to 400 μg/mL, have no significant toxic effect on mammalian cell viability, and Alshatwi et al. found human MSC cytotoxicity via Al2O3 nanoparticles at 40 μg/mL; in this study, aluminum content in all specimens was substantially below this threshold (maximum aluminum content within range of standard deviation; Figure 6: 4.24 μg/mL) [44,45]. In a clinical, in vivo study, Swiatkowska et al. described titanium levels ranging between 2.20 and 2.56 μg/L in blood and plasma, measured via ICP-MS, in well-functioning unilateral hip implants, indicating some level of titanium may be tolerable by patients, without eliciting a whole-body response [46].

Study limitations included difficulties with staining TiNT surfaces, especially at time- points greater than 4 h. Because of the nanotopography and voids, which were ideal for cell attachment, stain penetration of the cells was complicated and incomplete. While three naïve animals were included to assess background metal levels, inclusion of additional animals may have been beneficial to further study and quantify these levels. Additionally, a full assessment of metal levels of animal food, bedding, enrichment materials, etc. may have allowed further reduction and elimination of background aluminum, titanium, and vanadium levels in all animals [47,48]. Based on the results of this study, a full systematic characterization of this material's biocompatibility and toxicity properties should be conducted in accordance with applicable standards, such as ISO 10993-11 (Biological Evaluation of Medical Devices—Part 11: Tests for Systemic Toxicity).

5. CONCLUSIONS

Previous studies have demonstrated the biocompatibility of TiNT surfaces and correspond with the findings of this study showing TiNT surfaces support cell attachment and proliferation and do not initiate systemic effects in an in vivo model of intramedullary implantation. The presented in vitro data demonstrated that cells cultured on bare TI6Al4V surfaces were spindle shaped, while those cultured on TiNT surfaces demonstrated a more rounded appearance, which may confer benefit with respect to in vivo bone formation. In vivo data demonstrated that the accumulation of metal ions in filtering organs was largely similar between the two morphologies of TiNT surfaces and control bare Ti6Al4V surfaces, aside from an increased aluminum concentration in the lungs of rats with aligned TiNT implants. This increased aluminum concentration in lung specimens was not associated with alterations in overall health of the animal or observed pathology. As such, it can be concluded that at the time points studied, metal accumulation associated with TiNT surfaces was similar to control surfaces, and no systemic or local complications were observed in TiNT- or control-implanted animals.

Appendix A. Supplemental Methods and Results

Appendix A.1. Titanium Nanotube-Etched Specimen Fabrication

Titanium nanotube-etched specimens were prepared for in vitro and in vivo experiments, based on previous methods developed and tested by our group [17,20–22]. Titanium alloy (Ti-6Al-4V ELI) sheet (in vitro experiments) or wire (in vivo study) were polished with 600 grit abrasive sheet and deionized (DI) water, then rinsed in DI water and air-dried. All sample materials were cleaned with acetone just prior to etching, followed by air drying. On wire samples, the trocar tip was coated to prevent nanotube formation on the site of insertion. To etch, samples (sheet or wire) were suspended vertically toward one side of a glass beaker and then connected as the anode to a variable DC power supply. Additionally, in the beaker, a small diameter graphite rod was suspended, diametrically opposed to the sample, and connected to ground (cathode). An electrolyte solution (98 vol.% ethylene glycol, 2 vol.% deionized water, and 0.6 wt.% NH4F) was added to the beaker, with a final volume just below the electrical connections to the sample and graphite. Before adding NH4F to the electrolyte solution, it was first completely dissolved in DI water; following this addition, the electrolyte solution was mixed to ensure all constituents were fully combined. Then, the power supply was engaged (+60 VDC) and etching was allowed to proceed for 40 minutes. After etching, the power supply was disengaged and sample(s) were immediately removed from the beaker, followed by a 1-minute rinse under DI water and air drying. After drying, the coating on each trocar tip was removed after etching and sonication (aligned TiNT samples were sonicated for 2 min; trabecular TiNT samples were not sonicated). Sheet samples were then sectioned into coupons (10 mm × 10 mm); ires were net-shape and did not require sectioning. To convert the amorphous titania oxide layer to the crystalline anatase phase, which increases the hydrophilicity of the TiNT surfaces, samples were placed in a programmable annealing oven. The temperature was increased by 7.5 C per minute to a steady state of 450 C for a total heating time of three hours. Samples were removed after the oven was completely cooled, which was approximately 5 h. Specimens were stored in cushioned, sealed storage cases until use (Figure A1).

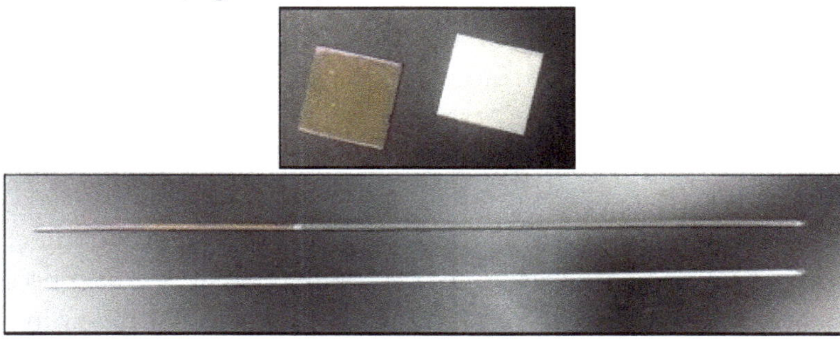

Figure A1. Digital photographs showing the final appearance of 10 mm × 10 mm coupon samples

Figure A1. Digital photographs showing the final appearance of 10 mm 10 mm coupon samples (top) and Kirschner wires (bottom). TiNT-etched surfaces appear golden and smooth, compared to (top) and Kirschner wires (bottom). TiNT-etched surfaces appear golden and smooth, compared to unmodified control surfaces.

Appendix A.2. In Vitro Experimentation to Assess Adherent Cell Count, Cell Equivalent Diameter, and Cell Eccentricity

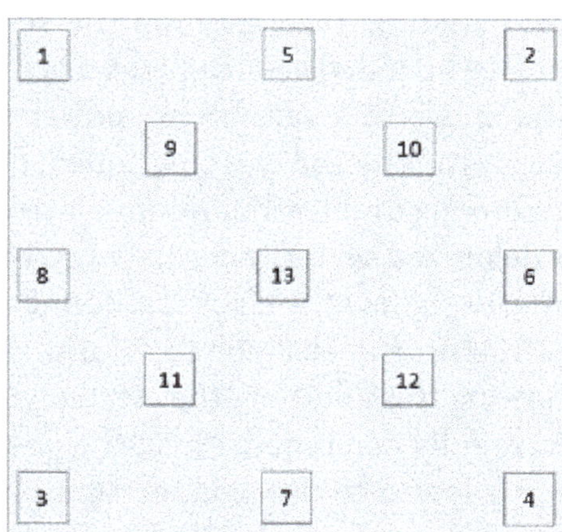

Figure A2. Imaging convention for each coupon in the experiment. Regions were chosen to reflect the corners (1–4), edges (5–8), central area (9–12), and absolute center (13) of each sample.

Appendix A.3. Metal Ion Analysis of Remote Organs and Whole Blood

Inductively coupled plasma mass spectrometry (ICP-MS; HP 4500, Agilent Technologies, Santa Clara, CA) was performed to quantify aluminum, titanium, and vanadium content in whole-blood and remote organ samples. Sample tubes, made of ultra-low leachable material to avoid metal contamination, were labeled and weighed to three significant figures before sample collection. At endpoint, whole blood (($\geq$2 mL) was collected, before proceeding with dissection of remote organs (i.e., spleen, liver, lungs, kidneys, brain). Organs were weighed before being placed into tubes, and each filled sample tube was weighed again to determine the actual/ net weight of the collected specimen.

All samples were transported and held on ice (~4 _C) until analysis. Samples were digested to dissolve the metals into an aqueous solution by placing a combination of highly pure solutions of nitric acid, hydrochloric acid, hydrogen peroxide, and water into the tubes. The tubes were heated at 95 _C for 2 h per cycle. Some of the samples went through multiple digestion cycles for thorough digestion. For ICP-MS analysis, samples were included in analytical batches of 20 samples, which includes a laboratory reagent blank (LRB) reference to the laboratory background value, a laboratory fortified blank (LFB) to measure accuracy of the analytical procedure, and a duplicate LFB to measure the precision of the analytical procedure. The final volume of digestion was set to 40 mL per specimen digested, and the digested solutions were further diluted by 10 times and analyzed for aluminum, titanium, and vanadium contents using a standard operating procedure (UL SOP# 114) based on the EPA guidelines (Method 200.8).

The total weight of each metal per specimen and concentration of each metal in a specimen were recorded; weights and concentrations were reported to three significant figures. Accuracy and precision data as part of the quality control carried through the analytical batches were also documented. Chain of custody forms containing sample descriptions, sampling dates, etc. were completed and retained.

Appendix A.4. Histologic Analysis of Bone-Implant Interface

Average cell count data of five cell types (foreign body giant/multinucleated, granulocyte, neutrophil, monocyte, lymphocyte) for the three implant groups (trabecular TiNT, aligned TiNT, control) in the three regions of interest (ROI; distal, midshaft, and proximal femora) have been provided (Table A1).

Table A1. Average histologic grade for three regions of interest in each treatment group *.

[illegible]	[illegible]	[illegible]	[illegible]	[illegible]	[illegible]	[illegible]
Distal	Trabecular TiNT	0.8 (1.0)	0.1 (0.0)	0.7 (1.0)	0.8 (1.0)	1.1 (1.0)
	Aligned TiNT	0.7 (1.0)	0.2 (0.0)	1.1 (1.0)	0.8 (1.0)	1.1 (1.0)
	Control	0.6 (0.5)	0.3 (0.0)	1.0 (1.0)	0.7 (1.0)	1.2 (1.0)
Midshaft	Trabecular TiNT	0.5 (0.5)	0.6 (1.0)	1.0 (1.0)	0.6 (1.0)	1.2 (1.0)
	Aligned TiNT	0.4 (0.0)	0.8 (1.0)	0.7 (1.0)	0.7 (1.0)	1.1 (1.0)
	Control	0.8 (1.0)	0.7 (1.0)	1.0 (1.0)	0.7 (1.0)	1.2 (1.0)
Proximal	Trabecular TiNT	0.8 (1.0)	0.6 (1.0)	0.9 (1.0)	0.8 (1.0)	1.4 (1.0)
	Aligned TiNT	0.7 (1.0)	0.4 (0.0)	1.1 (1.0)	0.9 (1.0)	1.3 (1.0)
	Control	0.7 (1.0)	0.7 (1.0)	1.0 (1.0)	0.8 (1.0)	1.4 (1.0)

* Standard deviation listed in parentheses.

6. CONCISE SUMMARY

Orthopedic implants requiring osseointegration are often surface modified; however, implants may shed these coatings and generate wear debris leading to complications. Titanium nanotubes (TiNT), a new surface treatment, may promote osseointegration. In this study, in vitro (rat marrow-derived bone marrow cell attachment and morphology) and in vivo (rat model of intramedullary fixation) experiments characterized local and systemic responses of two TiNT surface morphologies, aligned and trabecular, via animal and remote organ weight, metal ion, hematologic, and nondecalcified histologic analyses. In vitro experiments showed total adherent cells on trabecular and aligned TiNT surfaces were greater than control at 30 min and 4 h, and cells were smaller in diameter and more eccentric. Control animals gained more weight, on average; however, no animals met the institutional trigger for weight loss. No hematologic parameters (complete blood count with differential) were significantly different for TiNT groups vs. control. Inductively coupled plasma mass spectrometry (ICP-MS) showed greater aluminum levels in the lungs of the trabecular TiNT group than in those of the controls. Histologic analysis demonstrated no inflammatory infiltrate, cytotoxic, or necrotic conditions in proximity of K-wires. There were significantly fewer eosinophils/basophils and neutrophils in the distal region of trabecular TiNT-implanted femora; and, in the midshaft of aligned TiNT-implanted femora, there were significantly fewer foreign body giant/multinucleated cells and neutrophils, indicating a decreased immune response in aligned TiNT-implanted femora compared to controls.

REFERENCES

1. Branemark, P.I.; Adell, R.; Albrektsson, T. Osseointegrated titanium fixtures in the treatment of edentulousness. Biomaterials 1983, 4, 25–28. [CrossRef]
2. Albrektsson, T.; Brånemark, P.-I.; Hansson, H.-A.; Lindström, J. Osseointegrated Titanium Implants:Requirements for Ensuring a Long-Lasting, Direct Bone-to-Implant Anchorage in Man. Acta Orthop. Scand. 1981, 52, 155–170. [CrossRef] [PubMed]
3. Ionescu, F.; Reclaru, L.; Ardelean, L.C.; Blatter, A. Comparative Analysis of the Corrosion Resistance of Titanium Alloys Intended to Come into Direct or Prolonged Contact with Live Tissues. Materials 2019, 12, 2841. [CrossRef]
4. Ratner, B.D.; Schoen, F.J.; Lemons, J.E. Biomaterials Science: An Introduction to Materials in Medicine, 2nd ed.; Elsevier Academic Press: San Diego, CA, USA, 2004.
5. Goodman, S.B.; Gallo, J. Periprosthetic Osteolysis: Mechanisms, Prevention and Treatment. J. Clin. Med. 2019, 8, 2091. [CrossRef] [PubMed]
6. Goodman, S.B.; Gallo, J.; Gibon, E.F.; Takagi, M. Diagnosis and management of implant debris-associated inflammation. Expert Rev. Med. Devices 2019, 17, 41–56. [CrossRef]
7. Hergel, C.A.; Acar, Y.B.; Ates, M.; Kucukkeles, N. In-vitro evaluation of the effects of insertion and sterilization procedures on the mechanical and surface characteristics of mini screws. Eur. Oral Res. 2019, 53, 25–31. [CrossRef] [PubMed]
8. Wang, J.; Wang, L.; Fan, Y. Adverse Biological Effect of TiO2 and Hydroxyapatite Nanoparticles Used in Bone Repair and Replacement. Int. J. Mol. Sci. 2016, 17, 798. [CrossRef]
9. Hallab, N.J.; Jacobs, J.J. Biologic effects of implant debris. Bull. NYU Hosp. Jt. Dis. 2009, 67, 182–188. [PubMed]
10. Hallab, N.J. Biologic Responses to Orthopedic Implants: Innate and Adaptive Immune Responses to Implant Debris. Spine 2016, 41, S30–S31. [CrossRef] [PubMed]

11. Reddy, A.; Caicedo, M.S.; Samelko, L.; Jacobs, J.J.; Hallab, N.J. Implant debris particle size affects serum protein adsorption which may contribute to particle size-based bioreactivity differences. J. Long Term Eff. Med. 2014, 24, 77–88. [CrossRef] [PubMed]
12. Bjursten, L.M.; Rasmusson, L.; Oh, S.; Smith, G.C.; Brammer, K.S.; Jin, S. Titanium dioxide nanotubes enhance bone bondingin vivo. J. Biomed. Mater. Res. Part A 2009, 92, 1218–1224. [CrossRef]
13. Von Wilmowsky, C.; Bauer, S.; Roedl, S.; Neukam, F.W.; Schmuki, P.; Schlegel, K.A. The diameter of anodic TiO2 nanotubes affects bone formation and correlates with the bone morphogenetic protein-2 expression in vivo. Clin. Oral Implant. Res. 2012, 23, 359–366. [CrossRef] [PubMed]
14. Brammer, K.S.; Frandsen, C.J.; Jin, S. TiO2 nanotubes for bone regeneration. Trends Biotechnol. 2012, 30, 315–322. [CrossRef] [PubMed]
15. Xing, H.; Komasa, S.; Taguchi, Y.; Sekino, T.; Okazaki, J. Osteogenic activity of titanium surfaces with nanonetwork structures. Int. J. Nanomed. 2014, 9, 1741–1755. [CrossRef] [PubMed]
16. Rani, V.D.; Vinoth-Kumar, L.; Anitha, V.; Manzoor, K.; Deepthy, M.; Shantikumar, V.N. Osteointegration of titanium implant is sensitive to specific nanostructure morphology. Acta Biomater. 2012, 8, 1976–1989. [CrossRef] [PubMed]
17. Baker, E.A.; Vara, A.D.; Salisbury, M.R.; Fleischer, M.M.; Baker, K.C.; Fortin, P.T.; Roberts, R.V.; Friedrich, C.R. Titania nanotube morphologies for osseointegration via models of in vitro osseointegrative potential and in vivo intramedullary fixation. J. Biomed. Mater. Res. Part B Appl. Biomater. 2020, 108, 1483–1493. [CrossRef]
18. Wang, J.; Meng, F.; Song, W.; Jin, J.; Ma, Q.; Fei, D.; Fang, L.; Chen, L.; Wang, Q.; Zhang, Y. Nanostructured titanium regulates osseointegration via influencing macrophage polarization in the osteogenic environment. Int. J. Nanomed. 2018, 13, 4029–4043. [CrossRef] [PubMed]
19. Bai, L.; Liu, Y.; Du, Z.; Weng, Z.; Yao, W.; Zhang, X.; Huang, X.; Yao, X.; Crawford, R.; Hang, R.; et al. Differential effect of hydroxyapatite nano-particle versus nano-rod decorated titanium micro-surface on osseointegration. Acta Biomater. 2018, 76, 344–358. [CrossRef] [PubMed]
20. Shokuhfar, T.; Hamlekhan, A.; Chang, J.-Y.; Choi, C.K.; Sukotjo, C.; Friedrich, C. Biophysical evaluation of cells on nanotubular surfaces: The effects of atomic ordering and chemistry. Int. J. Nanomed. 2014, 9, 3737–3748. [CrossRef]

21. Shin, D.H.; Shokuhfar, T.; Choi, C.K.; Lee, S.-H.; Friedrich, C. Wettability changes of TiO2nanotube surfaces. Nanotechnology 2011, 22, 315704. [CrossRef]

22. Friedrich, C.R.; Kolati, M.; Moser, T.; Sukotjo, C.; Shokuhfar, T. Survivability of TiO2nanotubes on the surface of bone screws. Surf. Innov. 2014, 2, 60–68. [CrossRef]

23. Gupta, A.; Liberati, T.A.; Verhulst, S.J.; Main, B.J.; Roberts, M.H.; Potty, A.G.R.; Pylawka, T.K.; Iii, S.F.E.-A. Biocompatibility of single-walled carbon nanotube composites for bone regeneration. Bone Jt. Res. 2015, 4, 70–77. [CrossRef] [PubMed]

24. Back, D.A.; Pauly, S.; Rommel, L.; Haas, N.P.; Schmidmaier, G.; Wildemann, B.; Greiner, S.H. Effect of local zoledronate on implant osseointegration in a rat model. BMC Musculoskelet. Disord. 2012, 13, 42. [CrossRef] [PubMed]

25. Su, Y.; Komasa, S.; Sekino, T.; Nishizaki, H.; Okazaki, J. Nanostructured Ti6Al4V alloy fabricated using modified alkali-heat treatment: Characterization and cell adhesion. Mater. Sci. Eng. C 2016, 59, 617–623. [CrossRef] [PubMed]

26. Pauly, S.; Back, D.A.; Kaeppler, K.; Haas, N.P.; Schmidmaier, G.; Wildemann, B. Influence of Statins locally applied from orthopedic implants on osseous integration. BMC Musculoskelet. Disord. 2012, 13, 208. [CrossRef]

27. Virdi, A.S.; Irish, J.; Sena, K.; Liu, M.; Ke, H.Z.; McNulty, M.A.; Sumner, D.R. Sclerostin Antibody Treatment Improves Implant Fixation in a Model of Severe Osteoporosis. J. Bone Jt. Surg. Am. Vol. 2015, 97, 133–140. [CrossRef] [PubMed]

28. Betoni, W.; Queiroz, T.P.; Luvizuto, E.R.; Valentini-Neto, R.; Garcia-Júnior, I.R.; Bernabé, P.F.E. Evaluation of Centrifuged Bone Marrow on Bone Regeneration Around Implants in Rabbit Tibia. Implant. Dent. 2012, 21, 481–485. [CrossRef] [PubMed]

29. Freire, A.R.; Rossi, A.C.; Queiroz, T.P.; Gulinelli, J.L.; Souza, F.A.; Margonar, R.; Garcia-Júnior, I.R.; Hochuli-Vieira, E.; Okamoto, R. Histometric Analysis of Bone Repair in Bone-Implant Interface Using a Polylactic/Polyglycolic Acid Copolymer Associated With Implants in Rabbit Tibia. J. Oral Implant. 2012, 38, 449–457. [CrossRef]

30. Nich, C.; Marchadier, A.; Sedel, L.; Petite, H.; Vidal, C.; Hamadouche, M. Decrease in particle-induced osteolysis in ovariectomized mice. J. Orthop. Res. 2009, 28, 178–183. [CrossRef]

31. Popat, K.C.; Leoni, L.; Grimes, C.A.; Desai, T.A. Influence of engineered titania nanotubular surfaces on bone cells. Biomaterials 2007, 28, 3188–3197. [CrossRef] [PubMed]

32. Liu, X.; Zhou, X.; Li, S.; Lai, R.; Zhou, Z.; Zhang, Y.; Zhou, L. Effects of titania nanotubes with or without bovine serum albumin loaded on human gingival fibroblasts. Int. J. Nanomed. 2014, 9, 1185–1198. [CrossRef]

33. Minagar, S.; Wang, J.; Berndt, C.C.; Ivanova, E.P.; Wen, C. Cell response of anodized nanotubes on titanium and titanium alloys. J. Biomed. Mater. Res. Part A 2013, 101, 2726–2739. [CrossRef]

34. Rangamani, P.; Lipshtat, A.; Azeloglu, E.U.; Calizo, R.C.; Hu, M.; Ghassemi, S.; Hone, J.; Scarlata, S.; Neves, S.R.; Iyengar, R. Decoding Information in Cell Shape. Cell 2013, 154, 1356–1369. [CrossRef] [PubMed]

35. Wang, J.-X.; Fan, Y.-B.; Gao, Y.; Hu, Q.-H.; Wang, T.-C. TiO2 nanoparticles translocation and potential toxicological effect in rats after intraarticular injection. Biomaterials 2009, 30, 4590–4600. [CrossRef] [PubMed]

36. Fan, X.; Feng, B.; Liu, Z.; Tan, J.; Zhi, W.; Lu, X.; Wang, J.; Weng, J. Fabrication of TiO2 nanotubes on porous titanium scaffold and biocompatibility evaluation in vitro and in vivo. J. Biomed. Mater. Res. Part A 2012, 100, 3422–3427. [CrossRef] [PubMed]

37. Bayat, N.; Lopes, V.R.; Schölermann, J.; Jensen, L.D.; Cristobal, S. Vascular toxicity of ultra-small TiO2 nanoparticles and single walled carbon nanotubes in vitro and in vivo. Biomaterials 2015, 63, 1–13. [CrossRef] [PubMed]

38. Ames, B.N.; Lee, F.D.; Durston, W.E. An Improved Bacterial Test System for the Detection and Classification of Mutagens and Carcinogens. Proc. Natl. Acad. Sci. USA 1973, 70, 782–786. [CrossRef] [PubMed]

39. Naya, M.; Kobayashi, N.; Ema, M.; Kasamoto, S.; Fukumuro, M.; Takami, S.; Nakajima, M.; Hayashi, M.; Nakanishi, J. In vivo genotoxicity study of titanium dioxide nanoparticles using comet assay following intratracheal instillation in rats. Regul. Toxicol. Pharmacol. 2012, 62, 1–6. [CrossRef]

40. Neacsu, P.; Mazare, A.; Cimpean, A.; Park, J.; Costache, M.; Schmuki, P.; Demetrescu, I. Reduced inflammatory activity of RAW 264.7 macrophages on titania nanotube modified Ti surface. Int. J. Biochem. Cell Biol. 2014, 55, 187–195. [CrossRef]

41. Östberg, A.-K.; Dahlgren, U.; Sul, Y.-T.; Johansson, C.B. Inflammatory cytokine release is affected by surface morphology and chemistry of titanium implants. J. Mater. Sci. Mater. Electron. 2015, 26, 1–9. [CrossRef]

42. Latha, T.S.; Reddy, M.C.; Durbaka, P.V.R.; Muthukonda, S.V.; Lomada, D. Immunomodulatory properties of titanium dioxide nanostructural materials. Indian J. Pharmacol. 2017, 49, 458–464. [CrossRef]

43. Smith, B.S.; Yoriya, S.; Grissom, L.; Grimes, C.A.; Popat, K.C. Hemocompatibility of titania nanotube arrays. J. Biomed. Mater. Res. Part A 2010, 95, 350–360. [CrossRef] [PubMed]

44. Radziun, E.; Wilczyn´ska, J.D.; Ksia˛z˙ek, I.; Nowak, K.; Anuszewska, E.; Kunicki, A.; Olszyna, A.; Za˛bkowski, T. Assessment of the cytotoxicity of aluminium oxide nanoparticles on selected mammalian cells. Toxicol. Vitr. 2011, 25, 1694–1700. [CrossRef] [PubMed]

45. Alshatwi, A.A.; Subbarayan, P.V.; Ramesh, E.; Al-Hazzani, A.A.; AlSaif, M.A.; Alwarthan, A.A. Al2O3 Nanoparticles Induce Mitochondria-Mediated Cell Death and Upregulate the Expression of Signaling Genes in Human Mesenchymal Stem Cells. J. Biochem. Mol. Toxicol. 2012, 26, 469–476. [CrossRef] [PubMed]

46. Swiatkowska, I.; Martin, N.G.; Henckel, J.; Apthorp, H.; Hamshere, J.; Hart, A.J. Blood and plasma titanium levels associated with well-functioning hip implants. J. Trace Elements Med. Biol. 2020, 57, 9–17. [CrossRef]

47. Agency for Toxic Substances and Disease Registry (ATSDR). Toxicological Profile for Aluminum; Department of Health and Human Services, Public Health Service: Atlanta, GA, USA, 2008.

48. Agency for Toxic Substances and Disease Registry (ATSDR). Toxicological Profile for Vanadium; Department of Health and Human Services, Public Health Service: Atlanta, GA, USA, 2012.

49. Maerz, T.; Fleischer, M.; Newton, M.D.; Davidson, A.; Salisbury, M.; Altman, P.; Kurdziel, M.D.; Anderson, K.; Bedi, A.; Baker, K.C. Acute mobilization and migration of bone marrow-derived stem cells following anterior cruciate ligament rupture. Osteoarthritis Cartilage 2017, 25, 1335–1344. [CrossRef]

06

POTENTIAL SYNTHESIS OF THERANOSTIC BORON NITRIDE NANOTUBES (^{64}CU-BNNTS) THROUGH NEUTRON CAPTURE RADIOLABELING

1. INTRODUCTION: AN OVERVIEW

In the realm of nanotechnology, the synthesis of advanced materials for theranostic applications has garnered significant attention due to their potential in revolutionizing targeted diagnostics and therapies. Among these materials, Boron Nitride Nanotubes (BNNTs) have emerged as promising candidates owing to their unique structural, mechanical, and chemical properties. This paper explores the prospective production of theranostic Boron Nitride Nanotubes labelled with the radioisotope Copper-64 (64Cu-BNNTs) through a method involving neutron capture. The integration of radiolabeling techniques enhances the multi-functionality of BNNTs, enabling simultaneous imaging and therapeutic capabilities for precision medicine applications. This study delves into the synthesis process, characterization, and potential applications of these radiolabeled BNNTs, shedding light on their role in advancing the field of theranostics.

The discovery and development of new materials are often the catalysts for techno- logical advances, particularly when they can be applied to various areas of research. A historic milestone in the search for new materials occurred in 1995 when boron nitride nanotubes (BNNTs) emerged as a key material in nanotechnology science [1]. BNNTs have cylindrical structures (one-dimensional—1D), formed only by atoms of boron (B) and nitrogen (N), with diameters in the order of nanometers and lengths in the order of microns [2]. Similar to single-wall carbon nanotubes (SWCNT), depending on the angle θ in which the sheet is rolled, nanotube structures are formed with armchair ($\theta = 30°$), zig-zag ($\theta = 0°$) and chiral ($0 < /\theta/ < 30°$) forms [3]. Otherwise, the multi-walled BNNTs are formed by several sheets of hexagonal boron nitride (h-BN) concentrically wrapped. Both BNNTs have excellent chemical, physical and mechanical properties [4–8] and a band gap of approximately 5.5 eV [3]. Several methods have been used for the synthesis of BNNTs [5]; however, chemical vapour deposition (CVD) is the most used method and requires a simple apparatus which produces BNNTs with excellent structural quality [9].

The interest in BNNT-based technologies in recent years can be measured by the large number of scientific works that have been published, in addition to the increase in large-scale production [10]. Within the class of nanostructured materials, the BNNTs have great potential for several biomedical applications. Some early studies demonstrated that nanotubes did not show toxicity at concentrations below 50 µg/mL [11]. Furthermore, they favour the reuptake of molecules into the cell interior and can be functionalized with different biological protein epitopes [12]. Recently, BNNTs have been used as nanovectors for DNA, drugs and radioisotopes, and as boosters for biomaterials. In 2012, Soares et al. [13] used BNNTs radiolabeled with 99mTc to investigate the cell-distribution behaviour in vivo through a process of passive accumulation in solid tumors. Diverse studies applying BNNTs to cancer treatment have been reported. For example, when linked to target molecules, BNNTs could be used as therapeutic agents capable of killing cancer cells by boron neutron capture therapy. This medical approach is generally applied in brain cancer treatments, and it is based on the capture of the neutron reaction 10B (n,α) 7Li, where a 10B atom captures a low-energy thermal neutron and then decays to produce 4He (alpha particles) and 7Li, resulting in a dense ionizing radiation which is capable of destroying the cells where the reaction takes place [12]. Another potential application of BNNTs is in diagnostic medicine. In this sense, BNNTs doped with rare earth beta-emitters with short half-lives, such as 153Sm and 159Gd, can also be used as radioisotopes for imaging [14].

In this context, nanotechnology has revolutionized so-called traditional medicine by introducing novel concepts and methods that had never been imagined. Thus, nanomedicine has improved the diagnosis of various diseases through techniques based on magnetism or nuclear reactions with different electronic devices, using biosensors or radioisotope-doped nanomaterials. In this way, the study of a more accurate diagnostic method using novel technologies is as relevant a goal as the prevention and treatment of oncological diseases. Therefore, a class of new nanomaterials, in which boron nitride nanotubes (BNNTs) stand out, has been the target of studies that have led to an understanding of the correlation between their structure and properties, which enables their use in diagnostic medicine. Due to their empty internal spaces, BNNTs can be filled by different chemical species, such as enzymes, noble metals, rare earths, and radioisotopes, especially copper-64 (64-Cu), which allows this type of material to be applied as a biological marker and in diagnostic medicine. For instance, copper-64 (T1/2 = 12.7 h; β+, 0.653 MeV (17.8%); β-, 0.579 MeV (38.4%)) has decay characteristics that allow it to be applied to obtain images of positron emission tomography (PET-scan) and in cancer-directed radiotherapy. Copper, for instance, has already well-established coordination chemistry that allows its reaction with an extensive variety of chelating systems that could potentially be linked to peptides and other interesting biological molecules such as antibodies, proteins, and nanoparticles. Its specific half-life expands the ability to image molecules of various dimensions, mainly including the slower compensating proteins and nanoparticles. Due to the versatility of applications of 64-Cu, a significant increase in scientific and technical publications has been seen over the last 2 decades, mainly in PET-scan imaging, but also in targeted cancer radiotherapy.

Thus, this work aimed to synthesize and characterize 64Cu-BNNTs with appreciable properties that suggest numerous multifunctional applications, with advantages for cancer diagnosis and therapy, such as: (i) increased bioavailability; (ii) reduction in systemic ad- verse effects, thereby increasing patient comfort and adherence to treatment; (iii) improved osteogenic differentiation response promoted by the 64Cu-BNNTs system and targeting of tumor cells, among others. It is also important to mention that the combination of 64Cu-BNNTs has not yet been reported in the literature.

2. EXPERIMENTAL METHODS

2.1 Raw Materials

Copper (II) chloride dihydrate (99.999%), iron (III) oxide nano powder (<50 nm particle size) and amorphous boron powder (≥95%) were obtained from Sigma Aldrich Brazil-Ltda, Sao Paulo, Brazil (CAS Number 10125-13-0) and used as received.

2.2 Synthesis and Purification of Boron Nitride Nanotubes

BNNTs were processed from mixing amorphous boron and iron (III) oxide powder (ratio 0.02) in a horizontal tubular reactor. This reactor consisted of an alumina with an inlet and outlet for the flow of ammonia and nitrogen gases. The synthesis was carried out under a NH3/N2 atmosphere at a 150/20 sccm (standard cubic centimetres per minute) flow rate with a heating rate of 10 ◦C min−1 from room temperature up to 1200 ◦C. An isotherm was maintained for 2 h. After this step was completed, the reactor was cooled down to room temperature under a N2 atmosphere.

The synthesized BNNTs were purified using sulfuric and nitric acids in the ratio of 3:1, respectively. The reaction mixture was kept under stirring and reflux conditions at 80 ◦C for 2 h, followed by the filtration process. The resulting solid was washed with deionized water and oven-dried for 4 h at 110 ◦C. In this process, hydroxyl groups (-OH) were introduced into the structure of the tubes.

2.2.1 Activation Process of ^{64}Cu Radioisotope

The radioisotope 64Cu was obtained by neutron activation of the copper (II) chloride dihydrate sample in a nuclear research reactor (TRIGA Mark-1) at CDTN (Belo Horizonte, Brazil) by the neutron capture reaction 63Cu(n,γ)64Cu. The irradiation was performed on 20 mg samples over 8 h under a thermal neutron flux of 6.6 × 1011 cm−2s−1. The theoretical induced activities were estimated according to the research of Zangirolami et al. [15]. The calculations were carried out while considering the amount of Cu in the sample and using the thermal neutron capture cross-sections as a reference, in accordance with an IAEA (International Atomic Energy Agency) publication [16].

2.2.2 Incorporation of Cu and ^{64}Cu to the BNNT Samples

The BNNT (100 mg) sample was dispersed in anhydrous ethanol. With the aid of an autoclave with a polytetrafluoroethylene (PTFE) vessel, the Cu and 64Cu radioisotope were incorporated into the BNNTs. The incorporation reaction was carried out in an oven at a temperature of 180 ◦C for two hours. After this period, the material was cooled to room temperature and filtered. The radiochemical purity of the sample was assessed by gamma spectroscopy, using an HP-Ge detector (Ortec Ametek, Oak Ridge, TN, USA) with 25% efficiency, and analyzed using the Canberra Genie 2000 software, Meriden, CT, USA [17]. The evaluation of the specific activity was carried out using a CRC 15R activimeter that Nanomaterials 2021, 11, For Peer Review had been previously calibrated for copper-64 emission. Figure 1a,d schematica4lloyf s1h4ow all stages of 64Cu-BNNT production.

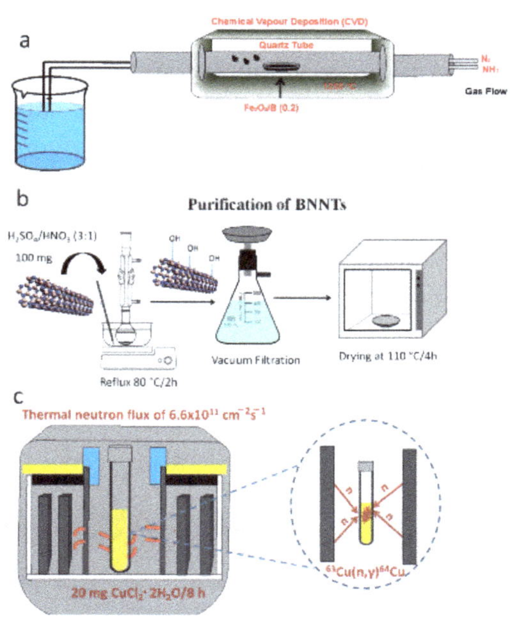

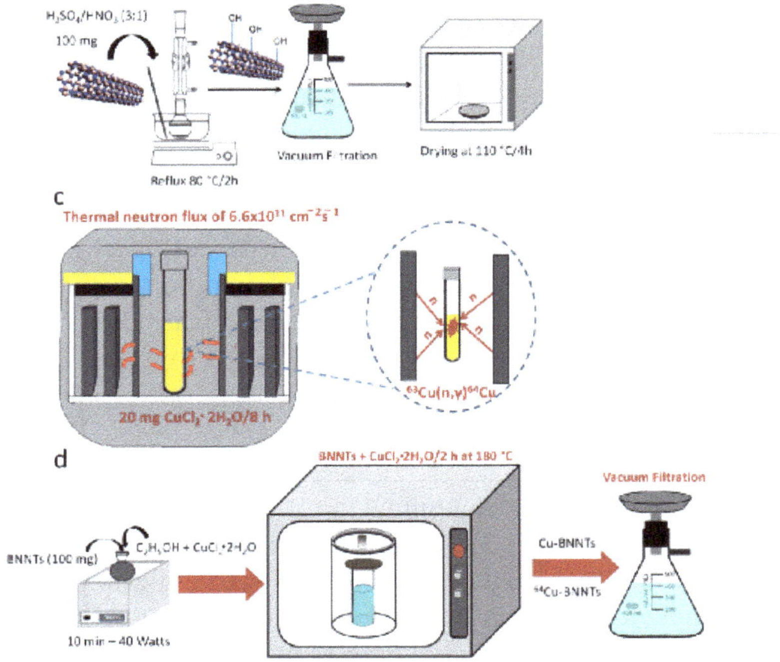

Figure 1. Schematic representation of synthesis of the BNNTS (a), purification process (b), activation process of CuCl2 • H O to obtainment of 64Cu radioisotope (c) and incorporation of Cu and 64Cu Figure 1. Schematic representation of synthesis of the BNNTS (a), purification process (b), activation into BNNTs samples (d).

3. CHARACTERIZATION

FTIR measurements of the BNNT and Cu-BNNT samples were performed with a Bruker model Vertex 70v instrument (Belo Horizonte, Brazil). The spectra were collected in ATR mode with 64 accumulations, a resolution of 4 cm^{-1}, and in the 4500–300 cm^{-1} region in transmission mode and then were systematically adjusted; baseline corrections were considered for this analysis. An ultima IV Rigaku Diffractometer with Cu-K radiation was employed to study the main crystalline phases in the synthesized BNNT and Cu-BNNT samples (Belo Horizonte, Brazil). The Bragg's angle values were measured in the 10–80 range, with a scanning rate of 0.02 min−1. XPS spectra were obtained using monochromatic Al K radiation (1486.6 eV) with an electron energy analyzer (Specs, Phoibos-150) that enabled high-energy resolution and an excellent signal-to-noise ratio (Belo Horizonte, Brazil). The signal of adventitious carbon (C 1 s at 284.6 eV) was used to correct the binding-energy scale of the survey and the high-resolution spectra. Highresolution spectra in the regions of interest were fitted assuming its shape as a convolution of Lorentzian and Gaussian functions of different components, and the background contribution was removed by the Shirley method [18,19]. SEM analysis was performed with a Carl Zeiss Field Emission Scanning Electron Microscope, model sigma VP (Belo Horizonte, Brazil), operating in vacuum with an electron-beam-acceleration voltage between 5 and 30 kV. The BNNT and Cu-BNNT powders were deposited directly onto the carbon tape. The Transmission Electron Microscopy (TEM) images were obtained on a FEI TEM-LaB6 TECNAI G2 microscope (Belo Horizonte, Brazil), with a tungsten-filament electron gun operating at 200 kV. Samples were dispersed in acetone for 30 min using a water bath sonicator and one drop was deposited onto a 200-mesh holey carbon–copper grid. The activity of the 64Cu-BNNTs after irradiation was obtained from the gamma spectrum, using an HP-Ge detector (Belo Horizonte, Brazil), with a nominal efficiency of 25%, and the Canberra Genie 2000 software.

4. RESULTS AND DISCUSSION

The FTIR spectrum was obtained in order to identify the vibrational modes in the BNNT samples (Figure 2). The absorption peaks between 3400 and 3200 cm–1could be attributed to the vibrational modes of the hydroxyl groups (-OH) from water molecules adsorbed on the sample surface [20,21]. However, it could also be attributed to the presence of copper hydroxide. The region between 2000 and 60 cm–1 has several peaks (Figure 2b,c). The ell-known longitudinal (LO) vibrations along the axis resonate sharply around 1369 cm^{-1}, and a second signal (1545 cm^{-1}) appears for tangential (T) circumferential in-plane modes (B-N). These T modes should be dependent on the diameter (curvature) but seem to only be visible in highly pure, crystalline BNNTs [22]. Another typical absorption peak for BNNTs is located around 790 cm^{-1} and is related to out-of-plane B-N-B bending (B-N-B) vibrations [20,21,23]. In both spectra, the peaks between the 1100 and 880 cm^{-1} regions give an account of the anti-symmetrical and symmetrical stretching vibrations of B-O bonds in BO3 and BO4 groups formed from B-OH, and peaks at 701.3, 685.8 and 453.3 cm^{-1} are assigned to the bend vibrations of B-O bonds in BO3 and BO_4 groups [24]. The peak at 426.0 cm^{-1} is assigned to the stretching vibrations Cu(II)-O of copper oxide CuO [25].

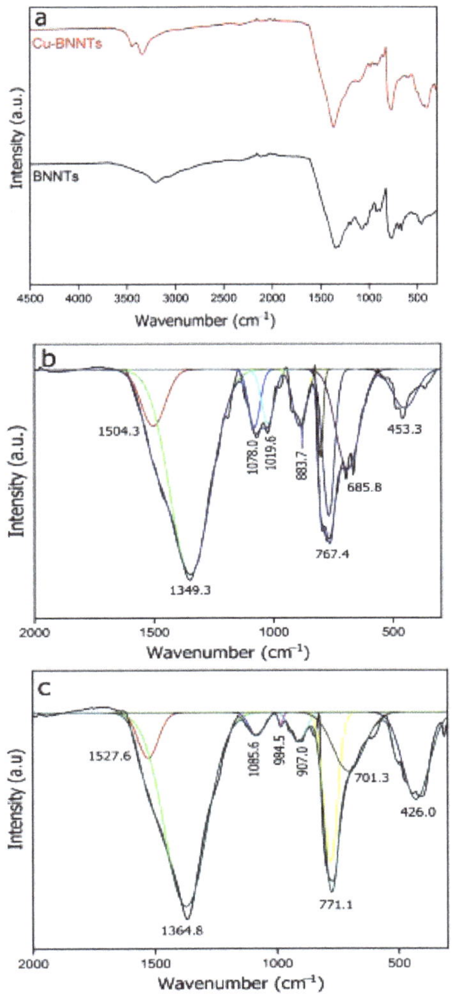

Figure 2. Infrared spectra of (a) BNNTs and Cu-BNNTs in the region between 4500 and 60 cm^{-1}. (a) Highlighted regions between 2000 and 60 cm-1 for (b) BNNTs and (c) Cu-BNNTs.

The XRD of BNNTs and Cu-BNNTs is shown in Figure 3. An intense peak close to 2θ = 26.65° (Figure 3a) corresponds to the plane (002) and is attributed to the main peak of the h-BN structure. Peaks assigned to h-BN are also observed at 2θ = 41.78°, 42.81°, 50.16°, 55.09°, 59.40° and 76.05° which correspond to the (100), (101), (102), (004), (103) and (110) planes, respectively [14,23,26]. After the introduction of Cu nanoparticles to the BNNTs, new diffraction peaks were observed (Figure 3b),so the region between 30° and 80° was highlighted. The presence of CuO and Cu_2O nanoparticles were identified at 36.89°, 39.71° and 65.3°, which may have occurred due to the exposure of the nanoparticles to the surrounding environment during characterization [27]. The characteristic diffraction peaks of copper nanoparticles located at 32.42◦ and 44.81◦ were observed. They correspond to the (110) and (200) crystallographic planes of face-center cubic (fcc), respectively [27,28]. Debye–Scherrer's equation, i.e., $D = 0.9 \times \lambda/(\beta \times \cos\theta)$, was used to calculate the size of copper nanoparticles, where D represents crystalline size, 0.9 is Scherrer's constant, λ is the wavelength of the X-ray, β is the full width at the half-maximum of the diffraction peak (FWHM) and θ represents Bragg's angle [27,29]. The calculations were performed using the mean values of the FWHM of the peaks, with 2θ of 32.4◦ and 44.8◦. The average size of the Cu nanoparticles was 16 nm. This is a dimension in which nanoparticles can penetrate tumor cells through the Increased Permeability and Retention Effect (EPR), thus they can be used as a theranostic nanomaterials.

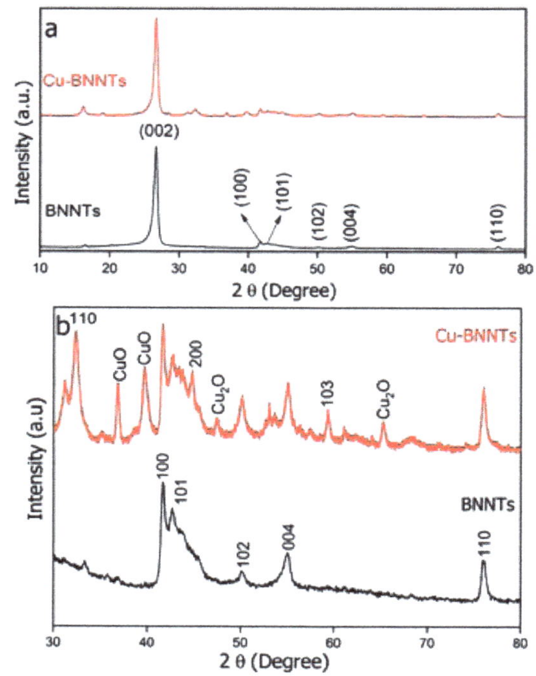

Figure 3. XRD of BNNTs and Cu-BNNTs in the region between 10° and 80° (a). Highlighted regions between 30° and 80° (b). (b).

Figure 4 shows the XPS survey of the samples. In all the survey spectra, only the presence of B (190.10 eV), N (398.12 eV), C (284.49 eV) and O (532.51 eV) was identified. The C and O signals were also identified on the BNNT surface. The presence of C is related to the surface contamination that usually occurs during the preparation process and to the exposure of the specimens to air, and the presence of O is due to the purification process and is commonly observed in XPS measurements. The presence of F (689.51 eV) and Si (102.53 eV) is due to residues from the vessel that was used for synthesis and the Cl (198.53 eV) comes from the reagent CuCl2 2H2O that was used for the synthesis of copper nanoparticles.

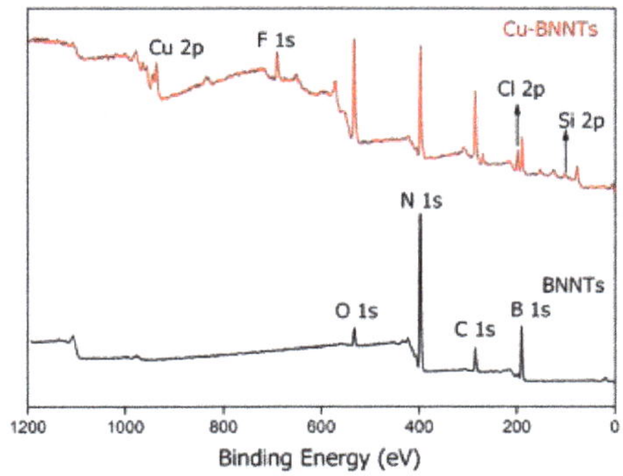

Figure 4. Survey XPS spectra for BNNTs and Cu-BNNTs.

Table 1 shows the data obtained from the XPS survey. The stoichiometry rate of boron and nitrogen atoms (B:N) is confirmed by the peak areas of the XPS survey spectra. The obtained value for the B:N ratio is 1.09. A similar result was obtained by Silva et al. and Juan Li et al. [21,30]. According to these studies, the non-stoichiometric BN nanotubes presentan excess of B atoms, and oxygen-doping in the h-BN network leads to the formation of ternary BN_xO_y species [31]. Hydroxyl groups increase in modulus to negative charges on the tubes' surfaces. As copper nanoparticles have positively charged surfaces, we believe that an electrostatic interaction occurs between the two nanostructures.

Table 1. Surface composition (at.%) and B: N ratio, as determined by XPS for Cu-BNNT sample.

O 1s	532.51	16.047
C 1s	284.49	26.010
N 1s	398.12	21.914
F 1s	689.51	4.179
Cu 2p	936.50	2.771
Cl 2p	198.53	3.004
B 1s	190.10	23.936
Si 2p	102.53	2.138
B:N	-	1.090

The B 1s, N 1s and O 1s core-level photoemission spectra for all samples are shown in Figure 5. The B 1s peak (Figure 5a,b) at 190.1 eV and N 1s peak (Figure 5c,d) at 398.0 eV correspond to the B–N bonding, matching the BE values reported for bulk h-BN [21]. The component at 188.3 eV (Figure 5a,b) is attributed to B bonded to Fe from the catalyst (Fe2B) [32]. A minor contribution of boron oxide (B_2O_3) was also identified at 192.4 eV (Figure 5a). After the synthesis of Cu nanoparticles, two new contributions were observed at 200.4 and 197.0 eV (Figure 5b); the first is attributed to Cl II and the second to the presence of metallic copper (Cu 0). In addition, the binding energies at 396.6 eV and 400.0 eV are attributed to B–N–B bonding [33] and O-B-N bonding [14], respectively. The O 1s spectra are also shown in Figure 5e,f. Both samples show peaks at 535.5, 532.9 and 530.2 eV, respectively. The first is associated with oxygen atoms bonded to O-H from the purification process, the second is characteristic of B–O bonds (B_2O_3) and the third is associated with oxygen atoms bonded to Fe-O (Fe_2O_3), which is related to the synthesis process [26].

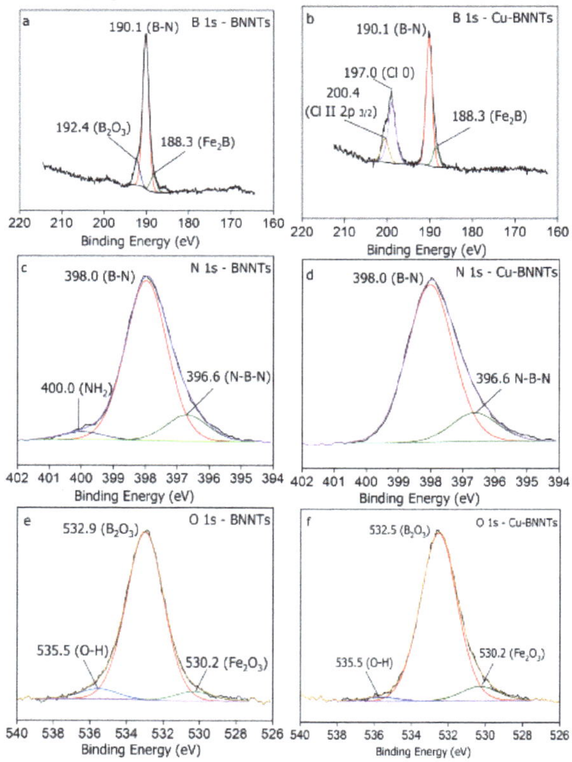

Figure 5. B 1s, N 1s and O 1s high-resolution XPS spectra for (a,c,e) BNNTs, (b,d,f) Cu-BNNTs.

Figure 6 depicts the high-resolution spectrum XPS for Cu-BNNTs. The dominant Cu $2p_{3/2}$ peaks at 933.8 and Cu $2p_{1/2}$ at 953.6 eV, which demonstrate the successful reduction of $CuCl_2$, can be assigned to metallic Cu(0), whereas the small peak at 936.2 eV can be assigned to Cu(II). Meanwhile, the peaks of Cu $2p_{1/2}$ at 955.9 eV, in combination with the satellite peak at 944.3 eV, are typical characteristics of CuO, implying the uniform surface oxidation of Cu nanoclusters that were exposed to air under ambient conditions [34,35].Cu(0) and Cu(I) are hard to differentiate since they have a ~0.3 eV separation in binding energy, whereas Cu(0) and Cu(II) have more than a 2eV separation. The Cu $2p_{5/2}$ peak around 958.1 eV indicates a Cu(II) oxidation state [36,37].

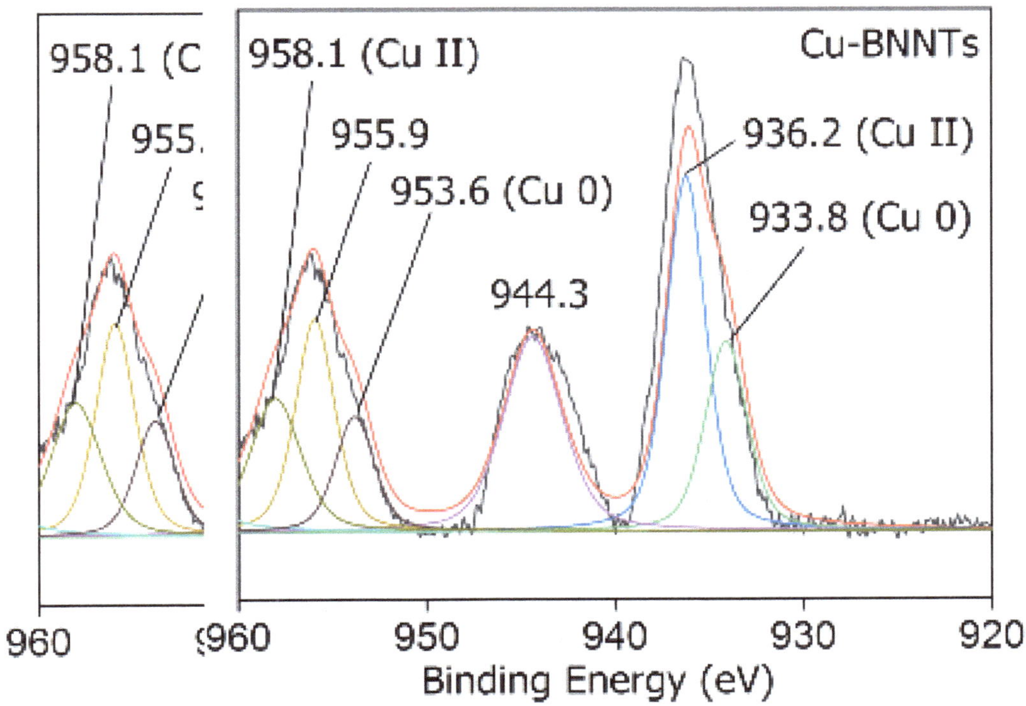

Figure 6. High-resolution spectrum XPS for Cu-BNNTs.

The morphological characteristics of the produced samples were studied using SEM and TEM microscopy, shown in Figures 7 and 8, respectively. It was observed in both samples that the produced BNNTs have non-uniform lengths and diameters. This feature is due to the nanotubes entanglement during their growth process, which is very common during the synthesis process. This nanotube characteristic was also observed in our previous work [23].

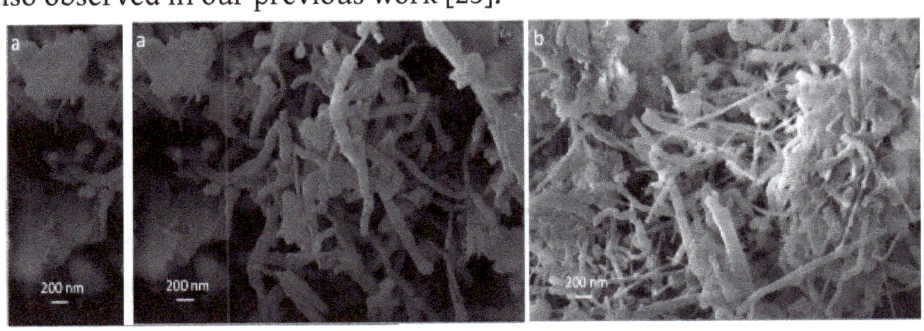

Figure 7. SEM images of BNNTs (a) and Cu-BNNTs (b).

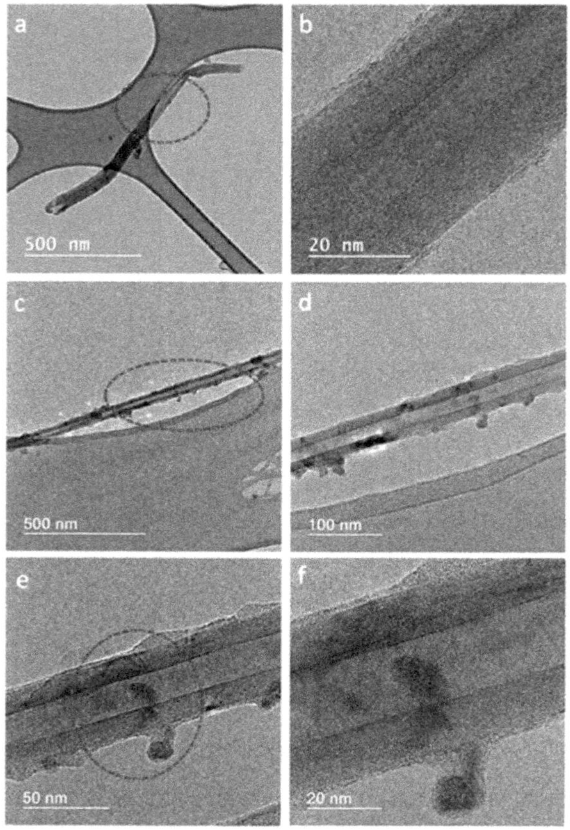

Figure 8. TEM images of BNNTs (a,b) and Cu-BNNTs (c–f).

TEM images are shown in Figure 8a–f. The highlighted regions in Figure 8a,c,e are shown in high-resolution images in Figure 8b,d,f. All images represent typical nanotubes with defined internal channels and external walls that are structurally well-organized [14]. When comparing Figure 8a,b with Figure 8c–f, particles with higher electron absorptions, appearing as darker sites isolated from each other, can be associated with the presence of an electron-conducting nanostructure, indicating the presence of metallic Cu [33]. It is expected that combining these metal nanostructures with BNNTs can improve the stability of the nanoparticles within the solution, allowing their application as theranostic nanomaterials. After increasing image resolution (Figure 8d–f) it was observed that Cu nanoparticles are present in great numbers on the tube surfaces and inside the channels. Referring to the scale bar, it is important to note that the size of nanoparticles is close to 16 nm, as determined by the Debye–Scherrer,s equation.

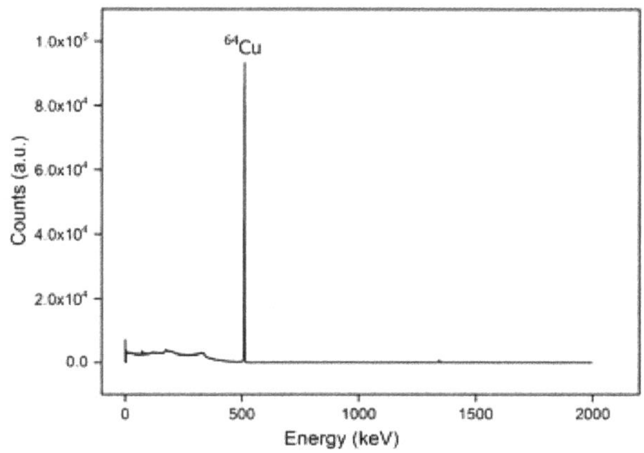

Figure 9. Gamma spectrum of ^{64}Cu-BNNTs

5. CONCLUSIONS

The ^{64}Cu nanostructures were incorporated within the BNNTs structure by a solvo-thermal method which produces stable and contaminant-free radioisotopes. Using XDR data and Debye–Scherrer's equation, it was possible to determine that metallic Cu nanoparticles have sizes of about 16 nm. The TEM images showed that the BNNTs are structurally well- organized, presenting Cu nanoparticles in their internal channels with an even distribution on their surfaces. The ^{64}Cu nanoparticles in the BNNTs also showed a pure photoemission peak of 511 keV, which is characteristic of gamma radiation. These results corroborate the fact that the studied system has high potential to be used in nuclear medicine as a theranostic material. However, this subject needs to be further explored.

6. CONCISE SUMMARY

In this work, the radioisotope 64Cu was obtained from copper (II) chloride dihydrate in a nuclear research reactor by neutron capture, (63Cu(n,γ)64Cu), and incorporated into boron nitride nanotubes (BNNTs) using a solvo-thermal process. The produced 64Cu-BNNTs were analyzed by TEM, MEV, FTIR, XDR, XPS and gamma spectrometry, with which it was possible to observe the formation of64Cu nanoparticles, with sizes of up to 16 nm, distributed through nanotubes. The synthesized of 64Cu nanostructures showed a pure photoemission peak of 511 keV, which is characteristic of gamma radiation. This type of emission is desirable for Photon Emission Tomography (PET scan) image acquisition, as well as its use in several cancer treatments. Thus, 64Cu-BNNTs present an excellent alternative as theranostic nanomaterials that can be used in diagnosis and therapy by different techniques used in nuclear medicine.

REFERENCES

1. Chopra, N.G.; Luyken, R.J.; Cherrey, K.; Crespi, V.H.; Cohen, M.L.; Louie, S.G.; Zettl, A. Boron Nitride Nanotubes. *Science* **1995**, *269*, 966–967. [CrossRef]
2. Terrones, M.; Romo-Herrera, J.M.; Cruz-Silva, E.; López-Urías, F.; Muñoz-Sandoval, E.; Velázquez-Salazar, J.J.; Terrones, H.; Bando, Y.; Golberg, D. Pure and dopedboronnitridenanotubes. *Mater. Today* **2007**, *10*, 30–38. [CrossRef]
3. Golberg, D.; Bando, Y.; Huang, Y.; Terao, T.; Mitome, M.; Tang, C.; Zhi, C. Boron nitride nanotubes and nanosheets. *ACS Nano* **2010**, *4*, 2979–2993. [CrossRef] [PubMed]
4. Chen, Y.; Zou, J.; Campbell, S.J.; le Caer, G. Boron nitride nanotubes: Pronounced resistance to oxidation. *Appl. Phys. Lett.* **2004**, *84*, 2430–2432. [CrossRef]
5. Zhi, C.; Bando, Y.; Tang, C.; Golberg, D. Boron nitride nanotubes. *Mater. Sci. Eng. R Rep.* **2010**, *70*, 92–111. [CrossRef]
6. Golberg, D.; Bando, Y.; Tang, C.; Zhi, C. Boron Nitride Nanotubes. *Adv. Mater.* **2007**, *19*, 2413–2432. [CrossRef]
7. Cohen, M.L.; Zettl, A. The physics of boron nitride nanotubes. *Phys. Today* **2010**, *63*, 34–38. [CrossRef]
8. Roy, S.; Zhang, X.; Puthirath, A.B.; Meiyazhagan, A.; Bhattacharyya, S.; Rahman, M.M.; Babu, G.; Susarla, S.; Saju, S.K.; Tran,M.K.; et al. Structure, Properties and Applications of Two-Dimensional Hexagonal Boron Nitride. *Adv. Mater.* **2021**, *2021*, 2101589. [CrossRef]
9. Ahmad, P.; Khandaker, M.U.; Amin, Y.M.; Khan, Z.R. Influence of growth duration on size and morphology of boron nitride nanotubes grown via chemical vapor deposition technique. *J. Phys. Chem. Solids* **2015**, *85*, 226–232. [CrossRef]
10. Guo, L.; Singh, R.N. Catalytic growth of boron nitride nanotubes using gas precursors. *Phys. E Low-Dimens. Syst. Nanostruct.* **2009**, *41*, 448–453. [CrossRef]
11. Ciofani, G.; Raffa, V.; Menciassi, A.; Cuschieri, A. Boron nitride nanotubes: An innovative tool for nanomedicine. *In Vitro* **2009**, *4*, 8–10. [CrossRef]
12. Ciofani, G.; Danti, S.; Genchi, G.G.; Mazzolai, B.; Mattoli, V. Boron nitride nanotubes: Biocompatibility and potential spill-over in nanomedicine. *Small* **2013**, *9*, 1672–1685. [CrossRef] [PubMed]

13. Soares, D.C.F.; Ferreira, T.H.; Ferreira, C.D.A.; Cardoso, V.N.; de Sousa, E.M.B. Boron nitride nanotubes radiolabeled with 99mTc: Preparation, physicochemical characterization, biodistribution study, and scintigraphic imaging in Swiss mice. *Int. J. Pharm.* **2012**, *423*, 489–495. [CrossRef] [PubMed]

14. da Silva, W.M.; Silva, R.H.d.A.e.; Cipreste, M.F.; Andrade, G.F.; Gastelois, P.L.; Macedo, W.A.d.; de Sousa, E.M.B. Boron nitride nanotubes radiolabeled with 153Sm and 159Gd: Potential application in nanomedicine. *Appl. Radiat. Isot* **2020**, *2020*, 157. [CrossRef] [PubMed]

15. Zangirolami, D.M.; Oliveira, A.H.; Ferreira, A.V. Thermal and Epithermal Neutron Fluence Rates in the Irradiation Facilities of the TRIGA IPR-R1 Nuclear Reactor. *Brazilian J. Phys.* **2010**, *2021*, 40. [CrossRef]

16. Mughabghab, S.F. Resonance Integrals and G-Factors. *Int. Nucl. Data Commun.* **2003**, *11376*, 7.

17. Hossain, I.; Sharip, N.; Viswanathan, K.K. Efficiency and resolution of HPGe and NaI(Tl) detectors using gamma-ray spectroscopy. *Sci. Res. Essays* **2012**, *7*, 86–89. [CrossRef]

18. Végh, J. The Shirley background revised. *J. Electron Spectros. Relat. Phenomena* **2006**, *151*, 159–164. [CrossRef]

19. Végh, J. The analytical form of the Shirley-type background. *J. Electron Spectros. Relat. Phenomena* **1988**, *46*, 411–417. [CrossRef]

20. Nazar, A.M.D.; Silva, W.M.; de Oliveira, E.F.; de Sousa, E.M.B. Boron nitride nanotubes decorated with magnetite nanoparticles for application as a multifunctional system in cancer treatment. *Nano-Struct. Nano-Objects* **2020**, *24*, 100616. [CrossRef]

21. da Silva, W.M.; Monteiro, G.A.A.; Gastelois, P.L.; de Sousa, R.G.; Macedo, W.A.d.; Sousa, E.M.B. Efficient sensitive polymer-grafted boron nitride nanotubes by microwave-assisted process. *Nano-Struct. Nano-Objects* **2018**, *15*, 186–196. [CrossRef]

22. Harrison, H.; Lamb, J.T.; Nowlin, K.S.; Guenthner, A.J.; Ghiassi, K.B.; Kelkar, A.D.; Alston, J.R. Quantification of hexagonal boron nitride impurities in boron nitride nanotubes: Via FTIR spectroscopy. *Nanoscale Adv.* **2019**, *1*, 1693–1701. [CrossRef]

23. da Silva, W.M.; Ribeiro, H.; Ferreira, T.H.; Ladeira, L.O.; Sousa, E.M.B. Synthesis of boron nitride nanostructures from catalyst of iron compounds via thermal chemical vapor deposition technique. *Phys. E Low-Dimens. Syst. Nanostruct.* **2017**, *89*, 177–182. [CrossRef]

24. Petrov, T.; Markova-Deneva, I.; Chauvet, O.; Nikolov, R.; Denev, I. Sem and ft-ir spectroscopy study of Cu, Sn and Cu-Sn nanoparticles. *J. Univ. Chem. Technol. Metall.* **2012**, *47*, 197–206.
25. Amelkovich, Y.A.; Nazarenko, O.B.; Sechin, A.I.; Visakh, P.M. Characterization of copper nanopowders after natural aging. *IOPConf. Ser. Mater. Sci. Eng.* **2015**, *81*, 012072. [CrossRef]
26. da Silva, W.M.; Ferreira, T.H.; de Morais, C.A.; Leal, A.S.; Sousa, E.M.B. Samarium doped boron nitride nanotubes. *Appl. Radiat. Isot* **2018**, *131*, 30–35. [CrossRef]
27. Chauhan, P.S.; Shrivastava, V.; Tomar, R.S. Biofabrication of copper nanoparticles: A next-generation antibacterial agent against wound-associated pathogens. *Turkish J. Pharm. Sci.* **2018**, *15*, 238–247.
28. Betancourt-Galindo, R.; Reyes-Rodriguez, P.Y.; Puente-Urbina, B.A.; Avila-Orta, C.A.; Rodríguez-Fernández, O.S.; Cadenas-Pliego, G.; Lira-Saldivar, R.H.; García-Cerda, L.A. Synthesis of copper nanoparticles by thermal decomposition and their antimicrobial properties. *J. Nanomater.* **2014**, *2014*, 10. [CrossRef]
29. Shirui, Z.; Yabin, D.; Mingyan, Y.; Xiaolong, G.; Xinwei, X.; Yupeng, J.; Yang, X. Microwave annealing effects on ZnO films deposited by atomic layer deposition. *J. Semicond.* **2014**, *35*,1–7.
30. Li, J.; Li, J.; Yin, Y.; Chen, Y.; Bi, X. Water-assisted chemical vapor deposition synthesis of boron nitride nanotubes and their photoluminescence property. *Nanotechnology* **2013**, *24*, 365605. [CrossRef]
31. Hubácek, M.; Sato, T.; Ishii, T. A Coexistence of Boron Nitride and Boric Oxide. *J. Sol.* **1994**, *109*, 384–390. [CrossRef]
32. Ferreira, T.H.; Marino, A.; Rocca, A.; Liakos, I.; Nitti, S.; Athanassiou, A.; Mattoli, V.; Mazzolai, B.; de Sousa, E.M.B.; Ciofani, G. Folate-grafted boron nitride nanotubes: Possible exploitation in cancer therapy. *Int. J. Pharm.* **2015**, *481*, 56–63. [CrossRef]
33. Monteiro, G.A.A.; de Sousa, R.G.; da Silva, W.M.; Gastelois, P.L.; Macedo, W.A.d.; de Sousa, E.M.B. Microwave radiation-assisted covalent functionalization of boron nitride nanotubes and their grafting with cationic thermo and pH-sensitive hydrogel. *Appl. Nanosci.* **2021**, *11*, 505–520. [CrossRef]
34. Jin, Z.; Liu, C.; Qi, K.; Cui, X. Photo-reduced Cu / CuO nanoclusters on TiO_2 nanotube arrays as highly efficient and reusable catalyst. *Sci. Rep.* **2017**, *7*, 39695. [CrossRef] [PubMed]

35. Ghosh, S.; Das, R.; Chowdhury, I.H.; Bhanja, P.; Naskar, M.K. RSC Advances. *RSC Adv.* **2015**, *5*, 101519–101524. [CrossRef]

36. Mondal, P.; Sinha, A.; Salam, N.; Roy, A.S.; Jana, N.R.; Islam, S.M. Enhanced catalytic performance by copper nanoparticle- graphene based composite. *RSC Adv.* **2013**, *3*, 5615–5623. [CrossRef]

37. Fernandez, V.; Kiani, D.; Fairley, N.; Felpin, F.X.; Baltrusaitis, J. Curve fitting complex X-ray photoelectron spectra of graphite- supported copper nanoparticles using informed line shapes. *Appl. Surf. Sci.* **2020**, *505*, 143841. [CrossRef]

38. Kim, J.Y.; Park, H.; Lee, J.C.; Kim, K.M.; Lee, K.C.; Ha, H.J.; Choi, T.H.; An, G.I.; Cheon, G.J. A simple Cu-64 production and itsapplication of Cu-64 ATSM. *Appl. Radiat. Isot* **2009**, *67*, 1190–1194. [CrossRef]

39. Pérot, B.; Jallu, F.; Passard, C.; Gueton, O.; Allinei, P.-G.; Loubet, L.; Estre, N.; Simon, E.; Carasco, C.; Roure, C.; et al. The characterization of radioactive waste: A critical review of techniques implemented or under development at CEA. *EPJ Nucl. Sci. Technol.* **2018**, *4*, 3. [CrossRef]

07 DEVELOPMENT OF A NEW NANOEMULSION FORMULATION CONTAINING METHOTREXATE FOR TREATING RHEUMATOID ARTHRITIS THROUGH SUBCUTANEOUS DELIVERY WITH SPECIFIC TARGETING AT THE TREATMENT SITE

1. INTRODUCTION: AN OVERVIEW

Rheumatoid arthritis (RA) is a chronic autoimmune disorder characterized by inflammatory joint damage, leading to pain, swelling, and impaired mobility. The development of effective and targeted therapeutic strategies is crucial for managing RA and improving patient outcomes. In this context, nanotechnology has emerged as a promising avenue, offering innovative solutions to enhance drug delivery efficiency and specificity. This chapter explores the development of a novel Methotrexate (MTX)-loaded nanoemulsion designed for the treatment of rheumatoid arthritis through site-specific targeting and subcutaneous delivery.

Rheumatoid arthritis (RA) is a systemic autoimmune disease characterized by chronic inflammation of the joints and associated tissues. Although the inflammation is primarily associated with the joints, other organs in the body are also affected [1,2]. RA has a prevalence of approximately 0.24%, with a higher prevalence in females than males [3]. There is no clear cause for the disease, but genetic factors, environmental factors, and lifestyle changes trigger the inflammatory response in the body. Hormonal imbalance alters cortisol and androgen levels triggered by stress and leads to inflammatory response [4,5]. It is believed that the influx of inflammatory cells (T cells and B cells) into the pannus and synovial fluid is the cause of the pathology of RA leading to complete tissue destruction [6]. Inflammation causes an increase in the production of cytokines, which results in joint damage, synovitis, and edema. Synovial fluid contains a large number of immune cells, most of which are monocytes and mast cells. In addition, there are a small number of adaptive immune cells, such as plasma cells, Th1 (T-helper type 1), Th17, and B cells, which are involved in the inflammatory process. Further, the joint fluid contains high concentrations of antibodies against citrullinated proteins, which stimulates the release of complement proteins, which in turn trigger the onset of the inflammatory response [7].

Treatment aims to reduce joint inflammation and pain as well as tissue abnormalities and deformities in order to restore life's quality. Corticosteroids, NSAIDs (non-steroidal anti-inflammatory drugs), and opioids are considered to be first-line drug therapies for inflammatory pain and swelling. These medications are given for a limited period and are intended to relieve the pain and swelling associated with inflammation. The prevention of further joint damage and reduction of deformity are the primary goals of therapy [8,9]. The nonbiologic drugs methotrexate, sulfasalazine, hydroxychloroquine, and leflunomide are the most commonly used disease-modifying antirheumatic drugs (DMARDs) [10]. Fewerpeople take gold salts, azathioprine, and cyclosporine, among other rare medications. In RA, biologic DMARDs consist of monoclonal antibodies and recombinant receptors that inhibit the production of cytokines that contribute to the prevention of inflammation [11,12].

Methotrexate is the most important component of DMARDs, and its structure is shown in Figure 1. It is a folate derivative that suppresses pyrimidine and purine production [12]. Methotrexate reduces inflammation in RA by reducing the proliferation of inflammatorycells and lymphocytes, resulting in a decrease in the number of active inflammatory cells. Methotrexate increases the level of intracellular adenosine, which interacts with certain cell-surface receptors to inhibit the synthesis of the chemokines and leukotriene B4 as well as the regulation of synovial collagenase genes [13].

The systemic side effects of methotrexate, which limit dose escalation, are a significant obstacle in treating RA with this drug. The use of encapsulated carriers for targeted delivery may reduce the need for excessive and frequent dosing. When it comes to the pathophysiology of RA, the lymphatic system plays an important role [14,15]. Anti-inflammatory drugs can be delivered through the lymphatic system to improve bioavailability and therapeutic efficacy while reducing adverse effects [16]. Drugs can be more effectively targeted to the lymphatic system when encapsulated in carriers ranging in size from micro- to nanoscale.

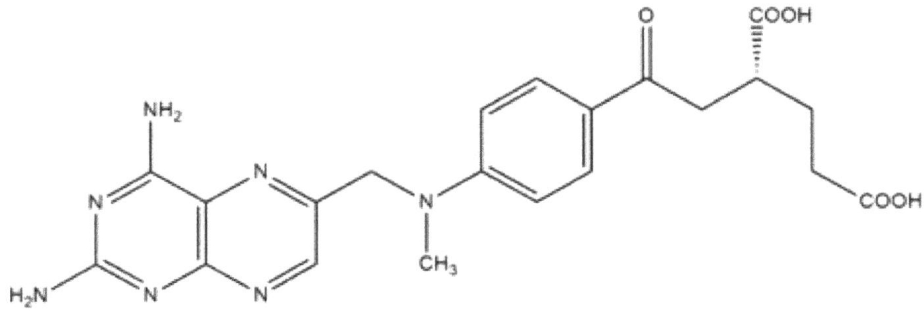

Figure 1. Chemical structure of methotrexate.

The use of encapsulated carriers, particularly nanoparticles, can contribute to delayed release while reducing the amount of drugs used [17,18].

When nanoparticle carriers are administered intravenously, they are consumed by reticular endothelial cells, preventing the drug from reaching the lymphatic system. Subcutaneous administration can be used for lymphatic targeting because of the permeability of capillaries, which allows the transport of smaller molecules [19]. Subcutaneous delivery of drug-loaded nano-molecules and their uptake into the lymph make it a suitable route that has a positive impact on the ongoing treatment profile of RA. Because of the above statement, it was planned to prepare and characterize a novel methotrexate-loaded nanoemulsion for the treatment of RA by targeted subcutaneous administration with improved bioavailability and therapeutic efficacy. The use of the nanoemulsion as a carrier can improve the bioavailability of methotrexate. The toxicity of the nanoemulsion as a carrier was studied using the cytotoxic and hemolysis assays to ensure that it was safe to use. In addition, CFA (complete Freund adjuvant)-induced arthritis in rats was used as a disease paradigm to evaluate the antirheumatic effect of methotrexate-loaded nanoemulsion.

2. RESULTS AND DISCUSSIONS

2.1 Particle Size Study

The size of the particles plays a significant influence in their uptake and retention in lymphoid tissue. When the particle size of different formulations was determined using a zetasizer by differential light scattering, a decrease in particle size was observed with the addition of a mixture of surfactant and co-surfactant. The optimized nanoemulsion loaded with methotrexate had a mean particle size of 87.89 ± 2.86 nm, whereas the particle size of the nanoemulsion without the drug was 76.39 ± 1.53 nm (Figure 2). The larger size of the nanoemulsion loaded with methotrexate compared to the nanoemulsion without the drug indicates that the drug was successfully loaded into the nanoemulsion. It has been documented that a nanoemulsion with a particle size smaller than 100 nm increases the transport of the drug through the lymphatic capillaries, which absorb the drug from the interstitial space [20]. Moreover, this particular size of the nanoemulsion can be easily absorbed and retained in the lymphoid tissue, resulting in improved therapeutic efficacy compared to conventional drugs. The PDI of the optimized nanoemulsion loaded with methotrexate was 0.27 (Figure 2). The nanoemulsion has smaller vesicles with a low PDI. The PDI reflects the potential of the size distribution of the vesicles in the nanoemulsion. In addition, the PDI is one of the most effective techniques with which to evaluate the homogeneity and stability of the nanoemulsion vesicles [21,22]. A higher PDI value of about 0.7 shows that the formulation has a very wide particle-size distribution, while values of 0.2 and below are mostly agreeable formulations. A PDI value of 0.3 and below is considered acceptable and indicates a uniform distribution of lipiodol vesicles [22]. The size distribution of the optimized nanoemulsion was in agreement with the particle size.

Zeta potential is an indicator of the stability of nanoemulsion. Determination of the zeta potential gave a value of -35.9 ± 0.73 mV for the optimized nanoemulsion loaded with methotrexate, while the zeta potential for nanoemulsion without drug was -33.9 ± 1.03 mV (Figure 3). The values of the zeta potential confirmed that the prepared nanoemulsion was negatively charged droplets with the capacity of higher lymphatic uptake and longer retention time. It measures the electrical repulsion force between the particles. It has been documented a higher zeta potential (>-30 mV) is advantageous to vesicle's physical stability as it avoids accumulation between vesicles owing to electrostatic repulsion resulting in a stable emulsion. The low zeta potential of emulsions leads to coagulation or flocculation of vesicles, resulting in poor physical stability [23–25]. According to these findings, we concluded that the selected composition of drug, oil, and surfactant makes the nanoemulsion electrically stable.

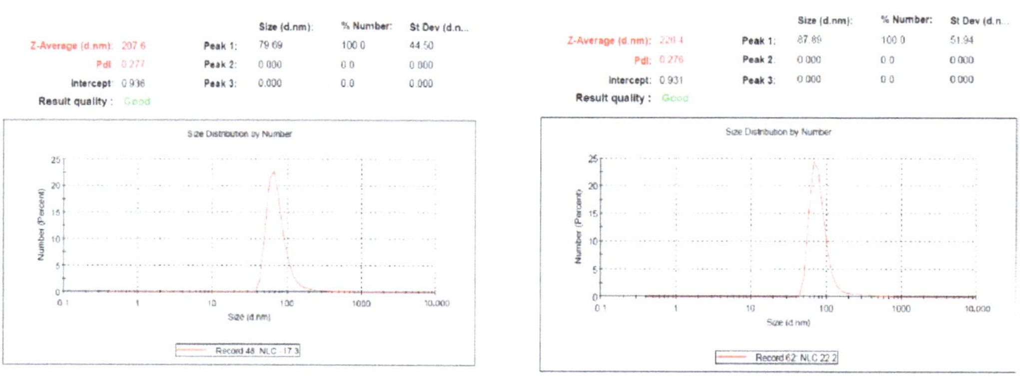

A B

Figure 2. Data of particle-size distribution and zeta potential of the prepared nanoemulsion; (A): Nanoemulsion without drug; (B): Methotrexate-loaded nanoemulsion.

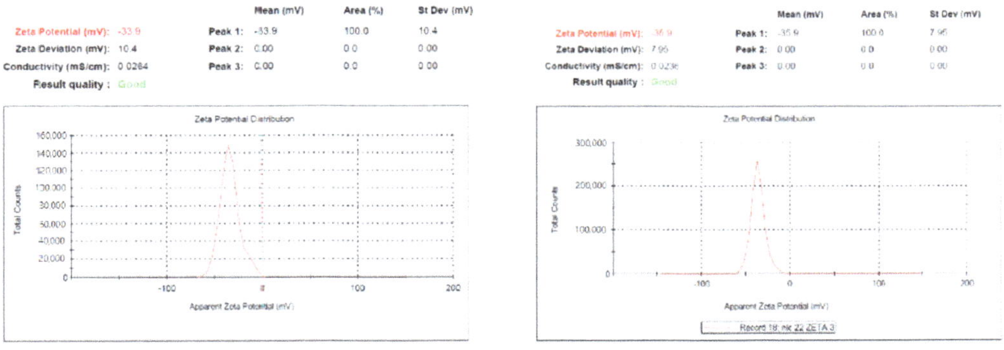

A B

Figure 3. Data of zeta potential. (A): Nanoemulsion without drug; (B): Methotrexate-loaded nanoemulsion.

2.2 Morphology of Particle

The SEM (scanning electron microscope) photograph of the methotrexate-loaded nanoemulsion shows that the particles have almost spherical morphology, are uniform and monodispersed (Figure 4). A nanoemulsion is a kinetically stable, dispersed system with particles of small droplet size. The spherical shapes indicate the stability of the system without aggregation and gravitational separation. This stability can be attributed to the presence of surface-active agents. The study conducted by Zhou et al. shows that the spherical nanoemulsion can rapidly penetrate the cell [26]. The results from SEM are consistent with the result of particle-size characterization.

The TEM (transmission electron microscopy) image of methotrexate-loaded nanoemul- sion shows a particle size of less than 100 nm with a spherical shape (Figure 5). The results of the particle size and SEM study support the findings of TEM. El-Refai et al. and Antil et al. documented a spherical and uniform distribution of vesicles in the nanoemul-sion prepared with sesame oil and Tween 80 in their TEM study [27,28]. The results of TEM are consistent with the study in which the anoemulsion was prepared with sesame oil and Tween 80.

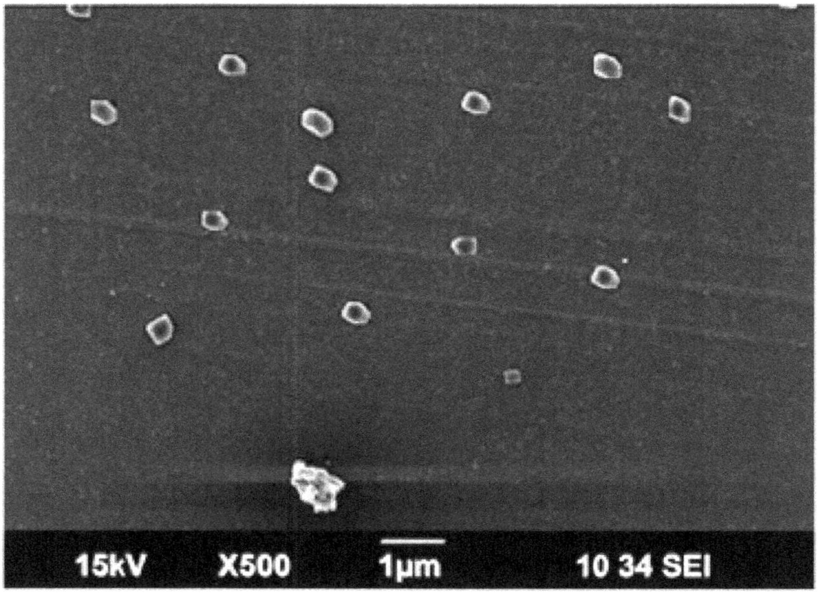

Figure 4. Representation of SEM findings of methotrexate-loaded nanoemulsion.

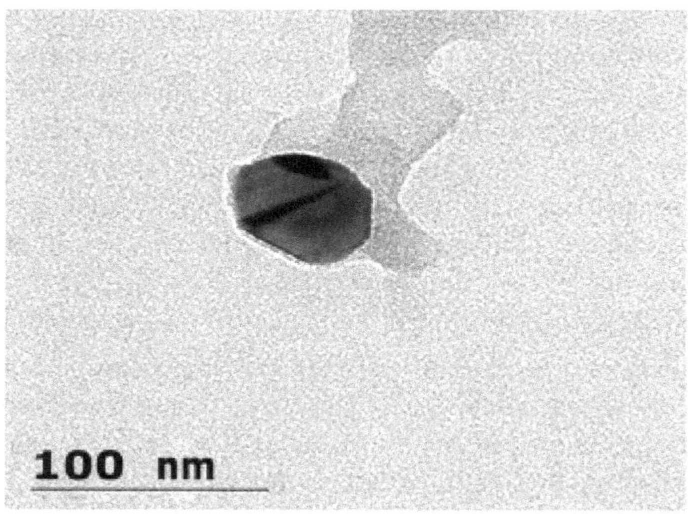

Figure 5. Representation of TEM findings of methotrexate-loaded nanoemulsion.

2.3. DSC Study

The DSC thermogram of methotrexate showed an endothermic peak at 104.74 °C, which is related to the melting point of the methotrexate, and a peak endothermic peak at 82.57 °C, which resembles the melting point of the methotrexate-loaded nanoemulsion (Figure 6). The DSC thermogram shows a sharp endothermic peak of methotrexate, proving its crystalline state. The endothermic peak of methotrexate did not appear in the DSC thermogram of the nanoemulsion, and the broadened thermal peak of the nanoemulsion confirms that methotrexate was fully incorporated into the vesicles. The different thermal peak of methotrexate and the nanoemulsion justifies that the drug is in an amorphous state. Moreover, as the temperature increases, there is an interaction between methotrexate and oil, which forms a complex that has a lower melting point (82.57 °C) than pure methotrexate [29,30].

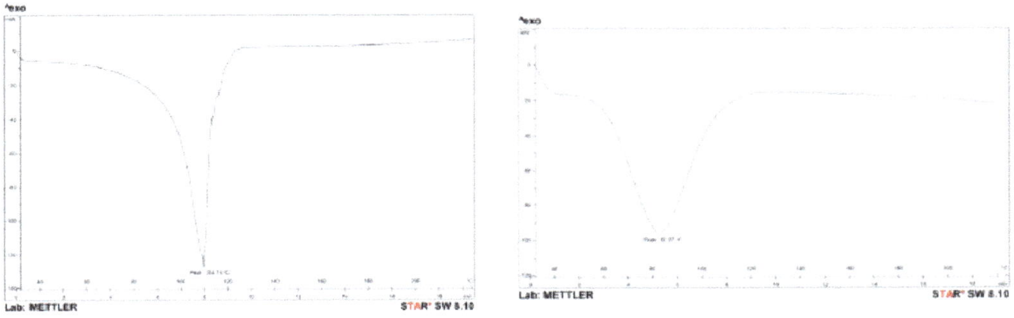

(A) (B)

Figure 6. Differential scanning colorimetry of pure-drug and methotrexate-loaded formulations: (A) Pure drug; (B) Methotrexate-loaded formulation.2.4. Entrapment Efficiency of Nanoemulsion To determine the quantity of drugs successfully incorporated into the nanoparticles,

2. 4 Entrapment Efficiency of Nanoemulsion

To determine the quantity of drugs successfully incorporated into the nanoparticles, the entrapment efficiency must be measured. The entrapment efficiency of the nanoemulsion was calculated using an indirect method by measuring the free drug available in the aqueous phase. The entrapment efficiency of the methotrexate-loaded nanoemulsion was found to be 87.25%, indicating high efficacy. The higher inclusion of methotrexate may be due to its stronger incorporation into the hydrophobic core. Antil et al. reported a 78% encapsulation efficiency of the metaxalone-loaded nanoemulsion containing sesame oil and Tween 80 [28]. The encapsulation-efficiency results are in agreement with the results of Antil et al.

2.5 In Vitro-Release Study

In vitro release from the methotrexate-loaded nanoemulsion was performed using a dialysis membrane that showed a biphasic release pattern. In this formulation, the drug was released in a burst at the beginning of the test, followed by a sustained release. The methotrexate-loaded nanoemulsion exhibited a 96.77% release after 72 h (Figure 7). The rapid release of the drug of 13.39 ± 1.14 from the nanoemulsion was observed during the first 2 h of the experiment. The suppressed drug release was observed after 2 h, indicating prolonged drug release from the nanoemulsion. The results are in agreement with other studies that addressed the biphasic release of methotrexate from the nanoemulsion. In this regard, Rashid et al. and Rathee et al. showed drug release of 72% after 20 h and 55% after 24 h, respectively [31–33]. Initially, the rapid release of the drug from the nanoemulsion could be caused by the methotrexate that is adsorbed on the droplet surface or dispersed in the surfactant. The sustained release indicates that the nano-sized drug is retained within the nanoemulsion droplet. In addition, the drug release in the nanoemulsion is also controlled by the interactions of the drug with the surfactants and its distribution between the aqueous and oil phases. The delayed drug release is of great interest for systemic delivery, especially for arthritis. The results suggest that the sustained release of the methotrexate-loaded nanoemulsion is a better formulation for the treatment of arthritis.

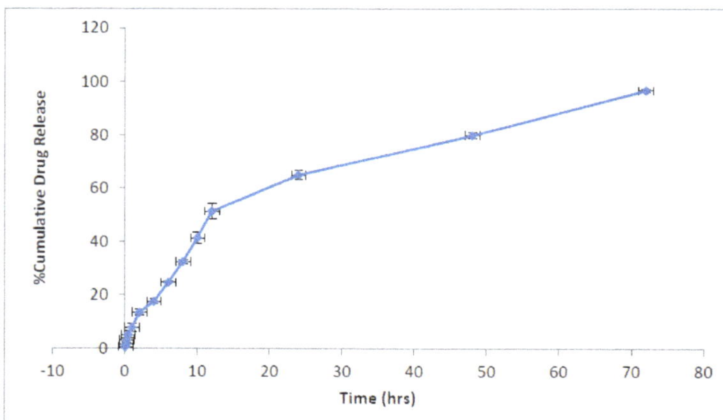

Figure 7. Graphical representation of drug release from the nanoemulsion loaded with methotrexate.

The order and mechanism of drug release from the nanoemulsion were investigated by applying the drug-release data to the zero-order, first-order, Higuchi and Korsmeyer–Peppas models. It was found that the drug release from the nanoemulsion was most closely associated with the Higuchi model, based on the R^2 values. The diffusion of methotrexate from the oily core and interface is retarded by the aqueous medium, which acts as a physical barrier to the release of the drug due to its poor solubility in water, a likely cause of prolonged release. It would appear that drug release is controlled by diffusion if the n value from the Korsmeyer–Peppas equation is used (Figure 8).

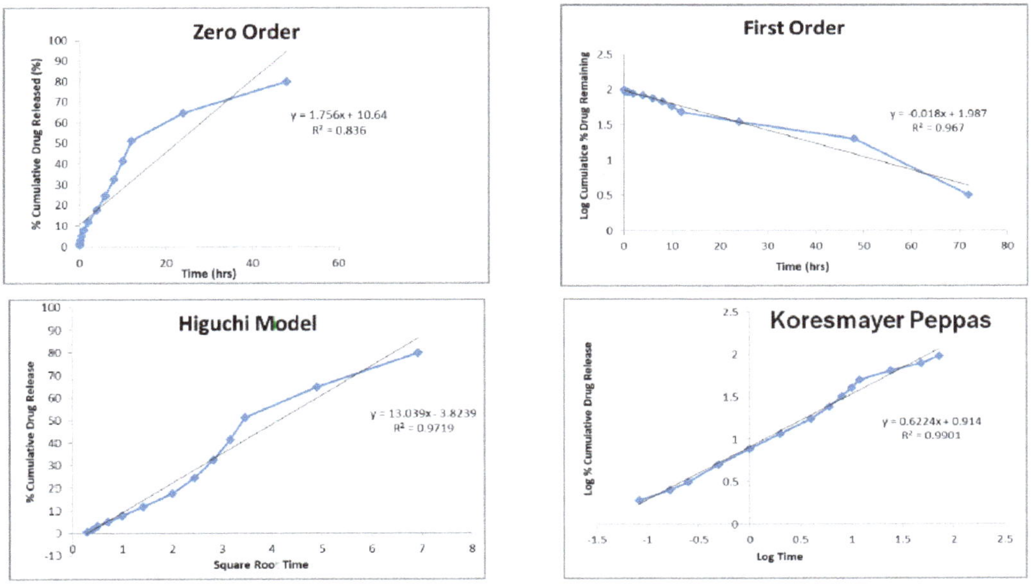

Figure 8. Graphical representation of kinetic-data analysis of methotrexate-loaded nanoemulsion.

2.6 Hemocompatibility Analysis of Nanoemulsion

The nanoemulsion loaded with methotrexate were tested and showed negligible hemolysis after a prescribed incubation period. The different concentrations of methotrexate- loaded nanoemulsion showed hemolysis ranging from 0.14 to 1.27% (Figure 9). The percentage of hemolysis in the samples does not increase significantly as the concentration of nanoemulsion increases. The marketed methotrexate drug solution exhibited 1.98% hemolysis. It has been documented that materials with hemolysis greater than 5% are considered hemolytic, while materials with hemolysis between 5 and 2% are termed as slightly hemolytic, and materials with less than 2% hemolysis are considered very hemocompatible [34]. The results indicate that the nanoemulsion is less than 2% hemolytic. This suggests that the prepared nanoemulsion is very hemocompatible and has no toxic effect on blood vessels. The nanoemulsion loaded with methotrexate showed less hemolysis compared to the marketed drug solution, indicating that the nanoemulsion is much safercompared to the conventional formulation.

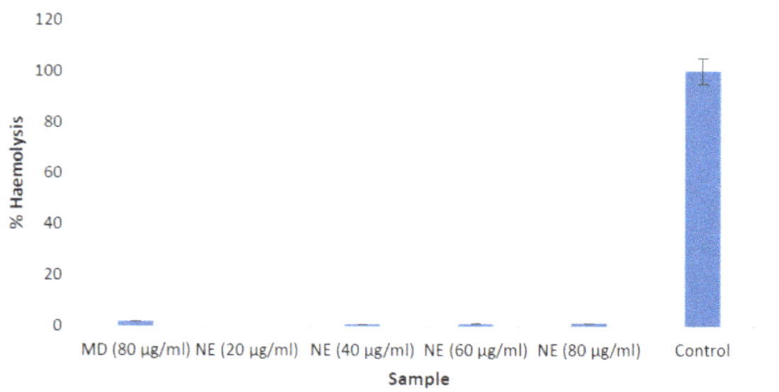

Figure 9. Percentage hemolysis of methotrexate-loaded nanoemulsion and pure methotrexate solution. NE: Methotrexate-loaded nanoemulsion; MD: Marketed drug.

2.7 Cytotoxic Assay

The MTT study was used to detect the living cell's metabolic activity in order to evaluate the cell viability of methotrexate-loaded nanoemulsion against the macrophagecell line RAW 264.7. The cell viability of RAW 264.7 cells at various concentrations of methotrexate-loaded nanoemulsion ranged from 95.79% to 60.21% (Figure 10). It was found that the cell viability of RAW 264.7 cells decreased in direct proportion to the concentration of methotrexate-loaded nanoemulsion. This indicates that the higher concentration of nanoemulsion induces mild cell cytotoxicity. The RAW 264.7 cell line exhibited a cell viability of less than 88% at concentrations of 6.25 and 12.5 µg/mL in the nanoemulsion.The results of hemolysis support the results of the cytotoxicity assay. In addition, the findings indicate that the higher dose of methotrexate-loaded nanoemulsion may cause mild toxicity to cells, while the lower dose is safe.

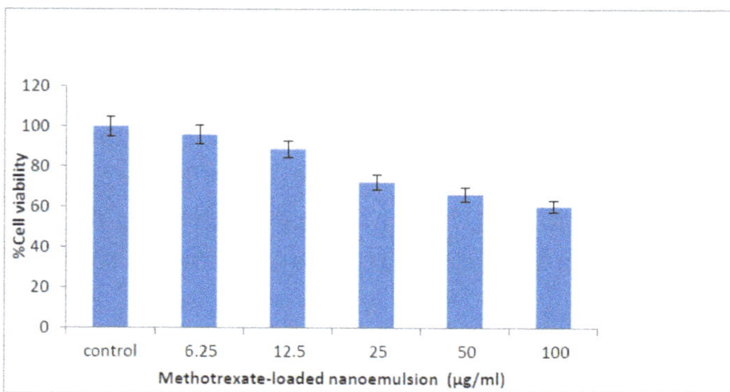

Figure 10. Graphical representation of in vitro cytotoxic effect of methotrexate-loaded nanoemulsion.

2.8 In Vivo Anti-Arthritic Activity of Nanoemulsion

The methotrexate-loaded nanoemulsion was intended to investigate the anti-arthritic efficacy by measuring the potential of the nanoemulsion to inhibit CFA-induced knee edema in rats. It showed a significant reduction in inflammation in the methotrexate- loaded nanoemulsion and pure methotrexate groups over 30 days compared to the control group. Figure 11 demonstrated a continuing increase in knee inflammation by the disease control animals. The animal treated with the methotrexate-loaded nanoemulsion (10 mg/mL per kg body weight) and marketed methotrexate significantly ($p < 0.05$) decreased the knee inflammation compared to control animals on days 6–30. The highest knee edema was noted during the 30 days of study in control animals (23.11 ± 0.14 mm), whereas the edema was controlled by the methotrexate-loaded-nanoemulsion-treated (8.12 ± 0.13 mm) and marketed-drug-treated (9.40 ± 0.39 mm) animals. The methotrexate- loaded nanoemulsion showed a greater reduction in swelling compared to the marketeddrug, suggesting that the methotrexate-loaded nanoemulsion has improved anti-arthritic potential compared to the conventional dosage form.

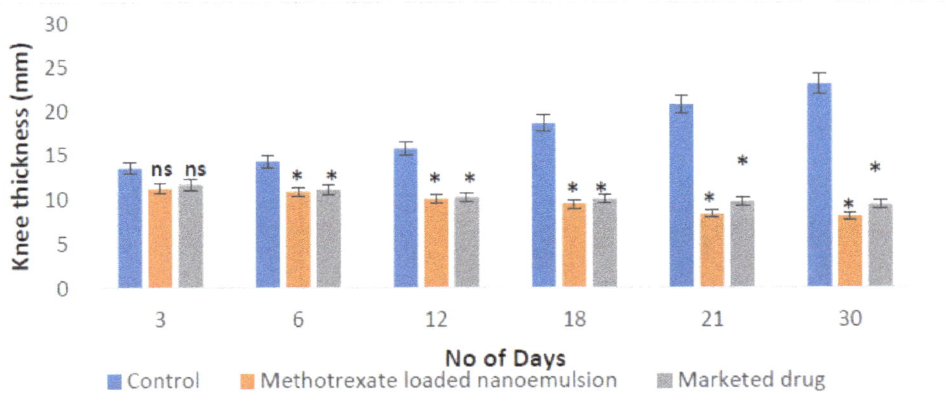

Figure 11. Knee-edema studies in complete-Freund-adjuvant-induced arthritis model. Results are shown as mean ± SEM (n = 6) and analyzed by one-way ANOVA followed by Tukey's test; ns: non-significant; * p < 0.05 significance difference to compared to control group.

The liver-protective properties of the methotrexate-loaded nanoemulsion were studied after the end of the experiment. When methotrexate is administered for a prolonged period or at higher doses, it causes liver toxicity, but in the present study, the formulations were administered to the rats only for a short time, so the possibility of liver toxicity was low. The function of hepatic marker enzymes SGOT (serum glutamic oxaloacetic transam- inase), SGPT (serum glutamic pyruvic transaminase) and ALP (alkaline phosphatase) were used to evaluate the cellular architecture of the animals with CFA-induced arthritis. SGOT, SGPT, and ALP were elevated in the control group of animals. Hepatic marker enzymes were significantly ($p < 0.05$) decreased in methotrexate-loaded-nanoemulsion- and marketed-drug-treated animals (Figure 12). It was observed that the methotrexate-loaded nanoemulsion exhibited greater liver-protective properties than the marketed drug solution. The results indicated that the CFA-induced animals experienced a significant effect on hepatic enzyme activity, while the methotrexate-loaded nanoemulsion and marketed methotrexate significantly decreased hepatic enzyme activity compared with the control group. The results suggest that the nanoemulsion loaded with methotrexate has better liver-protective properties compared to the marketed drug, as the function of liver marker enzymes such as SGOT, SGPT and ALP is much closer to the healthy liver function. This confirms that the methotrexate-loaded nanoemulsion enables targeted drug delivery with minimal access of the drug to the liver and also prevents the metabolism of the drug by the liver cell, resulting in improved therapeutic efficacy. The in vivo anti-arthritic study confirms the capability of the methotrexate-loaded nanoemulsion to lessen systemic toxicityas well as improve anti-arthritic efficacy compared to the usual dosage forms.

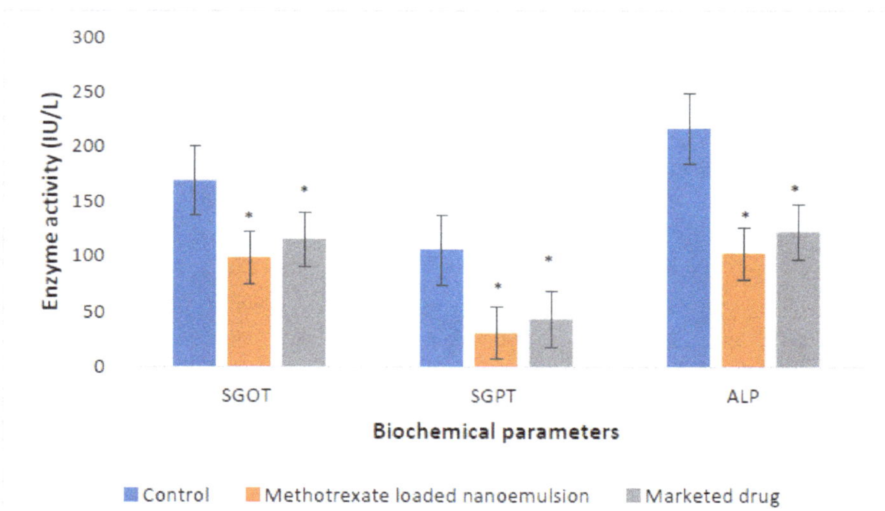

Figure 12. Liver-function tests for methotrexate-loaded nanoemulsion and pure methotrexate solution. Results are shown as mean ± SEM (n = 6) and analyzed by one-way ANOVA followed by Tukey's test; * p < 0.05 significance difference compared to control group.

A radiographic analysis of the animals on day 30 showed bone devastation and soft-tissue swelling along with joint-space narrowing in the CFA-induced control group, indicating subchondral erosion in the arthritic state. The methotrexate-loaded nanoemulsion and marketed drug solution showed no bone destruction and normal soft-tissue swelling in the animals (Figure 13). When RA occurs, radiography is an important diagnostic technique that can be used to determine the extent of arthritis. The early stages of arthritis are characterized by soft-tissue swelling and inflammation, whereas the later stages are characterized by subchondral erosions and narrowing of the joint space. The methotrexate- loaded nanoemulsion and marketed drug solution showed a significant reduction in jointdestruction and soft-tissue damage in the animals. The potent anti-arthritis effect of the methotrexate-loaded nanoemulsion was subsequently validated and confirmed by the radiographic study of the knee joints of the animals.

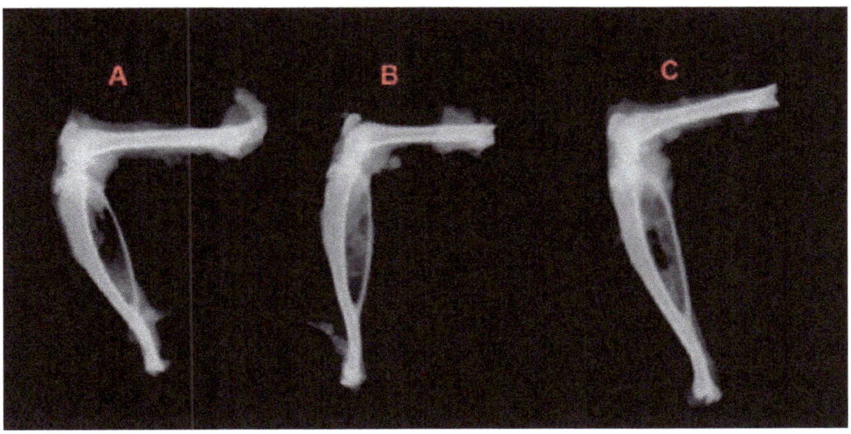

Figure 13. Radiographic analysis of the knees with the administration of the formulation. (A): FCA- induced arthritis, control group; (B): Methotrexate-loaded nanoemulsion; (C): marketed drug solution.

2.9 Stability Studies

To verify the stability properties of the methotrexate-loaded nanoemulsion, the stability studies of the nanoemulsion were carried out at refrigerated, room and elevated temperatures for three months. The particle size and entrapment efficiency changed significantly during storage at room temperature and elevated temperature. The possible mechanism for this was the polymorphic transformation of methotrexate. No changes in particle size and entrapment efficiency were observed at cooling temperatures (Figure 14). This indicates that the methotrexate-loaded nanoemulsion formulation was more stable under cooling temperature. Therefore, the methotrexate-loaded nanoemulsion is an excel- lent formulation for curing RA as it mitigates systemic side effects and improves treatment efficacy compared to the conventional dosage form.

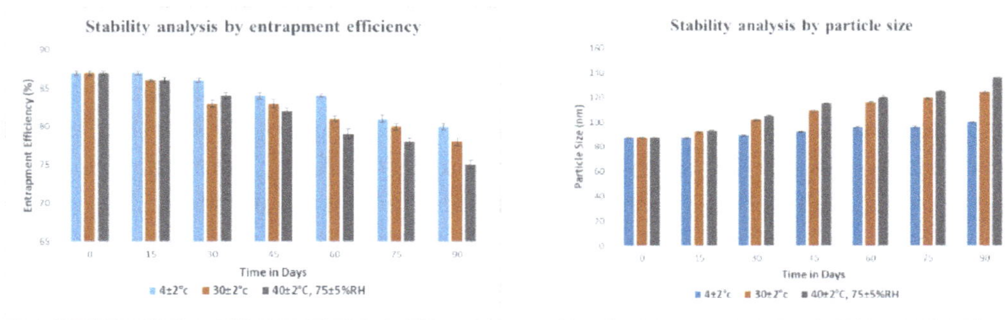

Figure 14. Graphical representation of stability study of methotrexate-loaded nanoemulsion according to particle size and entrapment efficiency

3. MATERIALS AND METHODS

3.1 Materials

The sample of the methotrexate injection was acquired from the leading IP Pharmacy center, AIMS, Kochi, Kerala, India. The surfactant, namely sesame oil, was purchased from the local grocery. Nice Chemicals, located in Ernakulam, Kerala, India, provided the Tween 80 for this study. The DMEM (Dulbecco's modified Eagle's medium) was procured from the well-known supplier thermo fisher scientific India Pvt Ltd., Ernakulum, Kerala, India.

3.2 Methodology

3.2.1 Formulation of Methotrexate-Loaded Nanoemulsion

Aqueous-phase titration was used to prepare the methotrexate-loaded nanoemulsion. In this procedure, the distilled water was gently added to the oil mixture and mixed dropwise with dynamic stirring until the resulting formulation appeared transparent and clear. Methotrexate solution at the concentration of 50 mg/mL was prepared and 0.2 mL solution was mixed into 3 mL of sesame oil, and then a mixture of 7 mL of Tween 80 and 2 mL of ethanol was slowly added while stirring. The resultant solution was vortexed for about 5 min using a vortex mixer. The aqueous phase, which consisted of distilled water, was added dropwise while constantly stirring at 500 rpm until a nanoemulsion was formed [35].

3.2.2. Particle Characteristics

The droplet size, polydispersity index (PDI), and zeta potential of the prepared nanoemulsion were determined using the Malvern Zetasizer (Nano ZS Malvern Instruments Ltd., Mavern, UK). After appropriate dilution, the nanoemulsions were sonicated and the droplet size was measured. The goal of this study was to assess the stability of the prepared nanoemulsion by measuring the droplet size, distribution, and zeta potential [35,36].

3.2.3 Particle Surface Morphology

The morphology and size were also determined by scanning electron microscopy (SEM) (VEGA 3, TESCAN, Brno-city, Kohoutovice, Czech Republic) and transmission electron microscopy (TEM) (TALOS, Thermo fisher scientific India Private Limited, Mumbai, India), respectively. The methotrexate-loaded nanoemulsion was diluted 10-fold with distilled water and dried at room temperature on 200 mesh film grids. For SEM analysis, samples were fixed in the sample holder with double-sided adhesive tape, and pictures were taken at a voltage of 5 kV. For TEM, samples were stained with a 2% phosphotungstic- acid solution and then dried for 2 min before being viewed under an electron microscope at 100 kV [37,38].

3.2.4 Differential Scanning Colorimetry (DSC)

The DSC (DSC 204F1 Phoenix, NETZSCH-Gerätebau GmbH, Selb, Germany) method was used to determine the change in physical properties and temperature of the optimized nanoemulsion. The sample was placed in an aluminum pan sealed with perforated lids and heated in the temperature range of 40–300 ◦C at a constant rate of 10 ◦C per minute. In addition, the inert atmosphere was created by a nitrogen purge of 50 mL/min. It assesses the stability of the formulation [39].

3.2.5 Entrapment Efficiency

In a slight modification of the method reported by Sarheed et al., a cellulose dialysis membrane technique was used to estimate the encapsulation of methotrexate in the nanoemulsion by assessing the free drug available in the aqueous phase. Before the experiment, the membrane was soaked overnight in phosphate-buffer solution (pH 7.4). By capping both ends, the nanoemulsion with 10 mg of drug incorporated was kept in the dialysis membrane. Phosphate-buffer solution and ethanol were mixed in a ratio of 80:20, and 100 mL was added into the receptor compartment. The dialysis membrane was placed in this 100 mL receptor compartment and shaken with a mechanical stirrer for 24 h. Aliquots were withdrawn from the receptor compartment and the amount of free drug crossing the dialysis membrane was measured by UV spectroscopy (UV-1700, Shimadzu, Kyoto, Japan) at 303.5 nm [40]. The following equation was used to measure the entrapment efficiency:

$$\text{Entrapment Efficiency} = \frac{\text{Quantification sample drug} - \text{Quantification of free drug in the solution}}{\text{Quantification sample drug}} \times 100$$

3.2.6 In Vitro Drug Release and Kinetic-Modeling Study

The cellulose dialysis membrane technique was used for in vitro drug release from the methotrexate-loaded nanoemulsion. The membrane was well cleaned and soaked overnight in phosphate buffer (pH 7.4) before use. Then, 10 mg of the drug was added to the membrane and both ends were tightly sealed. The dialysis membrane was kept in a beaker containing a phosphate-buffer solution with a pH of 7.4 and a temperature of 37 ± 1 ◦C throughout the study. The solution in the beaker was continuously stirred with a magnetic stirrer at 50 rpm. Samples were withdrawn from the beaker at specific intervals and the volume was maintained by adding the same amount of medium. The withdrawn aliquots were analyzed spectrophotometrically at 303.5 nm. Various kinetic models were developed to measure the release order and mechanism of release based on these results [41].

3.2.7 Hemocompatibility Analysis

The hemocompatibility study of the methotrexate-loaded nanoemulsion was performed to determine the suitability of the composition and the effect on red blood cells. To perform the test, a slight modification of the method reported by Roka and colleagues was employed. A volume of 5 mL of blood was drawn from healthy volunteers and an anticoagulant was added to prevent blood clotting. The blood sample was further diluted with a phosphate buffer with a pH of 7.4. The different concentrations of methotrexate- loaded nanoemulsion (20, 40, 60, and 80 µg/mL) and marketed drug (80 µg/mL) were added separately to the diluted blood samples and incubated for 24 h. The sample was then centrifuged at 3500 rpm for 10 min at 4 □C. The supernatant containing blood cells was placed in a microtiter plate, and absorbance was measured at 540 nm using an Elisa plate reader [42]. The percentage of hemolysis was calculated by using the following formula:

$$\%\text{Hemolysis} = \frac{\text{Blood cells in the supernatant of the sample solutions}}{\text{Blood cells in the distilled water}} \times 100$$

3.2.8 Cytotoxic Assay

MTT assay on RAW 264.7 (macrophage) cell lines was used for the evaluation of the in vitro cytotoxicity effect of the methotrexate-loaded nanoemulsion. The RAW 264.7 cell lines were cultured in DMEM (Dulbecco's modified Eagle's medium) media which was supplemented with 10% FBS, L-glutamine, sodium bicarbonate, and antibiotics at 37 ◦C maintaining humidification of 5% CO2. The different concentrations of 6.25 µg/L, 12.5 µg/L, 25 µg/L, 50 µg/L, and 100 µg/L of nanoemulsion were prepared by diluting with 5% DMEM. These samples were added to the cell culture and incubated at 37 ◦C in a humidity of 5% CO2. The normal saline solution was considered as the control solution for the measurement of cell viability.

The samples were incubated for 24 h, after which 30µL of reconstituted MTT solution was added to all test and control wells. The plate was shaken slowly and then incubated at 37 ◦C for 4 h to maintain the humidification of 5% CO^2 in the incubator. After completion of incubation, the supernatant was taken out, and added to 100µL of MTT solubilization solution. To solubilize the formazan crystals, the wells were carefully agitated by pipetting up and down. A microplate reader was used to assess the absorbance of samples at a wavelength of 570 nm, which signifies the optical density (OD) [43].

$$\%\text{ of viability} = \frac{\text{Mean OD of samples}}{\text{Mean OD of control}} \times 100$$

3.2.9 In Vivo Anti-Arthritic Activity

Approval for the animal study was obtained from the Central Animal Facility of the Amrita Institute of Medical Sciences (IAEC/2017/3/9) for the animal experiments. Experiments were conducted on adult male Sprague Dawley rats weighing about 200–250 g in order to determine the anti-arthritic effect. The animals were segregated into three groups of six animals each and kept in polypropylene cages at a temperature of 23 ± 2 °C. Intraarticular injection of 1 mg/mL CFA was injected to all groups of animals to induce the Chronic Arthritis condition. After that, methotrexate-loaded nanoemulsion (10 mg/mL) and marketed methotrexate drug (10 mg/mL) were administered subcutaneously once a week for 30 days to groups II and III, respectively, while no treatment was given to group I. Using a digital micrometer, the knee circumference (in mm) of each group was determined every day for 30 days. After 30 days of testing, retro-orbital blood was drawn from the rats and the serum was separated to assess the functions of liver enzyme markers (SGPT, SGOT, and ALP). In addition, radiographs of the knee joints of rats were performed to evaluate the effect of marketed methotrexate and methotrexate-loaded nanoemulsion on the severity of arthritis in FCA-induced rats [44,45].

3.2.10 Stability Studies

For three months, methotrexate-loaded nanoemulsion was stored at three different temperatures: refrigerator temperature (4 ± 2 °C), room temperature (30 ± 2 °C), and accelerated temperature (40 ± 2 °C, 75 ± 5% RH). Particle size (PS) and entrapment efficiency (EE) were evaluated at 15-day intervals over 90 days to check the changes in physical and chemical stability of the nanoemulsion [41].

4. CONCLUSIONS

In the treatment of RA, methotrexate is a DMARD that is considered a first-line therapy. There are only a small number of drug-delivery systems on the market that provide significant pharmacotherapy for rheumatoid arthritis. Methotrexate formulations currently on the market have been associated with systemic side effects. Therefore, a methotrexate-loaded subcutaneous nanoemulsion that delivers the drug exclusively to the lymphatic system was developed and studied in vitro and in vivo in order to minimize adverse effects and improve therapeutic efficacy. The nanoemulsion was found to encapsulate a significant amount of methotrexate with a smaller particle size and shape suitable for subcutaneous injection. Physicochemical studies confirmed that the methotrexate-loaded nanoemulsion exhibited homogeneous nanosized droplets and a stable formulation. In vitro drug-release studies confirmed the sustained release of the drug from the nanoemulsion. The results of hemolysis and cytotoxicity studies showed that the nanoemulsion is safer for systemic circulation at lower concentrations. The anti-arthritic study of the nanoemulsion in CFA- induced animals showed improved anti-arthritic activity compared to the marketed drug. The preliminary studies of the methotrexate-loaded nanoemulsion suggest that the formulation may be able to improve the lymphatic transport of drugs after systemic administration. In addition, this formulation may reduce drug doses while minimizing adverse effects on cell integrity. It can be concluded that methotrexate-loaded nanoemulsion is a superior formulation for the treatment of RA as it mitigates systemic side effects and improves treatment efficacy compared to the conventional dosage form. In the future, the drug-uptake study in the lymph node will demonstrate the ability of the nanoemulsion to deliver drugs subcutaneously into the lymphatic circulation.

5. CONCISE SUMMARY

Rheumatoid arthritis (RA) is a systemic, chronic autoimmune disease that causes disability due to progressive inflammation and destruction of the tissues around the joints. Methotrexate is mainly used to prevent the progression of joint destruction and reduce the deformity. The major challenge in treating RA with methotrexate is the systemic side effects that limit dose escalation. Hence, a novel formulation of a methotrexate-loaded nanoemulsion for subcutaneous administration was developed that aims to deliver methotrexate into the system via the lymph. The methotrexate-loaded nanoemulsion was prepared by using the aqueous-titration method. The prepared nanoemulsion was investigated for particle size, surface charge, surface morphology, entrapment efficiency, DSC (differential scanning colorimetry), drug release, hemocompatibility assay, and cytotoxicity, as well as anti-arthritic and stability studies. The vesicle size, zeta potential, PDI (polydispersity index), and entrapment efficiency of the optimized nanoemulsion were 87.89 ± 2.86 nm, 35.9 ± 0.73 mV, 0.27, and 87 ± 0.25%, respectively. The DSC study showed that the crystalline methotrexate was converted to an amorphous form and the drug was fully incorporated into the vesicles. After 72 h, the optimized nanoemulsion showed a drug release of 96.77 ± 0.63%, indicating a sustained-release dosage form. Cytocompatibility testing by MTT (3-(4,5-dimethylthiazol-2-yl)-2,5-diphenyl-2H-tetrazolium bromide) assay on macrophage cell lines showed that the nanoemulsion was non-toxic. The formulation showed significant anti-arthritic activity compared to the marketed drug solution. In addition, the nanoemulsion containing methotrexate remained stable for three months when stored at a low temperature. Since the nanoemulsion containing methotrexate has excellent physicochemical properties and lowers systemic side effects by targeted delivery, it is a desirable technology for subcutaneous drug delivery.

REFERENCES

1. Wong, S.H.; Lord, J.M. Factors underlying chronic inflammation in rheumatoid arthritis. Arch. Immunol. Ther. Exp. 2004, 52, 379–388.
2. Qiang, G.; Yuxiang, W.; Dan, X.; Johannes, N.; Nathan, J.P.; Jiake, X. Rheumatoid arthritis: Pathological mechanisms and modern pharmacologic therapies. Bone Res. 2018, 6, 1–14.
3. Cross, M.; Smith, E.; Hoy, D.; Carmona, L.; Frederick, W.; Theo, W. The global burden of rheumatoid arthritis: Estimates from the Global Burden of Disease 2010 study. Ann. Rheum. Dis. 2014, 73, 1316–1322. [CrossRef] [PubMed]
4. Allan, G. Epidemiology, Pathophysiology, and Diagnosis of Rheumatoid Arthritis: A Synopsis. Am. J. Manag. Care 2014, 20, S128-35.
5. Jefferies, W.M. The etiology of rheumatoid arthritis. Med. Hypothesis 1998, 51, 111–114. [CrossRef]
6. Kahlenberg, J.M.; David, A.F. Advances in the medical treatment of rheumatoid arthritis. Hand Clin. 2011, 27, 11–20. [CrossRef]
7. Chauhan, K.; Jagmohan, S.J.; Mohammed, A.A. Rheumatoid Arthritis, StatPearls [Internet]; StatPearls Publishing: Treasure Island, FL, USA, 2019.
8. Jacqueline, B.; Syed, A.A.R.; Ayman, M.S. Rheumatoid Arthritis: A Brief Overview of the Treatment. Med. Princ. Pract. 2018, 27, 501–507.
9. Rajitha, P.; Biswas, R.; Sabitha, M.; Jayakumar, R. Methotrexate in the treatment of psoriasis and rheumatoid arthritis: Mechanistic insights, current issues and novel delivery approaches. Curr. Pharm. Des. 2017, 23, 3550–3566. [CrossRef]
10. Amy, M.W. Diagonosis and management of Rheumatoid Arthritis. Am. Fam. Physician 2011, 84, 1245–1252.
11. Ram, M.; Anita, M.; Sujithra, R.M.; Aramya, A.R.; Mathew, J.; Parvathy, M. Comparative Effectiveness Research on Different Treatment Options for Rheumatoid Arthritis in Ayurveda. J. Altern. Complement. Med. 2014, 20, A75–A76.
12. Christine, T.N.P. Nanotherapeutic approaches for the treatment of rheumatoid arthritis. Wiley Interdiscip. Rev. Nanomed. Nanobiotechnol. 2011, 3, 607–619.
13. Cutolo, M.; Sulli, A.; Pizzorni, C.; Seriolo, B.; Straub, R. Anti-inflammatory mechanisms of methotrexate in rheumatoid arthritis. Ann. Rheum. Dis. 2001, 60, 729–735. [CrossRef] [PubMed]
14. Simon, S.; Michael, D. Inflammation and Lymphatic Function. Front. Immunol. 2019, 10, 308.

15. Schwartz, N.; Chalasani, M.L.S.; Li, T.M.; Feng, Z.; Shipman, W.D.; Lu, T.T. Lymphatic Function in Autoimmune Diseases. Front. Immunol. 2019, 10, 519. [CrossRef] [PubMed]

16. Zhang, X.Y.; Lu, W.Y. Recent advances in lymphatic targeted drug delivery system for tumor metastasis. Cancer Biol. Med. 2014,11, 247–254.

17. Yang, Q.; Laird, F. Drug Delivery to the Lymphatic System, Drug Delivery: Principles and Applications, 2nd ed.; Chapter 21; Wiley Publisher: Hoboken, NJ, USA, 2016.

18. Isabel, M.O.; Cristiana, G.; Rui, L.R.; Joaquim, M.O. Engineering nanoparticles for targeting rheumatoid arthritis: Past, present, and future trends. Nano Res. 2018, 11, 4489–4506.

19. Satheesh, J.; Shyam, S.R.; Jithan, A. Development of subcutaneous sustained release nanoparticles encapsulating low molecular weight heparin. J. Adv. Pharm. Technol. Res. 2015, 6, 58–64.

20. McLennan, D.N.; Porter, C.J.; Charman, S.A. Subcutaneous drug delivery and the role of the lymphatics. Drug Discov. Today Technol. 2005, 2, 89–96. [CrossRef]

21. Ghosh, V.; Saranya, S.; Mukherjee, A.; Chandrasekaran, N. Cinnamon oil nanoemulsion formulation by ultrasonic emulsification: Investigation of its bactericidal activity. J. Nanosci. Nanotechnol. 2013, 13, 114–122. [CrossRef]

22. Danaei, M.; Dehghankhold, M.; Ataei, S.; Hasanzadeh Davarani, F.; Javanmard, R.; Dokhani, A.; Khorasani, S.; Mozafari, M.R. Impact of Particle Size and Polydispersity Index on the Clinical Applications of Lipidic Nanocarrier Systems. Pharmaceutics 2018, 10, 57. [CrossRef]

23. Jang, J.H.; Jeong, S.H.; Lee, Y.B. Enhanced Lymphatic Delivery of Methotrexate Using W/O/W Nanoemulsion: In Vitro Characterization and Pharmacokinetic Study. Pharmaceutics 2020, 12, 978. [CrossRef] [PubMed]

24. Kaur, G.; Singh, P.; Sharma, S. Physical, morphological, and storage studies of cinnamon based nanoemulsions developed with Tween 80 and soy lecithin: A comparative study. Food Meas. 2021, 15, 2386–2398. [CrossRef]

25. Wik, J.; Bansal, K.K.; Assmuth, T.; Rosling, A.; Rosenholm, J.M. Facile methodology of nanoemulsion preparation using oily polymer for the delivery of poorly soluble drugs. Drug Deliv. Transl. Res. 2020, 10, 1228–1240. [CrossRef] [PubMed]

26. Zhou, H.; Yue, Y.; Liu, G.; Li, Y.; Zhang, J.; Gong, Q.; Yan, Z.; Duan, M. Preparation and Characterization of a Lecithin Nanoemulsion as a Topical Delivery System. Nanoscale Res. Lett. 2010, 5, 224. [CrossRef] [PubMed]

27. El-Refai, A.A.; Rabie, M.M.; El-Gammal, R.E.; Al-Saban, W.A. Nanoemulsion of Sesame Seeds Oil: Preparation, Evaluation and Stability. Asian J. Chem. 2019, 31, 3004–3008. [CrossRef]

28. Antil, S.; Gupta, C. Formulation and Evaluation of Nanoemulsion for Bioavailability Enhancement of Metaxalone. Int. J. Cur. Res. Rev. 2021, 13, 47–53. [CrossRef]

29. Jangde, R.; Elhassan, G.O.; Khute, S.; Singh, D.; Singh, M.; Sahu, R.K.; Khan, J. Hesperidin-Loaded Lipid Polymer Hybrid Nanoparticles for Topical Delivery of Bioactive Drugs. Pharmaceuticals 2022, 15, 211. [CrossRef]

30. Aswathy, K.N.; Asdaq, S.M.B.; Saritha, C.K.; Thomas, L.; Haridas, N.; Viswanad, V.; Sahu, R.K.; Fattepur, S.; Alamri, A.S.; Alsanie, W.F.; et al. Formulation and In-vitro Characterization of Fast-Disintegrating Herbal Extract Sublingual Immunotherapy Tablet for Peanut-Induced Allergic Asthma. Saudi J. Biol. Sci. 2022, 29, 1283–1297. [CrossRef]

31. Rashid, S.A.; Bashir, S.; Naseem, F.; Farid, A.; Rather, I.A.; Hakeem, K.R. Olive Oil Based Methotrexate Loaded Topical Nanoemulsion Gel for the Treatment of Imiquimod Induced Psoriasis-like Skin Inflammation in an Animal Model. Biology 2021, 10, 1121. [CrossRef]

32. Rathee, J.; Kanwar, R.; Kumari, L.; Pawar, S.V.; Salunke, D.B.; Mehta, S.K. Preparation of α-Tocopherol based nanoemulsion for efficacious delivery of Methotrexate. J. Dispers. Sci. Technol. 2022. [CrossRef]

33. Kelmann, R.G.; Kuminek, G.; Teixeira, H.F.; Koester, L.S. Carbamazepine parenteral nanoemulsions prepared by spontaneous emulsification process. Int. J. Pharm. 2007, 342, 231–239. [CrossRef]

34. Weber, M.; Steinle, H.; Golombek, S.; Hann, L.; Schlensak, C.; Wendel, H.P.; Avci-Adali, M. Blood-Contacting Biomaterials: In Vitro Evaluation of the Hemocompatibility. Front. Bioeng. Biotechnol. 2018, 6, 99. [CrossRef] [PubMed]

35. Akrawi, S.H.; Gorain, B.; Nair, A.B.; Choudhury, H.; Pandey, M.; Shah, J.N.; Venugopala, K.N. Development and Optimization of Naringenin-Loaded Chitosan-Coated Nanoemulsion for Topical Therapy in Wound Healing. Pharmaceutics 2020, 12, 893. [CrossRef]

36. Ali, H.H.; Hussein, A.A. Oral nanoemulsions of candesartan cilexetil: Formulation, characterization and in vitro drug release studies. AAPS Open 2017, 3, 4. [CrossRef]

37. Alhamdany, A.T.; Saeed, A.M.H.; Alaayedi, M. Nanoemulsion and Solid Nanoemulsion for Improving Oral Delivery of a Breast Cancer Drug:

Formulation, Evaluation, and a Comparison Study. Saudi Pharm. J. 2021, 29, 1278–1288. [CrossRef] [PubMed]

38. Saani, S.M.; Abdolalizadeh, J.; Heris, S.Z. Ultrasonic/sonochemical synthesis and evaluation of nanostructured oil in water emulsions for topical delivery of protein drugs. Ultrason. Sonochem. 2019, 55, 86–95. [CrossRef]
39. Rosso, A.; Lollo, G.; Chevalier, Y.; Troung, N.; Bordes, C.; Bourgeois, S.; Maniti, O.; Granjon, T.; Dugas, P.; Urbaniak, S.; et al. Development and structural characterization of a novel nanoemulsion for oral drug delivery. Colloids Surf. A Physicochem. Eng. Asp. 2020, 593, 124614. [CrossRef]
40. Sarheed, O.; Dibi, M.; Ramesh, K.V.R.N.S. Studies on the Effect of Oil and Surfactant on the Formation of Alginate-Based O/W Lidocaine Nanocarriers Using Nanoemulsion Template. Pharmaceutics 2020, 12, 1223. [CrossRef]
41. Vijaya Rani, K.R.; Rajan, S.; Bhupathyraaj, M.; Priya, R.K.; Halligudi, N.; Al-Ghazali, M.A.; Sridhar, S.B.; Shareef, J.; Thomas, S.; Desai, S.M.; et al. The Effect of Polymers on Drug Release Kinetics in Nanoemulsion In Situ Gel Formulation. Polymers 2022, 14, 427. [CrossRef]
42. Róka, E.; Ujhelyi, Z.; Deli, M.; Bocsik, A.; Fenyvesi, É.; Szente, L.; Fenyvesi, F.; Vecsernyés, M.; Váradi, J.; Fehér, P.; et al. Evaluation of the Cytotoxicity of □-Cyclodextrin Derivatives on the Caco-2 Cell Line and Human Erythrocytes. Molecules 2015, 20, 20269–20285. [CrossRef]
43. Sahlan, M.; Mahira, K.F.; Pratami, D.K.; Rizal, R.; Ansari, M.J.; Al-Anazi, K.M.; Farah, M.A. The cytotoxic and anti-inflammatory potential of Tetragonula sapiens propolis from Sulawesi on raw 264.7 cell lines. J. King Saud Univ.-Sci. 2021, 33, 101314. [CrossRef]
44. Cui, X.; Wang, R.; Bian, P.; Wu, Q.; Seshadri, V.D.D.; Liu, L. Evaluation of antiarthritic activity of nimbolide against Freund's adjuvant induced arthritis in rats. Artif. Cells Nanomed. Biotechnol. 2019, 47, 3391–3398.
45. Singh, V.S.; Dhawale, S.C.; Shakeel, F.; Faiyazuddin, M.; Alshehri, S. Antiarthritic Potential of Calotropis procera Leaf Fractions in FCA-Induced Arthritic Rats: Involvement of Cellular Inflammatory Mediators and Other Biomarkers. Agriculture 2021, 11, 68. [CrossRef]

08

THE INFLUENCE OF CITRATE AND LIPID-FUNCTIONALIZED MAGNETIC NANOPARTICLES ON DEHYDROPEPTIDE SUPRAMOLECULAR MAGNETOGELS: PROPERTIES, DESIGN AND DRUG RELEASE

1. INTRODUCTION: AN OVERVIEW

The utilization of magnetic nanoparticles in the realm of supramolecular magnetogels has garnered significant attention due to their potential impact on the properties, design, and controlled drug release mechanisms. In particular, the incorporation of citrate and lipid functionalizations adds a distinctive dimension to these magnetic nanoparticles, enhancing their efficacy and versatility. This study delves into the intricate interplay between citrate and lipid-functionalized magnetic nanoparticles within dehydropeptide supramolecular magnetogels, aiming to elucidate their properties, contribute to innovative design strategies, and unravel the intricacies of drug release mechanisms in this unique context. The exploration of such advanced nanomaterials holds promise for the development of novel therapeutic approaches, offering insights into the intersection of nanotechnology and supramolecular chemistry for improved biomedical applications.

Supramolecular magnetogels basically comprise two main components: the hydrogel and the magnetic nanoparticles. Following a stimulus, the self-assembly of the supramolecular hydrogelators is driven towards a kinetically-trapped intertwined fibrillar structure, such that solvent pocket microdomains are formed. This process takes place through the cooperative effect of different non-covalent intermolecular interactions: hydrogen bonding, van der Waals, electrostatic, and/or hydrophobic and aromatic interactions [1–7].

For example, dehydropeptide-based hydrogelators, such as the minimalist Cbz-L- Met-Z-ΔPhe-OH (see Structure S1 in Supplementary Materials), have been used due to the simplicity of producing biocompatible systems at low cost, and its promising properties for drug delivery [8–10]. The aforementioned dehydropeptide uses of a non-polar amino acid, methionine, to induce hydrophobic collapse, while the dehydrophenylalanine moiety provides a means for π-π interactions (like the N-capping group) and conformational constraints in the peptide backbone, meaning that it promotes the self-assembly into fibers and provides resistance to enzymatic degradation [8–10]. As assessed in this work, the hydrogel displays a moderate gelation kinetics and elastic modulus, which allows following the impact of the nanoparticles, so that diverse parameters can be optimized. Further, the gels can be formed in a skin pH range of 4-6 [11], thus being suitable for dermal application.

The retention or entrapment of nanoparticles during the gelation process allows for the tailoring and modulation of the matrix structure, along with the possibility of tuning different physical properties [3,12,13]. The modulation of the matrix structure also enables optimizing the loading of a wide variety of hydrophilic and hydrophobic drugs, reducing potential side effects, and enabling higher doses in therapy at the right location when using magnetogels for drug delivery. In this regard, taking advantage of the magnetic nanoparticles, the application of a magnetic field gradient allows the control and targeting of the nanosystem to a specific location [14–16], which can be further coupled with an alternating magnetic field (AMF), such that the nanoparticles can absorb energy and then release it as heat [17–20]. As a consequence, there is a synergistic effect that involves this magnetic hyperthermia and the subsequently associated enhancement of drug release [21–23], such that a much higher therapeutic efficiency can be attained. Along these lines, among transition metal ferrites, manganese ferrite nanoparticles exhibit very suitable magnetic properties, particularly in terms of the high magnetic susceptibility and in terms of magnetophoretic mobility, which render them as an appealing option to improve the supramolecular magnetogels for drug delivery applications [24,25]. Further, manganese ferrites display good biocompatibility and the large saturation magnetization reduces the required concentration of nanoparticles, thus averting side effects [8,25].

Supramolecular magnetogels endorsed with bare nanoparticles were demonstrated to be easily prone to aggregation, which led us to limit the concentration of nanoparticles that were encapsulated [8,9]. Alternatively, the use of a thick shell coating the nanoparticles was confirmed to hamper the gelation process, with the additional cost of requiring a larger concentration of hydrogelator to compensate for the bulkier nanocomposite [10]. On the other hand, forcing the supramolecular design strategies to conjugate the nanoparticles with the hydrogelator molecules has also been considered [25–27]. In general, these existing strategies for supramolecular magnetogels either imply complex fabrication steps and/or fail on the homogeneous incorporation of high concentrations of magnetic material and target drugs, which strongly limit their usability in theranostics. Furthermore, the current literature on supramolecular magnetogels lacks exploration of the composite functionalization and its effect on supramolecular magnetic gels properties. Consequently, the large amounts of hydrogelator employed and the aggregation of nanoparticles in most of the reported systems have demonstrated lower heating efficiency, hindering the sought on-demand controlled release of the payload.

Doxorubicin is one of the most commonly used chemotherapeutic drugs in a wide variety of cancers. However, its various side effects (e.g., neutropenia and heart failures) lead to a worsening of the patient's quality of life, which can be improved through incorporation with drug delivery system to increase the safety profile of the therapy. Considering this need, a useful strategy could be the encapsulation of doxorubicin in liposomal systems, such as the liposomal formulation Doxil®, or the use of citrate-stabilized nanoparticles, which strongly adsorb the positively charged drug molecules through electrostatic interactions [28].

Herein, supramolecular magnetogels endorsed with high nanoparticle concentration, employing a lipid coating (magnetoliposome-like structure) or citrate stabilization, are evaluated. Whereas the citrate stabilization provides electrostatic stabilization, the lipid coating assessment tackles the strategy, not only to ensure the steric stabilization of manganese ferrite nanoparticles, but also to provide the structure with enough hydrophobic cavities for an efficient drug loading and subsequent controlled drug diffusion. This option to obtain magnetogels systems, compared with the use of negatively-charged citrate-stabilized nanoparticles, demonstrates that the gelation becomes optimized, such that the concentration of stable magnetic nanoparticles within the gel is increased. As a consequence, besides improving the system stability and magnetic properties, the strategy demonstrates the role of the magnetic nanoparticles taking advantage of the hyperthermia they are responsible for, on drug release, using the antitumor drug doxorubicin.

2. MATERIALS AND METHODS

2.1 Synthesis Procedure of Magnetic Nanoparticles

2.2.1 Preparation of Citrate-Stabilized Manganese Ferrite Nanoparticles

A modified synthesis of citrate-stabilized nanoparticles was adapted from reference [29]. Trisodium citrate dehydrate (1 mmol) and NaOH (4 mmol) were added to 19 mL of ultrapure water at 100 °C. A 1 mL aqueous solution of $FeSO4 \cdot 7H2O$ (1.33 mmol) and $MnSO4 \cdot H2O$ (all reagents from Merck-Sigma, St. Louis, MO, USA) (0.66 mmol) was added, drop by drop, into the mixture under vigorous agitation. After

2 h, the solution was cooled down to room temperature, washed through magnetic decantation with water/ethanol 1:1, and dried at 100 ◦C.

2.1.2 Preparation of Lipid-Coated Manganese Ferrite Nanoparticles

Lauric acid (1 mmol) and NaOH (4 mmol) were added to 19 mL of ultrapure water at 100 ◦C. A 1 mL aqueous solution of FeSO4·7H2O (1.33 mmol) and MnSO4·H2O (0.66 mmol) was added, drop by drop, into the mixture under vigorous agitation. After 2 h, the solution was cooled down to room temperature, washed through repeated centrifugation with water, and dried at 100 ◦C. The stock solution was prepared by dispersion of the nanoparticles (4 mg) in 2 mL of 2 mM L-α-phosphatidylcholine (from egg yolk, egg-PC) (Merck-Sigma, St. Louis, MO, USA) solution through sonication at 190 W. The lipid-coated nanoparticles were then washed and purified with ultrapure water by magnetic decantation.

2.2 Self-Assembly of Magnetogels

2.2.1 Optimization of Hydrogel Gelation

Gelation optimization was carried out through turbidity measurements at 500 nm. The hydrogel and glucono-δ-lactone (GdL) concentrations were screened. The self-assembly was induced by dissolving the hydrogelator in basic pH through the addition of 2 v/v% NaOH (1 M) and, then, glucono-δ-lactone (GdL) was added to decrease the pH homogeneously. A solution of 0.05 wt% hydrogel was used to assess the fiber-catalyzed secondary nucleation. The aggregates fraction, f (t), was defined as follows:

$$f(t) = \frac{T_{obs} - T_{free}}{T_{agg} - T_{free}} \quad (1)$$

where Tt, Tf ree and T∞ stand for turbidity measured at 500 nm observed at time t, before the addition of GdL and when aggregated (turbidity at t = 5 h), respectively. An empirical exponential decay function was fitted according to Equation (2) [30]:

$$T(t) = \frac{T(\infty)}{\sqrt[v]{\left(1 + ve^{-k_{emp}(t-t_m)}\right)}} \quad (2)$$

where kemp is the rate constant (inverse of the relaxation time) of fibril formation and tm is the point of the maximum elongation rate. This constant rate includes various steps and its interpretation between different systems is misleading. As a result, other models were fitted to understand the influence of the parameters on the nucleation and elongation rates. Saitô's fractional aggregation model has been successful in the aggregation studies of β-amyloid, calcitonin, prion, and α-sinuclein [31–33]. At a concentration larger than critical micellar concentration ([M] » cmc), the aggregation can be described by a two-step mechanism scheme [33]:

$$n_0 M(monomers) \rightleftharpoons M_{n_0}(micelle) \xrightarrow{k_n} P_n$$
$$M + P_n \xrightarrow{k_e} P_{n+1}.$$

In this mechanism, kn and ke represent the average nucleation and growth rate constants, M is the monomer, Mn0 represents the micelle (precatalytic form of the monomer), Pn is the nucleus of fibril with n monomer molecules, and Pn+1 is the extended fibril with n + 1 monomer molecules; thus, P is both a product and catalyst in the growth step reaction. The fibril formation can be described according to the equation:

$$f(t) = \frac{\rho\left(e^{(1+\rho)k_s t} - 1\right)}{1 + \rho e^{(1+\rho)k_s t}} \quad (3)$$

where ks = ke [M] is the effective growth rate constant, ϱ = kn , and the initial condition is f (0) = 0. The secondary nucleation was evaluated through the aggregation models of Knowles et al. [30,34] and Cohen et al. [35]. The former model describes the concentration of monomer in the fibers according to the equation:

$$f(t) = 1 - e^{-C_+ e^{kt} + C_- e^{-kt} + k_n m_{tot}^{n_c - 1} k_-^{-1}} \quad (4)$$

where mtot is the total hydrogelator concentration and k = √2mtotk+ k $= \sqrt{2m_{tot}k_+k_-}$. of fibrils, at the beginning of the aggregation,. In the absence

$$C_\pm = \frac{\pm k_n m_{tot}^{n_c - 1}}{2k_-} \quad (5)$$

where k+ is the polymerization rate and k- is the secondary nucleation (fragmentation). The latter model describes the concentration of monomer in the fibers according to the equation:

$$f(t) = 1 - \left(\frac{B_+ + C_+}{B_+ + C_+ e^{\kappa t}} \frac{B_- + C_+ e^{\kappa t}}{B_- + C_+}\right)^{\frac{k_\infty^2}{\kappa \tilde{k} j_\infty}} - e^{-k_\infty t} \quad (6)$$

where k $= \sqrt{2k + j_2 m(0)^{n2+1}}$ is associated with the secondary pathways, k2 is the fibrilcatalyzed secondary nucleation, and $k_2 = k_-$ when $n_2 = 0$, $C_\pm = \pm \lambda^2/(2k^2)$ is related tion). The latter model describes the concentration of monomer in the fibers according to the equation:

$$f(t) = 1 - \left(\frac{B_+ + C_+}{B_+ + C_+ e^{\kappa t}} \frac{B_- + C_+ e^{\kappa t}}{B_- + C_+} \right)^{\frac{k_\infty^2}{\kappa k j_\infty}} - e^{-k_\infty t} \quad (6)$$

where $\kappa = \sqrt{2k_+ k_2 m(0)^{n_2+1}}$ is associated with the secondary pathways, k_2 is the fibril-catalyzed secondary nucleation, and $k_2 = k_-$ when $n_2 = 0$, $C_\pm = \pm\lambda^2/(2\kappa^2)$, $\lambda = \sqrt{2k_+ k_n m(0)^{n_c}}$ is related with the rate of formation of new aggregates through primary pathways, $B_\pm = (k_\infty \pm kj_\infty)/(2\kappa)$, $k_\infty = \sqrt{2\kappa^2/[n_2(n_2+1)] + 2\lambda^2/n_c}$, and $kj_\infty = \sqrt{k_\infty^2 - 4C_+C_-\kappa^2}$. The parameters n_c and n_2 describe the dependencies of the primary and secondary pathways, and $m(0)$ is the initial concentration of soluble monomers.

where k = $\sqrt{2k + j_2 m(0)^{n2+1}}$ is associated with the secondary pathways, k2 is the fibrilcatalyzed secondary nucleation, and k2 = k- when n2 = 0, C± = ± λ2/(2k2) is related with the rate of formation of new aggregates through primary pathways, B± = (k∞ ± kj∞)/(2k), k∞ = √ 2k2/[n2(n2 + 1)] + 2λ2/nc, and kj∞= √ 2k2∞/[n2(n2 + 1)] + 2λ2/nc . The parameters nc and n2 describe the dependencies of the primary and secondary pathways, and m(0) is the initial concentration of soluble monomers.

2.2.2 Development of Magnetogels

The prepared nanoparticles were added to the hydrogel solution at a final volume of 200 μL and at the required concentration from a starting solution at 2 wt%. All hydro- gel/magnetogel solutions were left standing at room temperature until the gel phase was attained. Here, the unit wt% stands for m/v%.

2.3 Spectroscopic Measurements

2.3.1 General Methods

Fluorescence measurements were carried out using a Horiba-Jobin Yvon Fluorolog 3 spectrofluorimeter (HORIBA Jobin Yvon IBH Ltd., Glasgow, UK), equipped with double excitation and emission monochromators, Glan-Thompson polarizers (HORIBA Jobin Yvon IBH Ltd., Glasgow, UK), and a temperature-controlled cuvette holder. Fluorescence emission spectra were corrected for the instrumental response of the system. The excitation of the hydrogelator was set at 280 nm, and the emission spectrum was collected between 290 nm and 600 nm with a slit of 6 nm in both excitation and emission. Absorption spectra were recorded in a Shimadzu UV-3600 Plus UV-Vis-NIR with the rate of formation of new aggregates through primary pathways, B± = (k∞ ± kj∞)/(2k), k∞ = 2k2/[n_2(n2 + 1)] + 2l2/nc, The parameters nc and n^2 describe the dependencies of the primary and secondary pathways, and m(0) is the initial concentration of soluble monomers.

2.2.2 Development of Magnetogels

The prepared nanoparticles were added to the hydrogel solution at a final volume of 200 μL and at the required concentration from a starting solution at 2 wt%. All hydro- gel/magnetogel solutions were left standing at room temperature until the gel phase was attained. Here, the unit wt% stands for m/v%.

2.3 Spectroscopic Measurements

2.3.1 General Methods

Fluorescence measurements were carried out using a Horiba-Jobin Yvon Fluorolog 3 spectrofluorimeter (HORIBA Jobin Yvon IBH Ltd., Glasgow, UK), equipped with double excitation and emission monochromators, Glan-Thompson polarizers (HORIBA Jobin Yvon IBH Ltd., Glasgow, UK), and a temperature-controlled cuvette holder. Fluorescence emission spectra were corrected for the instrumental response of the system. The excitation of the hydrogelator was set at 280 nm, and the emission spectrum was collected between 290 nm and 600 nm with a slit of 6 nm in both excitation and emission. Absorption spectra were recorded in a Shimadzu UV-3600 Plus UV-Vis-NIR spectrophotometer (Shimadzu Corporation, Kyoto, Japan). The fluorescence quantum yield, Φs, can be determined by Equation (7) (standard method) [36,37],

$$\Phi_s = \frac{(A_r F_s n_s^2)}{(A_s F_r n_r^2)} \Phi_r \tag{7}$$

where A is the absorbance at the excitation wavelength, F is the integrated emission area, and n is the refraction index of the solvents. Subscripts r and s refer to the reference and sample compound, respectively. The absorbance value at excitation wavelength was always less than 0.1 in order to avoid inner filter effects. L-Tryptophan in aqueous buffer solution (pH = 7.2) was used as a reference (Φr = 0.14 at 25 °C) [38].

2.3.2 Fluorescence Anisotropy Measurements

The steady-state fluorescence anisotropy values, r, provide information on the average microviscosity of the gel matrix where the fluorophore is localized and can be determined by Equation (8) [39],

$$r = \frac{I_{VV} - GI_{VH}}{I_{VV} + 2GI_{VH}} \quad (8)$$

where IVV and IVH are the intensities of the emission spectra obtained with vertical and horizontal polarization, respectively (for vertically polarized excitation light), IHV and IHH are the emission intensities obtained with vertical and horizontal polarization (for horizontally polarized excitation light), and G = IHV/IHH is the instrumental correction factor.

2.3.3 FRET Measurements

The drug incorporation into the magnetogels network was investigated by Förster Resonance Energy Transfer (FRET). FRET efficiency, ΦFRET, defined as the proportion of donor molecules that have transferred their excess energy to acceptor molecules, can be expressed by Equation (9) [39],

$$\Phi_{FRET} = 1 - \frac{I_{DA}}{I_D} \quad (9)$$

where IDA and ID are the donor integrated fluorescence intensities in the presence and absence of an acceptor, respectively. FRET efficiency can also be determined using the donor-acceptor intermolecular distance, RDA, and the Förster radius (critical diameter), R0, through Equation (10) [39],

$$\Phi_{FRET} = \frac{1}{1 + \left(\frac{R_{DA}}{R_0}\right)^6}. \quad (10)$$

The Förster radius, R0, the distance at which FRET efficiency is 50%, can be deter- mined by the spectral overlap, J(λ) between the donor fluorescence emission and the acceptor absorption, according to Equations (11) and (12) (with R0 in Å, λ in nm, ε A (λ) in M−1 cm−1) [39],

$$R_0 = 0.2108 \times \left[\kappa^2 \Phi_D n^{-4} J(\lambda)\right]^{1/6} \quad (11)$$

$$J(\lambda) = \int_0^\infty I_D(\lambda)\varepsilon_A(\lambda)\lambda^4 d\lambda \quad (12)$$

where κ2 = 2/3 is the orientational factor assuming random orientation of the dyes, ΦD is the donor fluorescence quantum yield in the absence of energy transfer, n is the refraction { ∞ index of the medium, ID (λ) is the fluorescence spectrum of the donor normalized so that $\int_0^\infty I_D(\lambda)\, d\lambda = 1$, and εA (λ) is the molar absorption coefficient of the acceptor.

2.4 Characterization Techniques

2.4.1 Scanning Transmission Electron Microscopy (STEM)

STEM images were recorded using a NanoSEM—FEI Nova 200 (FEI Company, Hillsboro, OR, USA), operating at 15 kV, coupled to an Electron Dispersive Spectroscopic analyzer (EDS) and Electron Backscatter Diffraction EDAX—Pegasus X4M analyser (AMETEK Inc., Berwyn, PA, US) and detection system (EBSD) at SEMAT (Serviços de Caracterização de Materiais, Guimarães, Portugal). After preparation of the hydrogel, a small portion of each sample was placed onto a TEM 400 mesh copper grid with Formvar/Carbon (ref. S162-4 from Agar Scientific), held by tweezers and the excess solution was cleaned. The processing of STEM images was performed using ImageJ software (National Institutes of Health (NIH), version 1.52p, Bethesda, MD, USA),which consisted of enhancing local contrast and adjusting brightness followed by a manual selection of fibers.

2.4.2 X-ray Diffraction

A conventional PAN'alytical X'Pert PRO diffractometer (Malvern Panalytical Ltd., Malvern, UK) was used for X-ray diffraction (XRD) analyses, operating with Cu Kα radiation, in a Bragg-Brentano configuration.

2.4.3 Raman Spectroscopic Measurements

Raman spectroscopy was used to assess the effect of nanoparticles in the secondary structure of the hydrogel fibers. Measurements were performed at room temperature with a Jobin Yvon T64000 triple Raman Spectrometer (HORIBA Jobin Yvon IBH Ltd., Glasgow, UK), equipped with a liquid nitrogen cooled charge couple device (CCD) detector, with a resolution better than 1 cm-1. The excitation line, 514.5 nm, of an argon ion laser was focused onto the sample using a ×50 objective (focused to ~1.5 μm of diameter) of an Olympus Microscope BHSM (Olympus Corporation, Tokyo, Japan) in a backscattering geometry. The spectra were acquired with a measured power of about 350 μW on the sample, with a spectral acquisition time of 45 s averaged over 10 scans, over the range 770–1800 cm-1.

2.4.4 Magnetic Properties

Magnetic measurements were performed using a SQUID magnetometer from Quantum Design (Quantum Design Inc., San Diego, CA, USA). The magnetization dependence with temperature in zero-field-cooling (ZFC) and field-cooling (FC) conditions was per- formed at 100 Oe in the 10–320 K range. Hysteresis loops were measured at different temperatures up to an external field of 50 kOe.

2.4.5 Rheology

The viscoelastic characterization of gels was performed with a stress-controlled rotational rheometer Anton Paar MCR300 (Anton Paar GmbH, Graz, Austria). Liquid samples were loaded into the Couette geometry of the rheometer and temperature was kept at 25 °C during testing. After a five hour rest period ensuring gel setting and structural equilibrium of samples, a sweep in the strain amplitude was performed from 0.001% to 500% to assess the linear regime of visco-elasticity and the large amplitude oscillatory strain (LAOS) regime.

2.5 Drug Release Assays

2.5.1 Incorporation of Doxorubicin

To study the incorporation and microenvironment of doxorubicin in gels through fluorescence spectroscopy, the drug was added to gel solutions prior to gelation, for a final concentration of 20 μM (to guarantee that fluorescence intensity is proportional to concentration). From the hydrogelator solution, 200 μL were transferred to a fluorescence microcuvette and left standing until the gel was formed.

2.5.2. Drug Release to pH = 7 Buffer

To assess doxorubicin release through fluorescence spectroscopy, gels (100 μL) loaded with 0.1 mM doxorubicin were prepared and left stabilizing overnight in Amicon® Ultra-0.5 mL centrifugal filters (MilliporeSigma, St. Louis, MO, USA) with a 0.1 μm pore size. Then, the filter tube was immersed in pH = 7 buffer (800 μL) to keep pH constant (besides neutralizing the gels), and left standing at room temperature, with or without an alternating magnetic field (AMF). The AMF was generated in a custom-designed solenoid device (800 turns per meter, length: 31 cm and internal diameter: 4.8 cm) by applying an alternating electric current. A magnetic field of 2.98 mT at 1000 kHz was used. Aliquots were taken and replaced with pH = 7 buffer, then fluorescence was measured to determine the concentration at each time point. Release profile assays were performed in triplicate.

3. RESULTS

3.1. Optimization of Hydrogel Gelation Kinetics

Turbidity kinetic assays were carried out to optimize hydrogel gelation, which is required to ensure quasi-homogeneous encapsulation of nanocomposites. Initially, the hydrogelator molecules were majorly organized in a mixture of micelles/aggregates and free monomer as suggested by fluorescence emissions at 450 nm and 360 nm, respectively (Figure 1A).

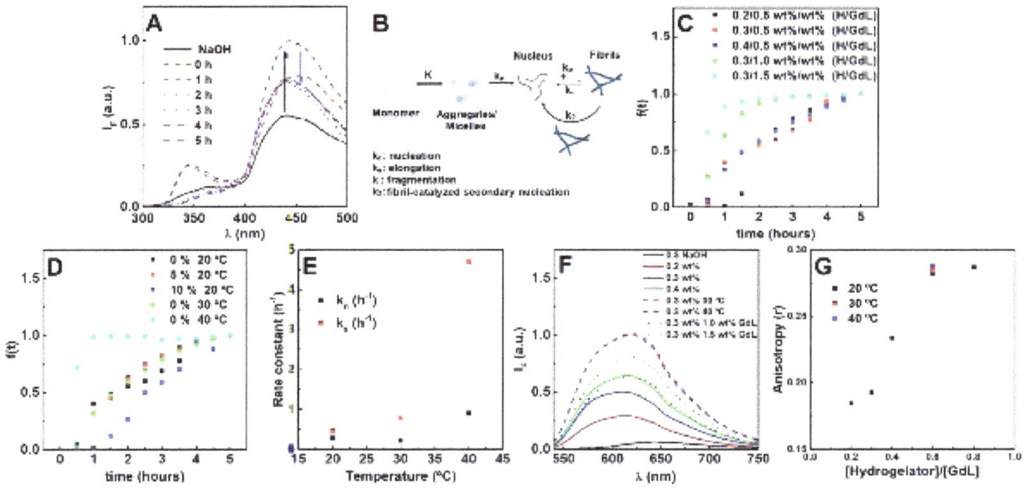

Figure 1. (A) Fluorescence emission spectra of the hydrogelator (0.3 wt%) over time after the gelation trigger (0.5 wt%). (B) Scheme of the evaluated pathways during the gelation process, where the fibril-catalyzed secondary nucleation contribution was found to be negligible. (C) Gelation kinetic profile dependence on hydrogelator and GdL concentration, (D) fibril concentration and temperature. (E) Average nucleation and elongation rate constants obtained from the Saitô's aggregation model fitting to turbidity profiles at increasing temperatures (hydrogel 0.3 wt%, GdL 0.5 wt%). Fluorescence emission (F) and anisotropy r (G) of Nile Red (2 μM) in hydrogels prepared at different hydrogelator-to-GdL ratios and temperatures (fixed hydrogel and GdL concentration).

Once gelation is initiated (after the addition of GdL), both emission bands increase, indicating the occurrence of a reorganization process. Here, the models of Knowles et al. [34] and Cohen et al. [35] (commonly used for amyloid aggregation) were also used as a strategy to assess the possible occurrence and influence of monomer independent (k-;fragmentation) and/or monomer dependent (k2; fibril-catalyzed secondary nucleation) secondary pathways (Figure 1B).

The gelation is characterized by a sigmoidal profile, which is a common feature of fibrillation processes comprising a nucleation and elongation phase (Figure 1C,D) [30]. Increasing the hydrogelator concentration at the same GdL (0.5 wt%) concentration resulted in an increase of the nucleation rate, while the elongation rate decreased. The increase in GdL concentration promoted both the nucleation and elongation phases. Hereby, the primary nucleation process is dependent on both the hydrogelator and GdL, while the elongation process is majorly affected by the GdL. The elongation phase dependence on GdL concentration (for a fixed hydrogelator concentration) demonstrates that its rate can be increased if more protons are made available over time, considering that GdL proton dissociation is also a kinetically-dependent process [40]. Such results evidence that an increase of hydrogelator concentration has to be accompanied by an increase of GdL concentration to keep the molar equivalents, thus favouring both nucleation and elongation. Notably, by increasing both hydrogelator and GdL concentrations, no major differences were obtained for the final pH, while increasing GdL alone strongly decreased pH.

A weak scaling of the half-time with the initial monomer concentration was obtained (-0.63), which is characteristic of monomer independent secondary pathways (such as fragmentation). Yet, the light scattering kinetic profiles cannot be satisfactorily matched using fixed kn /k- and k+ kn parameters in the Knowles' aggregation model [35]. The addition of pre-formed fibrils inhibited the average nucleation phase and enhanced the average elongation phase (Figure 1D), i.e., the secondary nucleation rate can be neglected. Further, increasing the hydrogelator decreased both secondary pathways rates, while GdL enhanced the monomer independent pathway and inhibited the monomer-dependent pathway. Nonetheless, self- assembly can be majorly attributed to the primary pathways. Temperature exponentially affected the average nucleation and elongation phase, which is associated with a faster GdL proton dissociation (Figure 1E) [40]. As a result, overall gelation can be enhanced by increasing both GdL and hydrogelator concentrations and preparing the gel at 30 or 40 ◦C, which favours the primary pathways (nucleation and elongation).

The dye Nile Red was used to evaluate the effect of preparation conditions on the microenvironment, as it is a solvatochromic probe that has almost negligible emission in water, but intensely emits fluorescence in non-polar environments [41–44], as observed after hydrogel formation (Figure 1F). Furthermore, the emission is accompanied by a blue-shift with a reduction of polarity [41–44]. Here, Nile Red is localized in a microenvironment with a polarity between acetone and ethanol [41]. The higher Nile Red fluorescence emission intensity on the gels prepared at 30 ◦C and 40 ◦C suggests that more hydrophobic regions were made available. The fluorescence anisotropy values reveal similar fluidity compared to the hydrogel prepared at room temperature. Although increasing GdL (fixed hydrogelator concentration at 0.3 wt%) also contributed for more hydrophobic regions, the microfluidity was lower than that obtained by increasing the hydrogelator concentration (fixed GdL concentration at 0.5 wt%). As such, increasing both GdL and hydrogelator concentrations (0.4 hydrogelator-to-GdL ratio or higher) promotes more hydrophobic regions with higher microviscosity.

3.2. Nanoparticles Characterization

Nanoparticles of manganese ferrite with different coatings were prepared using different synthetic methods, and are named, from now on, as citrate-stabilized or lipid-coated manganese ferrite nanoparticles. The X-ray diffraction (XRD) patterns of both samples present well-defined peaks (Figure 2A,B) characteristic of a crystalline structure, which was obtained without calcination. The diffraction peaks of the MnFe2O4 crystalline spinel structure are observed at 2θ= 29.7◦(2 2 52.7 ,(0 0 4) ◦42.5 ,(2 2 2) ◦36.5 ,(1 1 3) ◦34.9 ,(0◦ (4 2 2), 56.2◦ (3 3 3) and (,(1 1 5 74.0 ,(3 3 5) ◦73.1 ,(0 2 6) ◦70.1 ,(1 3 5) ◦65.0 ,(0 4 4) ◦61.8◦ (6 2 2), 78◦ (◦85.6 ,(4 4 4 1 3 7) ◦88.5 ,(2 4 6)) and (3 5 5), according to CIF file 2300618 (space group Fd3-m). Rietveld analysis was performed using the Full Prof software suite, confirming the spinel structure. As in previous works [45,46], we considered it to be important to use microabsorption correction [47], resulting in fits with good RF values. The calculated parameters offering a larger size of the crystalline domains in the nanoparticles that are citrate-stabilized. Additional diffraction peaks were observed for the lipid-coated nanoparticles, occurring at positions similar to those reported for layered manganese laurate [48], which can be ascribed to ordered lauric acid molecules at the surface of manganese ferrite.

The UV-visible absorption spectra of the prepared nanoparticles are represented in Figure 2C. From the Tauc plot (inset of Figure 2C), the optical band gap (E_g) between the citrate-stabilized and lipid-coated manganese ferrite nanoparticles was determined and a linear relation was obtained for an indirect semiconductor with a band gap of 1.13 eV and 1.19 eV, respectively, which are similar to the previous reported value of 1.08 eV [49].

The sedimentation profiles for bare, lipid-coated, and citrate-stabilized nanoparticles are displayed in Figure **2D**. The dependence of the sedimentation rate on nanoparticle concentration (obtained through fitting of a Becquerel function or compressed hyperbola) [50] is included (values are reported. The citrate-stabilized nanoparticles sedimentation profile suggests the occurrence of nanoparticles aggregation into stable agglomerates [51,52], which settle down at a faster rate than single nanoparticles. The lipid-coated nanoparticles show a sedimentation rate independent of the used concentration range (0.025–0.2 wt%). Hereby, the longer-term stability of the lipid-coated nanoparticles is expected to provide homogeneous gels at higher concentration of nanoparticles, compared to the ones that are citrate-stabilized. 1.19 eV, respectively, which are similar to the previous reported value of 1.08 eV [49].

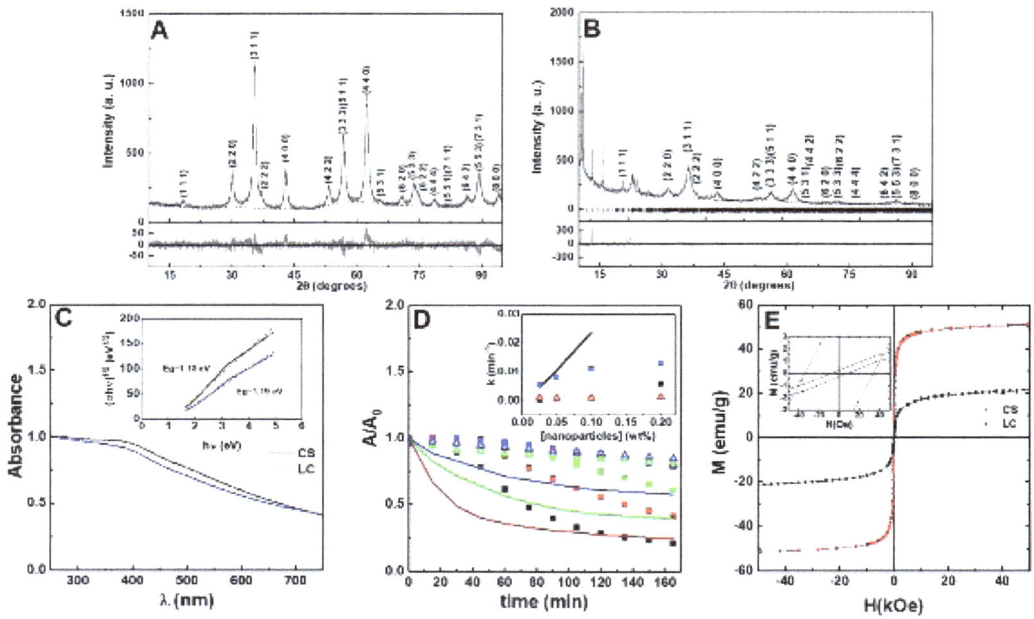

Figure 2. X-ray diffraction pattern of (A) citrate-stabilized and (B) lipid-coated manganese ferrite nanoparticles. Gray lines: experimental patterns; black lines: fitted patterns; dotted lines: fitted background; the vertical lines in (B) are the Bragg diffraction positions of manganese laurate. (C) Absorption spectra of citrate-stabilized (CS, black line) and lipid-coated (LC, blue line) $MnFe_2O_4$ nanoparticles. Inset: Tauc plot of citrate-stabilized and lipid-coated nanoparticles. (D) Sedimentation profiles of the citrate-stabilized (squares), lipid-coated (triangles), and bare (line) $MnFe_2O_4$ nanoparticles at 0.2 wt% (black), 0.1 wt% (red), 0.05 wt% (green), and 0.025 wt% (blue). Inset: sedimentation rate dependence on nanoparticle concentration. The citrate-stabilized aggregation rate is included (black squares). (E) Magnetization hysteresis loops of citrate-stabilized and lipid-coated manganese ferrite nanoparticles measured at room temperature (T = 300 K).

The magnetic hysteresis loops show that the saturation magnetization (emu/g) is higher for the citrate-stabilized nanoparticles than for the lipid-coated nanoparticles (Figure 2E), which can be explained by taking a different stoichiometry into account in the manganese ferrite in the two samples, as pointed out by the different lattice constant obtained in the X-ray diffraction analysis. A smaller size and a higher wt% of organic matter present in the second sample can also influence the final value of saturation magnetization. Consequently, considering that SAR $\propto$ Ms2 (SAR—specific absorption rate), the lipid-coated nanoparticles are expected to have lower heating efficiency than the citrate-stabilized nanoparticles [53,54]. The low Mr/Ms ratio, of around 0.1 is an indication that both types of nanoparticles display a super-paramagnetic behaviour [54].

3.3. Development of Magnetogels

An empirical equation was used to assist the estimation of the conditions required to maximize the homogeneity of the gel (Figure 3 displays plots of the obtained k_{emp}/k_{sed} for various nanoparticle concentration and GdL-to-hydrogelator concentration ratio. The estimation implies that the gelation conditions have to guarantee a $k_{emp}/k_{se}d > 41.7$, so that when gel fraction *f(t)* attains 0.9, the nanoparticles suspended fraction is also at 0.9 (if $v = 1$ and $a = 0.5$ is assumed).

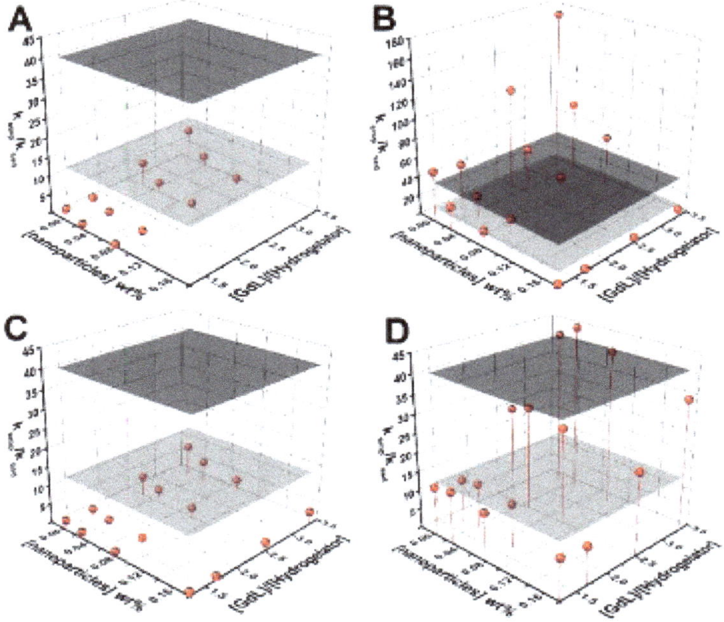

Figure 3. Surface plots of k_{emp}/k_{sed} dependence on nanoparticle concentration and the GdL-to-hydrogelator concentration ratio for: (A) bare nanoparticles, (B) the citrate-stabilized nanoparticles aggregation rate, (C) the sedimentation rate, and (D) lipid-coated nanoparticles. The gray planes define the estimated k_{emp}/k_{sed} required to guarantee that when the gel is about %10 of gelation completion, %90 of the nanoparticles remains in suspension ($\nu = 1$, $a = 0.5$, dark gray plane; □ = 0.5, a = 0.5, gray plane).

Gels prepared with 1.5 wt% GdL and 0.3 wt% hydrogelator retained the nanoparticles, though the pH value was lower than 4. Increasing the hydrogelator content to 0.5 wt% and reducing GdL to 1 wt% (has higher microviscosity) allowed the preparation of homogeneous magnetogels at 0.1 wt% of nanoparticles, with a pH of ~5. The 0.2 wt% content of nanoparticles can also be prepared at 0.5 wt% of hydrogelator and 1 wt% of GdL but preparation at 30 °C or 40 °C is required (yields k_{emp}/k_{sed} larger than 200). Nonetheless, the parameter v obtained from curve fitting was around 0.5, thus decreasing the required k_{emp}/k_{sed} to 14, which allowed the possibility of obtaining quasi-homogeneous gels at 0.1 wt% of nanoparticles at a [GdL]/[hydrogelator] ratio of 2.

3.4. Gels Microviscosity

The effect of the nanoparticles in the gels matrix microenvironment was studied using Nile Red as a fluorescence probe, while considering its sensitivity to polarity and viscosity [55,56]. In the 20–40°C temperature range, no major fluorescence emission decay changes were observed, which might be associated with the structure maintaining its integrity. A steep fluorescence emission decrease indicated that the phase transition occurs above 45°C for the hydrogel and the citrate-stabilized nanoparticle-containing magnetogels, while in the lipid-coated nanoparticle-containing magnetogels, it was shifted to 50 °C (Figure 4A). The fluorescence anisotropy increases at the phase transition temperature, which might be associated with a fluorescence emission lifetime decrease (Figure 4B) [57].

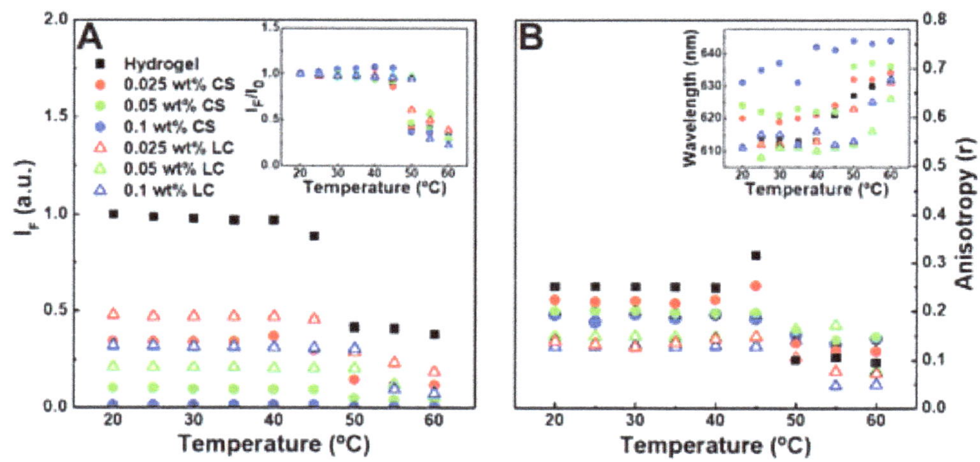

Figure 4. (A) Nile Red (2 μM) maximum fluorescence emission dependence on the temperature of hydrogel, citrate-stabilized (CS), and lipid-coated (LC) nanoparticles containing magnetogels. Inset: Normalized maximum fluorescence emission dependence on the temperature of the respective gels. (B) Nile Red fluorescence anisotropy dependence on temperature. Inset: Nile Red maximum emission wavelength dependence on temperature.

The results suggest that citrate-stabilized nanoparticles destabilized the hydrophobic domains, leading to a microviscosity reduction and polarity increase of the cavities where Nile Red is localized, which might occur through hydrogen bonding and ionic interaction between the nanoparticles and the fibers. The polarity in the hydrogel is similar to acetone, which was changed towards ethanol after addition of citrate-stabilized nanoparticles. Furthermore, after phase transition and at 0.1 wt% of citrate-stabilized nanoparticles, the Nile Red emission wavelength (640 nm) became close to the reported maximum wavelength in water (657 nm) [42]. The lipid-coated nanoparticles induced a lower fluorescence anisotropy than citrate-stabilized nanoparticles and one that is similar to the reported anisotropy values of Nile Red in mixed vesicles and micelle membranes [44], thus suggesting that lipid-fiber domains are formed with a polarity similar to the fibers and a viscosity near that of membranes.

3.5. Gels Secondary Structure

The Raman spectra of the hydrogels and magnetogels (0.1 wt% of nanoparticles) were obtained to assess influence of nanoparticles in the secondary structure, which are displayed in Figure 5. Reported Raman shifts of the phenylalanine phenyl ring and methionine side chain (CH_3 deformation at 1440 cm^{-1} and CH_2 wagging at 1320 cm^{-1}) are also displayed [58,59]. Structural changes upon the addition of lipid-coated nanoparticles were suggested by the appearance of a band at 982 cm^{-1}. Gaussian curves were fitted to the major phenyl ring signal at around 1003 cm^{-1}. A blue shift and decreasing cross-section were observed with an increasing nanoparticle concentration. The latter effect can be associated with an increasing exposure of the aromatic rings to a more hydrated environment [60].

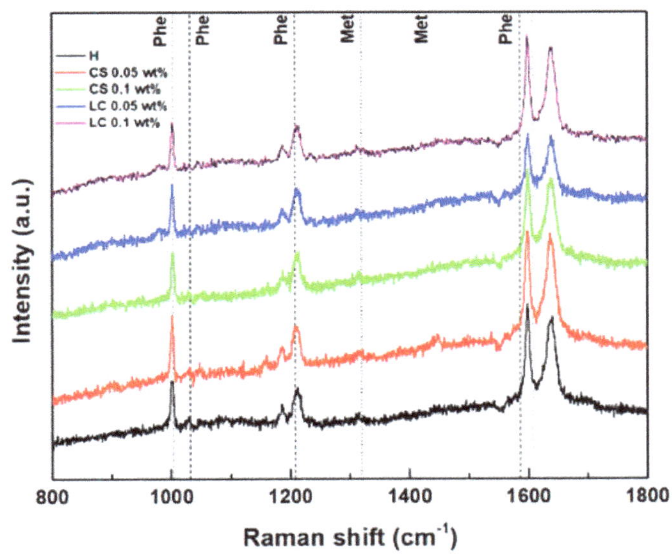

Figure 5. Raman scattering spectra of the hydrogel and magnetogels containing lipid-coated (LC) and citrate-stabilized (CS) nanoparticles. Vertical lines represent the reported Raman shifts of methionine (Met) and phenylalanine (Phe).

The Amide I region (1580–1700 cm^{-1}) arises from the C=O stretching vibration, which is sensitive to changes in backbone peptide conformation [61]. Deconvolution of the amide I band in its sub-bands is correlated with various secondary structure contributions [60,61,62]. A major contribution at 1637 cm^{-1} is common to all systems, which can be associated with a major β-sheet content [61,62]. The band at 1338 cm^{-1} also confirms the predominance of the β-sheet in all gels [61].

3.6. Rheological Properties

Large amplitude oscillatory shear strain sweeps (LAOSS) were carried out to assess the effect of the nanoparticles on the gels structure. Further, the effect of temperature at 37 °C on the hydrogel shear elastic and loss moduli was also assessed (Figure S6), the decrease of which evidences the phase transition behaviour observed in the microviscosity studies. The addition of nanoparticles reduced the elasticity of gels, as previously observed for other systems [10,63]. Interestingly, the lipid-coated nanoparticles induced a lower elasticity than the citrate-stabilized nanoparticles, which was similar to their influence on the gels' microviscosity. Further, the strain at which G′ crosses G″ increased in the presence of citrate-stabilized nanoparticles (at 0.05 wt%). As a result, the profiles suggest that different structures were obtained for each nanoparticle content, compared to the bare gel. The lipid-coated nanoparticle-containing magnetogel displays the Payne effect characteristic of viscoelastic matrices reinforced by solid fillers, i.e., a local maximum of G″ concomitant with a significant decrease of G′, which can be associated with the breakage and recovery of weak interaction bonds linking adjacent clusters, aggregation/disaggregation of nanoparticles, or molecular disentanglement [64].

3.7. Electron Microscopy

Figure 6 displays the STEM images of the citrate-stabilized and lipid-coated nanoparticles in solution and incorporated into the hydrogel matrix (magnetogels) prepared at 0.5 wt% of hydrogelator. The negative charge of the citrate molecules stabilizing the nanoparticles ensured there were well dispersed nanoparticles with an average size of 2.6 ± 14.4 nm, though some aggregates were observed (Figure 6A). Alternatively, the lipid-coated nanoparticles, with an average size of 8.9 ± 2.1 nm, displayed a tendency to form spherical aggregates, which can be associated with the dynamic membrane coating being prone to self-assembly (Figure 6B). The hydrogel network is also displayed for a matter of comparison with the magnetogels. The preparation conditions used here afforded a network comprised of thin and thick short fibers with a cross-section of 21.3 ± 3.4 nm and 48.4 ± 13.8 nm, and with an average length of 1030 ± 389 nm.

Furthermore, different effects on the magnetogels microstructure were observed when using the two types of nanoparticles. The citrate-stabilized nanoparticles are randomly distributed within the hydrogel matrix as aggregates. This stems from the fact that the magnetogels based on N-protected peptides lacked a cationic group in the hydrogelator structure electrostatically interacting with these negatively-charged stabilized nanoparticles and fixing them in the matrix. Nevertheless, the lipid-coated nanoparticles displayed an affinity towards the fibers surface, rendering them more adequate to avoid any potential leaking of nanoparticles. Furthermore, upon gelation, the lipid-coated nanoparticle aggregates become destabilized, as no aggregates were observed (see Figure 6E,F).

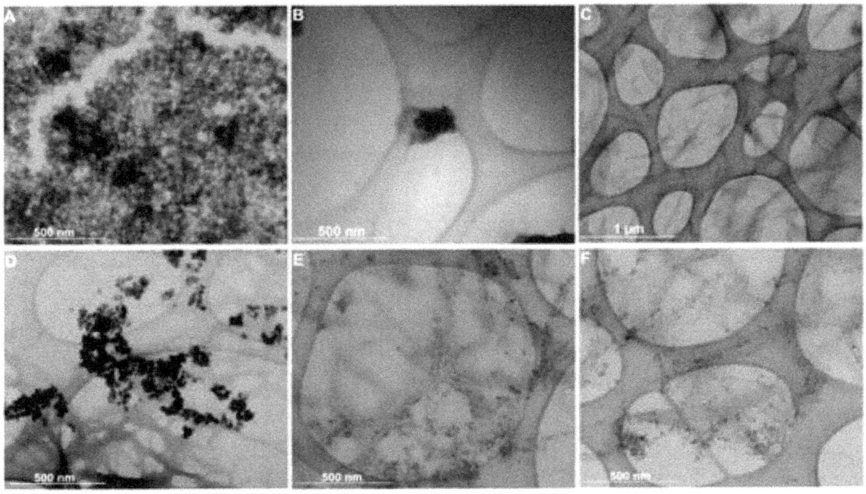

Figure 6. Scanning transmission electron microscopy (STEM) images of (A) citrate-stabilized nanoparticles, (B) lipid-coated nanoparticles, and (C) the hydrogel structure. Magnetogels containing (D) citrate-stabilized nanoparticles and (E, F) lipid-coated nanoparticles prepared at 0.5 wt% of hydrogelator, 1 wt% of GdL, and 0.05 wt% of nanoparticles.

3.8. Hyperthermia Studies

The calorimetric approach was carried out to assess the magnetic nanoparticle hyperthermia effect in gels, while considering the medical threshold limit of $H_0 f_0 \leq 10 \times 5^9$ A m^{-1} s^{-1} [65,66], or $H_0 f_0 \leq 10 \times 4.85^8$ A m^{-1} s^{-1} [66], depending on the area exposed. The increase in temperature over time, when the nanoparticles are dispersed in aqueous solution and in gels while applying an alternating magnetic field, is displayed in Figure 7.

In water solution, the citrate-stabilized nanoparticles attained higher temperatures than the lipid-coated nanoparticles, which could be associated with the different stoichiometry in the spinel ferrites belonging to the two samples, with the different average sizes, and with the fact that the lipid-coated nanoparticles are more likely to aggregate in aqueous solution. In both cases, the increments in temperature decreased when the magnetic nanoparticles were incorporated in the gels, likely because some of the heat generated was employed for inducing local changes in the gel structure.

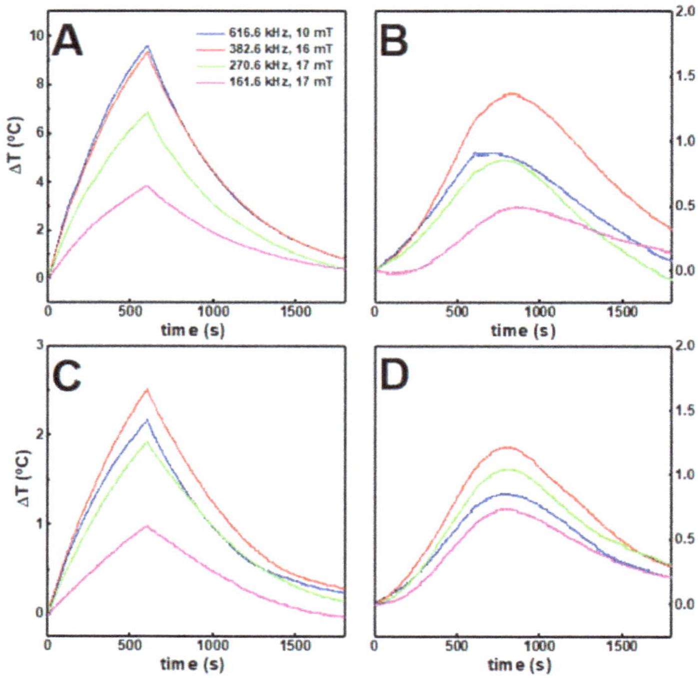

Figure 7. The temperature variation over time of (A) citrate-stabilized and (B) lipid-coated nanoparticles at 1 mg/mL (0.1 wt%) in water and in gels (C,D) respectively) under different magnetic field strengths and frequencies.

Furthermore, nanoparticles at high concentration within the gels underwent stronger magnetic dipolar interactions, which has a detrimental effect in the heat delivery capacity [67,68]. The heating efficiency was quantitatively evaluated through the intrinsic loss power (ILP) (see Table S6 in Supplementary Materials). While the ILP decreases for the citrate-stabilized nanoparticles around 80% when incorporated into the gels, for the lipid-coated nanoparticles no major changes were obtained. Accordingly, despite being less efficient than citrate-stabilized nanoparticles, the lipid-coated nanoparticles keep a similar heat delivery capacity when they are incorporated in the gels.

3.9. Drug Release Assays

3.9.1. Incorporation of Doxorubicin

Previously, supramolecular dehydropeptide-based hydrogels have shown promising results as drug delivery nanosystems [8]. Here, the nanoparticles concentration effect over doxorubicin incorporation is assessed. FRET (Förster Resonance Energy Transfer) process from the emissive moieties of the hydrogel aromatic moieties (acting as the energy donors) to doxorubicin (acting as the energy acceptor) allowed us to follow the encapsulation of doxorubicin owing to the overlap between the drug absorption band and hydrogel fluorescence emission.

Fluorescence spectra of hydrogel and magnetogels at 0.025 wt% of citrate-stabilized and lipid-coated nanoparticles, with and without doxorubicin, are displayed in Figure 8. The absence of doxorubicin results in a strong fluorescence emission (Figure 8A) of the aggregates associated with the stacking of the aromatic rings (λ_{max} ~ 450 nm).

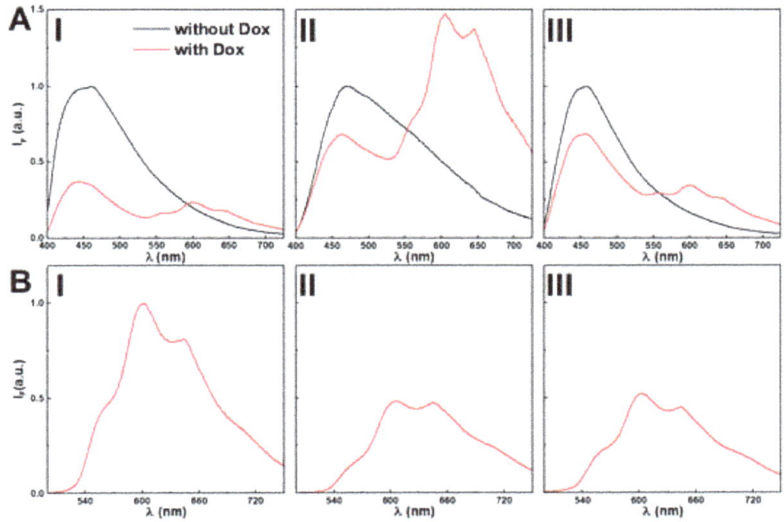

Figure 8. (A) Fluorescence emission spectra (λexc = 375 nm) of hydrogel (I) and magnetogels containing citrate-stabilized (II) and lipid-coated (III) manganese ferrite nanoparticles (0.025 wt%) incorporating the doxorubicin and comparison with the plain gels. (B) Fluorescence spectra (λ_{exc} = 480 nm) of directly-excited doxorubicin in gels, (I) and magnetogels containing citrate-stabilized (II) and lipid-coated (III) manganese ferrite nanoparticles (0.025 wt%) incorporating the doxorubicin and comparison with the plain gels

The addition of citrate-stabilized nanoparticles induced a red-shift (Figure 8B), which was modulated by increasing the nanoparticle concentration (see Figure S9 in Supplementary Materials). The lipid-coated nanoparticles induced a thinning of the emission band shape, suggesting that the variety of microenvironments in the vicinity of the fibers was restrained.

Doxorubicin displays reverse solvatochromism, making the assignment of its location ambiguous [69]. Further, the assays with Nile Red demonstrated that a higher concentration of citrate-stabilized nanoparticles was observed to increase the polarity of the gel hydrophobic cavities. Thus, the decreasing fluorescence emission ratio of doxorubicin between the peaks at λ = 560 nm and λ = 600 nm upon addition of citrate-stabilized nanoparticles might be associated with an increasingly acidic region and its location in cavities, which are destabilized by the increasing concentration of nanoparticles [69]. Nonetheless, the possibility of aggregation is not excluded, considering that at 10 μM, a fraction of 47% is dimerized [70,71], and the obtained spectra in gels resemble the fluorescence emission spectrum of a doxorubicin concentrated solution (0.1 mM) at pH = 7. The lipid-coated nanoparticles do not show the same effects, as the wavelength and emission ratio remain mostly unchanged, i.e., no major changes are induced in the doxorubicin location microenvironment. The fluorescence quenching by increasing nanoparticle concentration further suggests its proximity to the nanoparticles.

The hydrogelator fluorescence quantum yield at a nanoparticle concentration of 0.025 wt% was determined through Equation (7). The calculated FRET efficiencies (□FRETFRET), Förster radius (R_{00}), and donor-acceptor distances (R_{DA} are presented in Table 1. As reported in other systems [8,63], the distances between the fiber aromatic groups and doxorubicin remain similar with or without nanoparticles, and suggests a host-guest type interaction [8,63].

Table 1. Förster Resonance Energy Transfer (FRET) efficiencies (ΦFRET), fluorescence quantum yields of the donor (hydrogel aggregates) (Φ^0_D), donor-acceptor distances (Φ^0_D) and steady-state fluorescence anisotropy (ɤ) values for gels with an incorporated drug. Anisotropy value of doxorubicin in glycerol (at 25 °C): r = 0.285.

H	-	0.66	0.012	2.1	1.9	0.17
CS	0.025	0.33	0.001	1.7	1.4	0.16
	0.05	-	-	-	-	0.15
	0.1	-	-	-	-	0.14
LC	0.025	0.33	0.010	2.3	2.0	0.13
	0.05	-	-	-	-	0.11
	0.1	-	-	-	-	0.11

[a] *Relative to L-tryptophan in aqueous buffer solution, pH = 7.2 (Fr = 0.14 at 25 C) [39]. The error rate is about 10%.*

Here, similarly to the obtained results with Nile Red, the lipid-coated nanoparticles induced a lower anisotropy, while the fluorescence emission spectrum remains similar, suggesting the occurrence of lipid-fiber domains. Overall, the high anisotropy values suggest that doxorubicin has affinity towards the gel network fibers and the nanoparticles affect the arrangement of the fibers.

3.9.2. Doxorubicin Release Assays

In a previous drug release assay of hydrogels, a mixed behavior of gel erosion and drug diffusion was observed [8]. Doxorubicin release from the hydrogel and magnetogels to pH = 7 buffer (to keep pH conditions constant and neutralize gels) was assessed.

The release profiles of the hydrogel and both limiting conditions (0.1 wt% of nanoparticles) of magnetogels are displayed in Figure 9A. Similarly to previous results for dehydropeptide hydrogels (without nanoparticles) and other magnetogels [62], an initial burst release occurs, followed by a slow release phase. Moreover, both systems display a low drug release profile that can be associated with the strong interactions established between the drug and gels components. Such profiles are useful for therapeutic applications, as the systems can be loaded with high amounts of chemotherapeutic drugs and ensure a prolonged and controlled release of a therapeutically relevant dose in the target site.

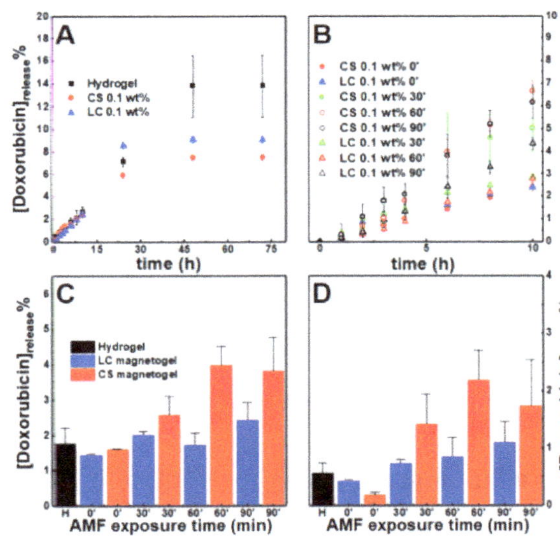

Figure 9. Cumulative doxorubicin percentage release profiles from the hydrogel and gels containing citrate-stabilized (CS) and lipid-coated (LC) nanoparticles (0.1 wt%) (A) without an externally applied alternating magnetic field (AMF) and (B) comparison of the magnetogels release when an AMF is applied for 30, 60, and 90 min between t = 4 h and t = 6 h. (C) Cumulative doxorubicin released at t = 6 h. (D) Variation of doxorubicin percentage released between t=4 h and t = 6 h, that is, for a period of 2 h.

Quantitative analysis of the cumulative drug release profiles without a magnetic field was carried out through the fitting of various mathematical models [72,73,74,75].

Overall, the obtained Gompertz model fitting suggests that both nanoparticles limited the maximum drug quantity released, while the Korsmeyer-Peppas mechanistic model ($0.45 < n < 0.89$) points to a mechanism that combines diffusion and erosion drug release (non-Fickian release) [72,73,74]. The latter contribution was also evidenced by the good fitting of the first-order model (associated with the diffusion of water-soluble drugs in porous matrices) and the Hixson-Crowell model (describes systems where changes in the surface area and diameter occur over time, but the initial geometrical shape is kept constant).

An alternating magnetic field (AMF) was applied for 30, 60, and 90 min, 2.98 mT at 1000 kHz after 4 h, which resulted into an increase of cumulative doxorubicin release comparatively to the systems that were not subjected to the AMF (Figure 9B,C). To remove the contribution from preparation anomalies, the enhancement of drug release was further evidenced by comparing the amount of drug release in the period between t = 4 h and t = 6 h (Figure 9D), that is, during two hours. The enhancement is larger in the citrate-stabilized nanoparticles than lipid-coated nanoparticles, which might be associated with the higher heating efficiency. However, the enhanced drug release displays a non-linear relation with the AMF exposure time. This can be associated with the nanoparticles being distributed as aggregates, as opposed to the lipid-coated nanoparticles that incrementally enhanced drug release, as the sample was exposed to longer periods. Further, gels subjected to the AMF for 30 min revealed a decrease of fluorescence anisotropy to 0.12 for the citrate-stabilized nanoparticles, while in the lipid-coated ones, it remained close to 0.15, i.e., the citrate-stabilized nanoparticles heating induced an irreversible collapse of the gel network that led to an increased variance of drug released, while the other nanoparticles might have not affected the network structure.

As a result, the incremental drug release enhancement and retention of the lipid-coated nanoparticles developed here indicate that this supramolecular system architecture is a suitable approach for controlling drug release, since the heating-induced effect displayed improved reproducibility. However, lipid-coated nanoparticles with a higher heating efficiency and gels that can homogeneously include more nanoparticles are required in future developments to improve the percentage of drug released upon the AMF-trigger, as suggested from the higher triggered release from citrate-stabilized nanoparticle-containing gels.

4. CONCLUSIONS

Pursuing the effect of citrate and lipid-functionalized nanoparticles in the development of supramolecular magnetogels, the gelation of the hydrogel Cbz-L-Met-Z-ΔPhe-OH was systematically optimized by using kinetic models to prepare homogeneous magnetogels, while considering both the kinetics of gelation and sedimentation of nanoparticles. Lipid-coated nanoparticles formed lipid-fiber domains and increased the gel irreversible phase transition temperature. The heating efficiency of lipid-coated nanoparticles was improved (maximum heat and reproducibility) when they were incorporated in the gels, while a detrimental effect was obtained for citrate-stabilized nanoparticles. Further, the former did not produce major changes in doxorubicin encapsulation, while the latter increased the micropolarity of its location and induced aggregation. The magnetogels revealed similar doxorubicin release profiles and AMF-trigger was stronger in the citrate-stabilized nanoparticles, though the triggered release was more reproducible in the lipid-coated nanoparticle-containing gels.

Overall, the lipid-coated nanoparticles displayed promising results for future developments of supramolecular magnetogels aiming at the control of drug release. This was mainly associated with the improved nanoparticle distribution (along the hydrogel fibers), unaffected heating efficiency upon gelation, and reproducible triggered drug release. On the other hand, despite the higher heating efficiency of the negatively-charged citrate-stabilized nanoparticles in solution, they are prone to aggregation upon gelation, which is reflected in a decreased heating efficiency, and local inhomogeneous distribution consequently leading to less reproducibility in drug release.

Hereby, this work reveals that negatively-charged stabilized and lipid-coated nanoparticles affect the final gel architecture differently and, thus, also affect its properties and the encapsulation of drugs in different ways. Further, comparison of both systems points out that on-demand drug release in dehydropeptide-based supramolecular magnetogels can be optimized by developing nanoparticles that can adsorb onto hydrogel fibers, while providing domains that improve or do not affect drug encapsulation (lipid-coated nanoparticles). In particular, the unaffected drug encapsulation and reproducible release from the lipid-coated nanoparticle-containing gels, upon application of AMF, is anticipated to potentiate the supramolecular magnetic gels in drug delivery towards on-demand drug release.

Future developments will be focused on improving heating efficiency of the lipid-coated nanoparticles and the synergy between hyperthermia and triggered drug release, without inducing a major collapse of the hydrogel.

5. CONCISE SUMMARY

Currently, the nanoparticle functionalization effect on supramolecular peptide-based hydrogels remains undescribed, but is expected to affect the hydrogels' self-assembly and final magnetic gel properties. Herein, two different functionalized nanoparticles: citrate-stabilized (14.4 ± 2.6 nm) and lipid-coated (8.9 ± 2.1 nm) magnetic nanoparticles, were used for the formation of dehydropeptide- based supramolecular magnetogels consisting of the ultra-short hydrogelator Cbz-L-

Met-Z-ΔPhe- OH, with an assessment of their effect over gel properties. The lipid-coated nanoparticles were distributed along the hydrogel fibers, while citrate-stabilized nanoparticles were aggregated upon gelation, which resulted into a heating efficiency improvement and decrease, respectively. Further, the lipid-coated nanoparticles did not affect drug encapsulation and displayed improved drug release reproducibility compared to citrate-stabilized nanoparticles, despite the latter attaining a stronger AMF-trigger. This report points out that adsorption of nanoparticles to hydrogel fibers, which display domains that improve or do not affect drug encapsulation, can be explored as a means to optimize the development of supramolecular magnetogels to advance theranostic applications.

REFERENCES

1. Ma, M.; Kuang, Y.; Gao, Y.; Zhang, Y.; Gao, P.; Xu, B. Aromatic-aromatic interactions induce the self-assembly of pentapeptidic derivatives in water to form nanofibers and supramolecular hydrogels. J. Am. Chem. Soc. 2010, 132, 2719–2728. [CrossRef] [PubMed]
2. Sutton, S.; Campbell, N.L.; Cooper, A.I.; Kirkland, M.; Frith, W.J.; Adams, D.J. Controlled release from modified amino acid hydrogels governed by molecular size or network dynamics. Langmuir 2009, 25, 10285–10291. [CrossRef] [PubMed]
3. Bhattacharya, S.; Samanta, S. Soft-nanocomposites of nanoparticles and nanocarbons with supramolecular and polymer gels and their applications. Chem. Rev. 2016, 116, 11967–12028. [CrossRef] [PubMed]
4. Jian, H.; Wand, M.; Dong, Q.; Li, J.; Wang, A.; Li, X.; Ren, P.; Bai, S. Dipeptide self-assembled hydrogels with tunable mechanical properties and degradability for 3D bioprinting. ACS Appl. Mater. Interfaces 2019, 11, 46419–46426. [CrossRef] [PubMed]
5. Fichman, G.; Gazit, E. Self-assembly of short peptides to form hydrogels: Design of building blocks, physical properties and technological applications. Acta Biomater. 2014, 10, 1671–1682. [CrossRef] [PubMed]
6. Yuan, D.; Bing, X. Heterotypic supramolecular hydrogels. J. Mater. Chem. B 2016, 4, 5638–5649. [CrossRef]
7. Draper, E.R.; Adams, D.J. Low-molecular-weight gels: The state of the art. Chem 2017, 2, 716–731. [CrossRef]
8. Veloso, S.R.S.; Magalhães, C.A.B.; Rodrigues, A.R.O.; Vilaça, H.; Queiroz, M.J.R.P.; Martins, J.A.; Coutinho, P.J.G.; Ferreira, P.M.T.; Castanheira, E.M.S. Novel dehydropeptide-based magnetogels containing manganese

ferrite nanoparticles as antitumor drug nanocarriers. Phys. Chem. Chem. Phys. 2019, 21, 10377–10390. [CrossRef]

9. Veloso, S.R.S.; Ferreira, P.M.T.; Martins, J.A.; Coutinho, P.J.G.; Castanheira, E.M.S. Core-shell magnetic-plasmonic nanoparticles enclosed in a biocompatible dehydropeptide-based hydrogel containing lysine. In Proceedings of the Fourth International Conference on Applications of Optics and Photonics, Lisbon, Portugal, 3 October 2019.
10. Carvalho, A.; Gallo, J.; Pereira, D.M.; Valentão, P.; Andrade, P.B.; Hilliou, L.; Ferreira, P.M.T.; Bañobre-Lopez, M.; Martins, J.A. Magnetic dehydrodipeptide-based self-assembled hydrogels for theragnostic applications. Nanomaterials 2019, 9, 541. [CrossRef]
11. Ali, S.; Yosipovitch, G. Skin pH: From basic science to basic skin care. Acta Derm. Venereol. 2013, 93, 261–267. [CrossRef]
12. Bonhome-Espinosa, A.; Campos, F.; Rodriguez, I.; Carriel, V.; Marins, J.; Zubarev, A.; Duran, J.; Lopez-Lopez, M. Effect of particle concentration on the microstructural and macromechanical properties of biocompatible magnetic hydrogels. Soft Matter 2017, 13, 2928–2941. [CrossRef] [PubMed]
13. Antman-Passig, M.; Shefi, O. Remote magnetic orientation of 3D collagen hydrogels for directed neuronal regeneration. Nano Lett. 2016, 16, 2567–2573. [CrossRef] [PubMed]
14. Kim, H.; Jo, A.; Baek, A.; Lim, D.; Park, S.; Cho, S.; Chung, J.; Yoon, J. Synergistically enhanced selective intracellular uptake of anticancer drug carrier comprising folic acid-conjugated hydrogels containing magnetite nanoparticles. Sci. Rep. 2017, 7, 1–10. [CrossRef] [PubMed]
15. Testa-Anta, M.; Liébana-Viñas, S.; Rivas-Murias, B.; Rodríguez-González, B.; Farle, M.; Salgueiriño, V. Shaping iron oxide nanoparticles for magnetic separation applications. Nanoscale 2018, 10, 20462–20467. [CrossRef]
16. Munaweera, I.; Shi, Y.; Koneru, B.; Saez, R.; Aliev, A.; Di Pasqua, A.; Balkus, K. Chemoradiotherapeutic magnetic nanoparticles for targeted treatment of nonsmall cell lung cancer. Mol. Pharm. 2015, 12, 3588–3596. [CrossRef]
17. Hervault, A.; Thanh, N. Magnetic nanoparticle-based therapeutic agents for thermo-chemotherapy treatment of cancer. Nanoscale 2014, 6, 11553–11573. [CrossRef]
18. Akbarzadeh, A.; Samiei, M.; Davaran, S. Magnetic nanoparticles: Preparation, physical properties, and applications in biomedicine. Nanoscale Res. Lett. 2012, 7, 144. [CrossRef]

19. Veloso, S.R.S.; Ferreira, P.M.T.; Martins, J.A.; Coutinho, P.J.G.; Castanheira, E.M.S. Magnetogels: Prospects and main challenges in biomedical applications. Pharmaceutics 2018, 10, 145. [CrossRef]

20. Zhang, Z.Q.; Song, S.C. Thermosensitive/superparamagnetic iron oxide nanoparticle-loaded nanocapsule hydrogels for multiple cancer hyperthermia. Biomaterials 2016, 106, 13–23. [CrossRef]

21. Ramos-Docampo, M.; Fernández-Medina, M.; Taipaleenmäki, E.; Hovorka, O.; Salgueiriño, V.; Stadler, B. Microswimmers with heat delivery capacity for 3D cell spheroid penetration. Acs Nano 2019, 13, 12192–12205. [CrossRef]

22. Bhardwaj, A.; Parekh, K.; Jain, N. In vitro hyperthermic effect of magnetic fluid on cervical and breast cancer cells. Sci. Rep. 2020, 10, 15249. [CrossRef] [PubMed]

23. Guisasola, E.; Asín, L.; Beola, L.; de la Fuente, J.; Baeza, A.; Vallet-Regí, M. Beyond Traditional Hyperthermia: In Vivo Cancer Treatment with Magnetic-Responsive Mesoporous Silica Nanocarriers. Acs Appl. Mater. Interfaces 2018, 10, 12518–12525. [CrossRef] [PubMed]

24. Amiri, M.; Gholami, T.; Amiri, O.; Pardakhti, A.; Ahmadi, M.; Akbari, A.; Amanatfard, A.; Salavati-Niasari, M. The magnetic inorganic-organic nanocomposite based on ZnFe2O4-Imatinib-liposome for biomedical applications, in vivo and in vitro study. J. Alloy. Compd. 2020, 849, 156604. [CrossRef]

25. Colombo, M.; Carregal-Romero, S.; Casula, M.F.; Gutiérrez, L.; Morales, M.P.; Böhm, I.B.; Heverhagen, J.T.; Prosperi, D.; Parak, W.J. Biological applications of magnetic nanoparticles. Chem. Soc. Rev. 2012, 41, 4306–4334. [CrossRef]

26. Gila-Vilchez, C.; Mañas-Torres, M.; Contreras-Montoya, R.; Alaminos, M.; Duran, J.; de Cienfuegos, L.; Lopez-Lopez, M. Anisotropic magnetic hydrogels: Design, structure and mechanical properties. Philos. Trans. R. Soc. A 2019, 377, 20180217. [CrossRef]

27. Yang, Z.; Gu, H.; Du, J.; Gao, J.; Zhang, B.; Zhang, X.; Xu, B. Self-assembled hybrid nanofibers confer a magnetorheological supramolecular hydrogel. Tetrahedron 2007, 63, 7349–7357. [CrossRef]

28. Nawara, K.; Romiszewski, J.; Kijewska, K.; Szczytko, J.; Twardowski, A.; Mazur, M.; Krysinski, P. Adsorption of doxorubicin onto citrate-stabilized magnetic nanoparticles. J. Phys. Chem. C 2012, 116, 5598–5609. [CrossRef]

29. Hui, C.; Shen, C.; Yang, T.; Bao, L.; Tian, J.; Ding, H.; Li, C.; Gao, H.J. Large-scale Fe3O4 nanoparticles soluble in water synthesized by a facile method. J. Phys. Chem. C 2008, 112, 11336–11339. [CrossRef]

30. Pagano, R.; López Medus, M.; Gómez, G.; Couto, P.; Labanda, M.; Landolfo, L.; D'Alessio, C.; Caramelo, J. Protein fibrillation lag times during kinetic inhibition. Biophys. J. 2014, 107, 711–720. [CrossRef]

31. Wang, S.; Chen, Y.; Chen, P.; Liu, K. A kinetic study on the aggregation behavior of β-amyloid peptides in different initial solvent environments. Biochem. Eng. J. 2006, 29, 129–138. [CrossRef]

32. Morris, A.; Watzky, M.; Finke, R. Protein aggregation kinetics, mechanism, and curve-fitting: A review of the literature. Biochim. Biophys. Actaproteins Proteom. 2009, 1794, 375–397. [CrossRef] [PubMed]

33. Kamihira, M.; Naito, A.; Tuzi, S.; Saitô, H.; Nosaka, A. Conformational transitions and fibrillation mechanism of human calcitonin as studied by high-resolution solid-state 13C NMR. Protein Sci. 2000, 9, 867–877. [CrossRef] [PubMed]

34. Knowles, T.; Waudby, C.; Devlin, G.; Cohen, S.; Aguzzi, A.; Vendruscolo, M.; Terentjev, E.; Welland, M.; Dobson, C. An analytical solution to the kinetics of breakable filament assembly. Science 2009, 326, 1533–1537. [CrossRef] [PubMed]

35. Cohen, S.; Linse, S.; Luheshi, L.; Hellstrand, E.; White, D.; Rajah, L.; Otzen, D.; Vendruscolo, M.; Dobson, C.; Knowles, T. Proliferation of amyloid-□42 aggregates occurs through a secondary nucleation mechanism. Proc. Natl. Acad. Sci. USA 2013, 110, 9758–9763. [CrossRef]

36. Crosby, G.A.; Demas, J.N. The measurement of photoluminescence quantum yields—Review. J. Phys. Chem. 1971, 75, 991–1024. [CrossRef]

37. Fery-Forgues, S.; Lavabre, D. Are fluorescence quantum yields so tricky to measure? A demonstration using familiar stationery products. J. Chem. Educ. 1999, 76, 1260–1264. [CrossRef]

38. Kirby, E.P.; Steiner, R.F. Influence of solvent and temperature upon the fluorescence of indole derivatives. J. Phys. Chem. 1970, 74, 4480–4490. [CrossRef]

39. Valeur, B. Molecular Fluorescence—Principles and Applications, 2nd ed.; Wiley-VCH: Weinheim, Germany, 2001.

40. Cardoso, A.; Alvarez, A.; Cattoz, B.; Griffiths, P.; King, S.; Frith, W.; Adams, D. The influence of the kinetics of self-assembly on the properties of dipeptide hydrogels. Faraday Discuss. 2013, 166, 101–116. [CrossRef]

41. Greenspan, P.; Fowler, S. Spectrofluorometric studies of the lipid probe Nile Red. J. Lipid Res. 1985, 26, 781–789.

42. Ira; Krishnamoorthy, G. Probing the link between proton transport and water content in lipid membranes. J. Phys. Chem. B 2001, 105, 1484–1488. [CrossRef]

43. Coutinho, P.J.G.; Castanheira, E.M.S.; Rei, M.C.; Oliveira, M.E.C.D.R. Nile Red and DCM fluorescence anisotropy studies in C12E7/DPPC Mixed Systems. J. Phys. Chem. B 2002, 106, 12841–12846. [CrossRef]

44. Hungerford, G.; Castanheira, E.M.S.; Real Oliveira, M.E.C.D.; Miguel, M.G.; Burrows, H. Monitoring ternary systems of C12E5/water/tetradecane via the fluorescence of solvato-chromic probes. J. Phys. Chem. B 2002, 106, 4061–4069. [CrossRef]

45. Rodrigues, A.R.O.; Matos, J.O.G.; Nova Dias, A.M.; Almeida, B.G.; Pires, A.; Pereira, A.M.; Araújo, J.P.; Queiroz, M.J.R.P.; Castanheira, E.M.S.; Coutinho, P.J.G. Development of multifunctional liposomes containing magnetic/plasmonic MnFe2O4/Au core/shell nanoparticles. Pharmaceutics 2019, 11, 10. [CrossRef] [PubMed]

46. Cardoso, B.D.; Rodrigues, A.R.O.; Almeida, B.G.; Amorim, C.O.; Amaral, V.S.; Castanheira, E.M.S.; Coutinho, P.J.G. Stealth magnetoliposomes based on calcium-substituted magnesium ferrite nanoparticles for curcumin transport and release. Int. J. Mol. Sci. 2020, 21, 3641. [CrossRef] [PubMed]

47. Pitschke, W.; Hermann, H.; Mattern, N. The influence of surface roughness on diffracted X-ray intensities in Bragg-Brentano geometry and its effect on the structure determination by means of Rietveld analysis. Powder Diffr. 1993, 8, 74–83. [CrossRef]

48. Lisboa, F.S.; Gardolinski, J.E.F.C.; Cordeiro, C.S.; Wypych, F. Layered metal laurates as active catalysts in the methyl/ethyl esterification reactions of lauric acid. J. Braz. Chem. Soc. 2012, 23, 46. [CrossRef]

49. Rodrigues, A.R.O.; Ramos, J.M.F.; Gomes, I.T.; Almeida, B.G.; Araújo, J.P.; Queiroz, M.J.R.P.; Coutinho, P.J.G.; Castanheira, E.M.S. Magnetoliposomes based on manganese ferrite nanoparticles as nanocarriers for antitumor drugs. RSC Adv. 2016, 6, 17302–17313. [CrossRef]

50. Foderà, V.; Donald, A. Tracking the heterogeneous distribution of amyloid spherulites and their population balance with free fibrils. Eur. Phys. J. E 2010, 33, 273–282. [CrossRef]

51. Hsiung, C.; Lien, H.; Galliano, A.; Yeh, C.; Shih, Y. Effects of water chemistry on the destabilization and sedimentation of commercial TiO2 nanoparticles: Role of double-layer compression and charge neutralization. Chemosphere 2016, 151, 145–151. [CrossRef]

52. Phenrat, T.; Saleh, N.; Sirk, K.; Tilton, R.; Lowry, G. Aggregation and sedimentation of aqueous nanoscale zerovalent iron dispersions. Environ. Sci. Technol. 2007, 41, 284–290. [CrossRef]

53. Obaidat, I.; Issa, B.; Haik, Y. Magnetic properties of magnetic nanoparticles for efficient hyperthermia. Nanomaterials 2015, 5, 63–89. [CrossRef] [PubMed]

54. Smit, J. Magnetic Properties of Materials; McGraw Hill: New York, NY, USA, 1971.

55. Swain, J.; Mishra, A. Nile red fluorescence for quantitative monitoring of micropolarity and microviscosity of pluronic F127 in aqueous media. Photochem. Photobiol. Sci. 2016, 15, 1400–1407. [CrossRef] [PubMed]

56. Swain, J.; Mishra, J.; Ghosh, G.; Mishra, A. Quantification of micropolarity and microviscosity of aggregation and salt-induced gelation of sodium deoxycholate (NaDC) using Nile red fluorescence. Photochem. Photobiol. Sci. 2019, 18, 2773–2781. [CrossRef] [PubMed]

57. Marchi, M.; Bilmes, S.; Negri, R. Microviscosity in the cavities of titania gels studied by steady-state fluorescence anisotropy. Langmuir 1997, 13, 3665–3673. [CrossRef]

58. Hernández, B.; Pflüger, F.; Kruglik, S.; Ghomi, M. Characteristic Raman lines of phenylalanine analyzed by a multiconformational approach. J. Raman Spectrosc. 2013, 44, 827–833. [CrossRef]

59. Noothalapati, H.; Iwasaki, K.; Yoshimoto, C.; Yoshikiyo, K.; Nishikawa, T.; Ando, M.; Hamaguchi, H.; Yamamoto, T. Imaging phospholipid conformational disorder and packing in giant multilamellar liposome by confocal Raman microspectroscopy. Spectrochim. Acta A Mol. Biomol. Spectrosc. 2017, 187, 186–190. [CrossRef]

60. Kurouski, D.; Van Duyne, R.; Lednev, I. Exploring the structure and formation mechanism of amyloid fibrils by Raman spectroscopy: A review. Analyst 2015, 140, 4967–4980. [CrossRef]

61. Kuhar, N.; Sil, S.; Verma, T.; Umapathy, S. Challenges in application of Raman spectroscopy to biology and materials. RSC Adv. 2018, 8, 25888–25908. [CrossRef]

62. Maiti, N.; Apetri, M.; Zagorski, M.; Carey, P.; Anderson, V. Raman spectroscopic characterization of secondary structure in natively unfolded proteins: α-Synuclein. J. Am. Chem. Soc. 2004, 126, 2399–2408. [CrossRef]

63. Veloso, S.R.S.; Martins, J.A.; Hilliou, L.; Amorim, C.O.; Amaral, V.S.; Almeida, B.G.; Jervis, P.J.; Moreira, R.; Pereira, D.M.; Coutinho, P.J.G.; et al.

Dehydropeptide-based plasmonic magnetogels: A supramolecular composite nanosystem for multimodal cancer therapy. J. Mater. Chem. B 2020, 8, 45–64. [CrossRef]

64. Fan, X.; Xu, H.; Zhang, Q.; Xiao, D.; Song, Y.; Zheng, Q. Insight into the weak strain overshoot of carbon black filled natural rubber. Polymer 2019, 167, 109–117. [CrossRef]

65. Mehdaoui, B.; Meffre, A.; Carrey, J.; Lachaize, S.; Lacroix, L.; Gougeon, M.; Chaudret, B.; Respaud, M. Optimal size of nanoparticles for magnetic hyperthermia: A combined theoretical and experimental study. Adv. Funct. Mater. 2011, 21, 4573–4581. [CrossRef]

66. Umut, E.; Cos¸kun, M.; Pineider, F.; Berti, D.; Güngünes¸, H. Nickel ferrite nanoparticles for simultaneous use in magnetic resonance imaging and magnetic fluid hyperthermia. J. Colloid Interface Sci. 2019, 550, 199–209. [CrossRef] [PubMed]

67. Mehdaoui, B.; Tan, R.; Meffre, A.; Carrey, J.; Lachaize, S.; Chaudret, B.; Respaud, M. Increase of magnetic hyperthermia efficiency due to dipolar interactions in low-anisotropy magnetic nanoparticles: Theoretical and experimental results. Phys. Rev. B 2013, 87, 174419. [CrossRef]

68. Abenojar, E.; Wickramasinghe, S.; Bas-Concepcion, J.; Samia, A. Structural effects on the magnetic hyperthermia properties of iron oxide nanoparticles. Prog. Nat. Sci-Mater. 2016, 26, 440–448. [CrossRef]

69. Karukstis, K.; Thompson, E.; Whiles, J.; Rosenfeld, R. Deciphering the fluorescence signature of daunomycin and doxorubicin. Biophys. Chem. 1998, 73, 249–263. [CrossRef]

70. Tasca, E.; Alba, J.; Galantini, L.; D'Abramo, M.; Giuliani, A.; Amadei, A.; Palazzo, G.; Giustini, M. The self-association equilibria of doxorubicin at high concentration and ionic strength characterized by fluorescence spectroscopy and molecular dynamics simulations. Colloids Surf. A 2019, 577, 517–522. [CrossRef]

71. Changenet-Barret, P.; Gustavsson, T.; Markovitsi, D.; Manet, I.; Monti, S. Unravelling molecular mechanisms in the fluorescence spectra of doxorubicin in aqueous solution by femtosecond fluorescence spectroscopy. Phys. Chem. Chem. Phys. 2013, 15, 2937. [CrossRef]

72. Dash, S.; Murthy, P.N.; Nath, L.; Chowdhury, P. Kinetic modelling on drug release from controlled drug delivery systems. Acta Pol. Pharm. 2010, 67, 217–223.

73. Ghitman, J.; Stan, R.; Ghebaur, A.; Cecoltan, S.; Vasile, E.; Iovu, H. Novel PEG-modified hybrid PLGA-vegetable oils nanostructured carriers for improving performances of indomethacin delivery. Polymers 2018, 10, 579. [CrossRef]

74. Kini, S.; Bahadur, D.; Panda, D. Mechanism of anti-cancer activity of benomyl loaded nanoparticles in multidrug resistant cancer cells. J. Biomed. Nanotechnol. 2015, 11, 877–889. [CrossRef] [PubMed]

75. Joy, M.; Iyengar, S.; Chakraborty, J.; Ghosh, S. Layered double hydroxide using hydrothermal treatment: Morphology evolution, intercalation and release kinetics of diclofenac sodium. Front. Mater. Sci. 2017, 11, 395–408. [CrossRef]

09 NOVEL NANO-FORMULATIONS AND NANOMEDICINES ARE GAINING PROMINENCE IN THE REALM OF OCULAR DRUG DELIVERY, SHOWCASING PROMISING APPLICATIONS

1. INTRODUCTION: AN OVERVIEW

In recent years, the field of nanomedicine has witnessed remarkable advancements, particularly in the development of nano-formulations tailored for ocular drug delivery. The intricacies and challenges associated with delivering therapeutic agents to the eye have spurred innovative approaches, and nano-sized formulations have emerged as promising solutions. These nanomedicine exhibit unique properties that can enhance drug bioavailability, prolong therapeutic effects, and overcome barriers associated with traditional ocular drug delivery methods. This introduction delves into the burgeoning realm of emerging nano-formulations designed for ocular applications, exploring their potential to revolutionize the landscape of ocular drug delivery and address longstanding challenges in treating various ocular disorders. Vision stems from integrated and intricate communication between the eyes and brain. Visual impairment has a deteriorating effect on the life of an individual both physically and mentally. The global prevalence of ocular diseases based on 2015 statistics is an estimated 216.6 million people suffering from moderate to severe visual impairment and 188.5 million people affected by mild visual impairment. In the upcoming years, growth in the aging population will cause a spike in the number of cases of visual impairment, making it a major public health concern [1]. Common eye diseases include glaucoma, age-related macular degeneration (AMD), and diabetic retinopathy [2]. The delivery of therapeutics to the anterior segment of the eye is challenging, owing to the presence of anatomical and physiological barriers. Briefly, anatomical barriers are divided into static and dynamic barriers. Static barriers include corneal epithelium and blood-aqueous barrier, whereas dynamic barriers comprise tear drainage and conjunctival blood and lymph flow. On the other hand, physiological barriers include blinking action, tear film turnover, and nasolacrimal drainage. For drug delivery in the posterior segment, the major limitation is due to the presence of a blood-retinal barrier. Topical administration, like eye drops, is the most common and simple route for drug delivery to the anterior segment. However, the aforementioned problems result in poor bioavailability of topical applications, which has urged researchers to shift focus to novel nanotherapeutics for ophthalmic diseases [1,3]. Topical applications do not reach the posterior segment of the eyes, which makes intravitreal administration (IVT) of a drug the common route of administration. Although IVT injection efficiently delivers drugs to the posterior segment, it can be difficult to achieve for small and large bioorganic compounds like proteins and DNA, and some drugs need repeated injections to reach a therapeutic dose. Multiple injections can increase the risk of complications and possibly damage the eyes. Therefore, it is not a preferable option in treating ocular disease for most patients due to the lack of patient compliance [4].

In the past decade, clinical pharmacologists have focused on developing nanomedicine that can overcome barriers to provide a sustained and targeted release with minimal toxicity. The development of lipid-based nanocarriers has emerged into pharmaceutical industries and attracted attention from formulation scientists for its high drug solubility, substance release, and specific targeting. Organic nanoparticles, such as liposomes, polymeric nanoparticles, and dendrimers, are fabricated and developed to improve therapeutic drug delivery for the ocular disease [5,6]. Inorganic nanoparticles have garnered significant attention for their theranostic properties, substance release, and specific targeting. In this review, we aim to provide an insight into current treatment modalities based on nanotechnology for drug delivery in the eye. Finally, we will summarize the current nanoformulations under clinical trials and the potential formulations available on the market.

2. NANOTECHNOLOGY IN OCULAR DISEASES THERAPY

Nanotechnology combines science and technology with the ability to manipulate structures and properties at the nanoscale size range, which lies between 1 and 100 nm. The ability to manipulate molecules on such a fine scale has application in many areas of basic science and in clinical treatment, diagnosis, and managing disease, which is known as "nanomedicine" [6]. Nanotechnology has been introduced since the 1980s in drug delivery systems to treat ocular disease [7]. The development of nanoformulation produces an ability to overcome ocular barriers, improve drug residence time on the cornea surface, increase permeability and bioavailability of the drug, reduce degradation of unstable drug and be well tolerated by the patient compared to the conventional drug [8]. Organic and inorganic nanoparticles (NPs) provide a new tool to address the unmet clinical needs, especially in the ophthalmic fields, with a significant improvement in drug delivery [5,6]. Therefore, various nanoparticle formulations have been fabricated, including lipid-based nanoparticles, nanosuspension, nanoemulsion, and metal-based nanoparticles.

3. ORGANIC NANOPARTICLES

The group of organic nanocarriers comprises liposomes, niosomes, dendrimers, solid-lipid NPs, polymers and protein/peptide-based NPs, which are fabricated from proteins, lipids, carbohydrates or other organic compounds (Figure 1). The use of organic nanoparticles seems to offer several advantages to facilitate drug or gene delivery to the ocular surfaces.

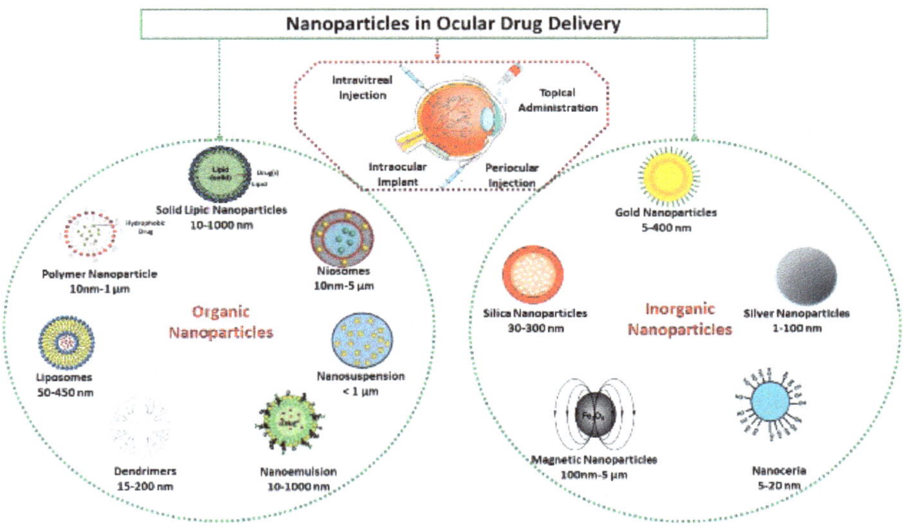

Figure 1. Schematic drawings for potential organic and inorganic nanoparticles for ophthalmic drug delivery

3.1. Liposomes

The liposome is a spherical-shaped nanoparticle with an aqueous core composed of one or more phospholipid bilayers, which resemble a cell membrane. Base on size and lamellarity, liposomes can be classified into three types: multilamellar vesicles, small unilamellar vesicles, and large unilamellar vesicles. Liposomes are widely used as nanocarriers in drug delivery systems and provide a significant advantage due to their size, biocompatibility, biodegradable, low toxicity, and potential to encapsulate hydrophilic, lipophilic, and amphiphilic drugs [9]. Due to their physicochemical and biophysical properties, liposomes can deliver drugs at different concentrations [10]. The liposome surface can be modified for targeted delivery, improving drug absorption to the posterior segment of the eyes, and the positive charge of the liposome can increase drug permeation to the cornea, prolonging drug release, which results in reducing multiple drug administration [7,10]. Law et al. [11] investigated acyclovir (ACV) embedded positively and negatively charged liposomes to target corneal penetration and absorption improvement. In an in vivo study, ACV-loaded liposomes were administered topically to the cornea surface, and the result suggested greater corneal permeation ability of ACV-loaded positively charged liposomes in comparison to negatively charged and plain ACV solution. The drug Edaravone loaded in submicron-sized liposomes was developed by Shimazaki et al. [12] for protection from light-induced retinal damage in mice. Their outcome demonstrated the eyedrop application of Edaravone encapsulated liposomes successfully inhibited light-induced reactive species oxygen production without any toxicity, making it a promising therapeutic potential in dry age-related macular degeneration management. Natarajan et al. [13] explored the therapeutic effect of liposome formulation to deliver latanoprost to treat glaucoma through subconjunctival injection administration. The liposomal formulation increased sustainability and drug targeting in the subconjunctival space. Along with this, latanoprost-encapsulated liposomes displayed an extended reduction of intraocular pressure (IOP) up to 50 days in rabbit eyes as compared to the conventional topical applications of eyedrops [13]. These promising effects have made liposomes an essential formulation in ophthalmic drug delivery. Bochot et al. [14] studied a novel approach to design an ocular drug delivery system that is based on the dispersion of PEGylated liposomes into a thermosensitive gel, which is made of a copolymer of ethylene oxide and propylene oxide (poloxamer 407). This approach offers new potentialities for ocular drug delivery.

3. 2 Niosomes

Niosomes were first widely used in the cosmetic industry, but later their potential in drug delivery was studied for different routes of administration [15,16]. Niosomes are bilayer structured vesicles with a submicron size of 10 nm to over 5 µm composed of amphiphilic nonionic surfactants in the aqueous phase [7,17,18]. The essential properties of these vesicle nanocarriers offer the ability to entrap both lipophilic and hydrophilic drugs like liposomes but with the additional advantages of chemical stability, minimum toxicity due to its nonionic surface, and improved permeation of drug in the corneal cells [19]. The small size of niosomes offers an essential property to overcome the ocular drainage mechanism and produce better drug retention on the ocular surface [20]. Furthermore, compared to liposomes, niosomes are convenient for handling and storage and cost less to make this vesical more preferable to produce large-scale in the pharmaceutical industry [15]. Gugleva et al. [21] reported that there was a high encapsulation efficiency of doxycycline hyclate in niosomes prepared by sorbitan monostearate (span60) and cholesterol formulation, which produced a prolonged drug-release rate and was well tolerated in mice eye. Abdelbary and El-Gendy [22] achieved an encapsulation efficiency of gentamycin sulfate in niosomes that consisted of polyoxyethylene sorbitan monostearate, also called Tween-60 dicetyl phosphate (DCP), and cholesterol with an effect of controlled drug release from the ophthalmic delivery system comparable to the drug solution. Zeng et al. [23] prepared hyaluronic acid-coated niosomes for improving the ocular bioavailability of the tacrolimus drug. The combination of niosomes with hyaluronic acid coating on tacrolimus improved ocular adhesion increased precorneal drug retention, and enhanced permeation of the transcorneal area compared to noncoated niosomes.

3.3 Solid Lipid Nanoparticles

Recently, solid lipid nanoparticles (SLNs) have been explored as a novel delivery system that includes liposome nanoparticles as a colloidal lipid nanocarrier [24]. SLNs offer essential characteristics such as small size, large surface area, controlled drug release, excellent physical stability, and prevention of drug degradation. SLNs typically range from 50 to 1000 nm in size, are made of biodegradable, biocompatible materials, and are stabilized in an aqueous surfactant solution [25,26]. Recently, SLNs have attracted great attention as potential candidates for drug delivery to ocular surfaces due to their capability of increasing corneal absorption and thus improving bioavailability with sustained and controlled drug release properties for hydrophilic and lipophilic drugs [27,28]. A previous study incorporated cyclosporine-A into solid lipid nanoparticles for ocular delivery and showed that the nanoparticle achieved higher concentration in aqueous and vitreous humor due to the controlled and substance release effects of SNLs as compared to a cyclosporine-A solution [29]. Cavalli et al. [30] evaluated solid lipid nanoparticles for the delivery of the tobramycin (TOB) drug. The formulation was applied topically to New Zealand albino rabbits, and the results indicated prolonged preocular drug retention, which is associated with increased ocular bioavailability. Attama et al. [31] investigated the ocular permeation and sustained release effect of diclofenac sodium by synthesized SNL combined with phospholipid and a homolipid obtained from goat fat. It was found that drug-loaded with SNL produced sustained release, and permeation studies of diclofenac from SNL through human cornea construct in cell culture revealed the improvement of cornea permeation potential [31].

3.4 Polymeric Nanoparticles

Polymeric nanoparticles are spherical-shaped solid colloidal particles composed of biocompatible and biodegradable polymers in a size range from 10 nm to 1000 nm [32]. Some polymeric NPs that are obtained from natural polymers include chitosan, alginate, albumin, gelatin, and dextran, while some are obtained from synthetic sources such as polylactic acid (PLA), polyglycolide (PGA), poly(L-lysine), polyaspartic acid, poly-alkyl cyanoacrylate, and polyethyleneimine (PEI) which are biodegradable synthetic polymers. Based on preparation methods, polymeric nanoparticles can be fabricated into nanospheres and nanocapsules [25,33]. Nanosphere polymer has a polymeric matrix core structure, in which drugs can either be entrapped inside a matrix or attached to the polymer surface. Nanocapsules are formed with a liquid phase in the core, surrounded by a solid polymeric wall [6]. These polymer characteristics make them potential nanocarriers as it is possible to conjugate a drug molecule at the surface or encapsulate it inside and deliver it to a specific target site of action at a high concentration with low systemic toxicity [32]. Poly lactic-co-glycolic acid (PLGA) nanoparticles, a biodegradable copolymer, have been used in ophthalmic drug delivery due to their high potential to entrap hydrophilic/hydrophobic drug molecules with prolonged drug release [34]. Glycol chitosan cerium oxide nanoparticles have been studied in treating dry eye diseases in which cerium oxide was entrapped in glycol chitosan. These studies have shown the ability of the drug to enter the corneal cells and conjunctiva cells and promoting SOD2 expression [35]. Sharma et al. [36] found that chloro trimethyl-ammonium methyl methacrylate (Eudragit RS 100/RL 100) polymeric nanoparticles containing amikacin sulfate formulation produced high corneal adherence, which led to an increase in drug retention in the cornea, and drug controlled release efficiency was observed during in vivo studies. Another interesting research conducted by Li et al. [37] reported the efficiency of the combination of hydroxypropyl-beta-cyclodextrin (HP-□-CD) and PLGA (PLGA) nanoparticles for triamcinolone acetonide encapsulation in ophthalmic delivery. The in vitro drug release study demonstrated a higher rate of drug release from hydroxypropyl-beta-cyclodextrin (HP-□-CD) /PLGA nanoparticles compared to the suspension of drug-loaded PLGA without □-CD. Moreover, the significantly improved drug penetration of HP-□-CD /PLGA in the transcorneal area led to increased drug concentration in the aqueous humor, which makes an HP-□-CD/PLGA nanoparticle-loaded triamcinolone acetonide a potential formulation in eye drops [37].

3.5 Dendrimers

Dendrimers, a name derived from the Greek etymology meaning "tree", are monodisperse macromolecules with well-defined size and a molecular structure composed of three major parts: (i) an inner core located inside the central molecule of the dendrimer, (ii) a highly branching unit forming a multilayer that links to the core called generation, and (iii) an outer surface with numerous valent functional groups [38]. Dendrimers have gained a prominent interest in the pharmaceutical application for drug delivery due to their tree-like branched structure with many covalent bonds allowing them to attach multiple functional groups, which can incorporate drugs inside dendrimers and conjugate them to the surface through covalent bonds. Their advantages include hydrophilic or lipophilic properties that offer control insolubility, and with their global shape, dendrimers can entrap small drug molecules inside their branches [39,40]. This unique architecture of dendrimers makes them an ideal nanocarrier for ophthalmic drug delivery with several purposes, such as improving permeability, water-solubility, biocompatibility, and bioavailability [41]. The most commonly used dendrimers-based nanocarriers are polyamidoamine (PAMAM) dendrimers that are hyperbranched [40]. It has been shown that polycationic PAMAM dendrimers are cytotoxic, and their toxicity increases as the generation increases, but hydroxy-terminated G4 PAMAM dendrimers are considered nontoxic due to their near- neutral surface charge that significantly reduces nonspecific retention and interactions in the tissues [42]. In other studies, PEGylation of PAMAM dendrimers has been shown to increase circulation time and biodistribution as well as reduce cytotoxicity of these dendrimers [43]. Mishra and Jain [44] have evaluated the intraocular pressure-lowering potential of acetazolamide (ACZ)-loaded poly-(propylene imine) dendrimer nanostructures for topical ocular administration. Acetazolamide-loaded dendrimer formulation was topically applied to adult male New Zealand albino rabbit eyes. The results showed that the drug-loaded dendrimer formulation caused a reduction in IOP for an extended time, approximately 4 h, compared to ACZ solution, which only showed a reduction effect up to 2 h [44]. In another study, PAMAM dendrimer was synthesized with primary amine and carboxylate surface groups to obtain better drug corneal permeation and properties during in vitro release. A dendrimer-based formulation was used for puerarin drug delivery and successfully tested on the albino rabbit model for ocular hypertension and cataracts glaucoma treatment [45]. Marano and coworkers [43] have designed lipophilic amino-acid dendrimer for anti-vascular endothelial growth factor (VEGF) oligonucleotide (ODN-1) retina delivery to treat choroidal neovascularization. It was reported that dendrimer conjugated ODN-1 significantly inhibits choroidal neovascularization (CNV) progression, and no side effects were observed during examination [46]. Given all the benefits of dendrimers, these formulations have the potential to improve the efficacy of drugs in ophthalmology.

3.6 Nanosuspensions

Poor solubility and low availability are the main concerns for the formulation of drugs in pharmaceutical industries as conventional formulations cannot overcome these problems, and over 40% of newly discovered drugs are reported to have low water solubility. Many strategies have been employed to improve poor drug solubility, but due to limitations in these strategies, scientists have found nanosuspensions to be a promising additional approach in drug delivery. Nanosuspensions are colloidal dispersions in nanoparticle size which have a surfactant as a stabilizer. Administration of nanosuspension offers benefits in sustained release of poorly soluble drug in the ocular surface, which leads to the increase of retention time [47–49]. Nanosuspensions have been reported in oral and topical formulations to enhance ocular drug bioavailability. In previous studies, a nanosuspension of ibuprofen sodium salt (IBU) sodium salt-coated chloro trimethyl-ammonium methyl methacrylate (Eudragit RS 100) polymer tested on rabbit eye by topical administration demonstrated penetration improvement to the anterior segment of the eye, extended drug release, and increased level of drug in the aqueous humor [50]. Some glucocorticoids used topically to treat eye inflammation, such as dexamethasone, hydrocortisone, and prednisolone, have been formulated in nanosuspension, showing an effect in prolonged drug absorption with higher drug bioavailability in ocular drug delivery, thus reducing frequent drug administration [51,52].

3.7 Nanoemulsion

Nanoemulsion is a colloidal system composed of two liquid phases, oil dispersed in water or water dispersed in oil, in which the nanoemulsion droplets with size 20–500 nm are stabilized by surfactants such as cetyltrimethyl ammonium bromide and sodium dodecyl sulfate [53,54]. The small size of droplets offers a high surface area with a significant potential to provide better efficacy in ophthalmic drug delivery with its benefits of improving ocular permeability and high bioavailability [55]. Nanoemulsion is considered to have low surface tension and greater drug spreading on the cornea, which can mix properly with the precorneal constituents. This enhances the drug contact time in the corneal epithelium [56]. The study by Ismail et al. [57] on anti-glaucoma drug dorzolamide hydrochloride nanoemulsion showed promising effects on ocular therapy. The drug for- mulation offered prolonged effect with a quick onset of action and thus reduced frequent administration of eye drops. Tayel et al. [58] developed terbinafine hydrochloride-loaded nanoemulsion gels. The formulation that was developed indicated extended drug residence time and bioavailability improvement. Clinical research conducted by Mahboobian et al. [59] evaluated the uptake of acyclovir nanoemulsion in the bovine cornea following transcorneal permeation. The results showed increased cornea permeation of acyclovir through the corneal membrane cells, and no irritation occurred on the rabbit eye studied, which indicates that the formulation can be considered safe for treating ocular infection [59]. Table 1 Summarizes the applications of organic nanoparticles for drug delivery to ocular surfaces.

Table 1. Applications of organic nanoparticles for ophthalmic drug delivery.

Liposomes	Acyclovir	Topical	Rabbit	Prolong drug penetration	[11]
	Edaravone	Topical	Rabbit	Decrease progression of dry AMD	[12]
	Latanoprost	Subconjunctival injection	Mice	Sustained drug release Reduce intraocular pressure	[13]
Niosomes	Doxycycline hyclate		Rabbit	Prolong drug release rate	[21]
	Tacrolimus	Topical	Rabbit	Increased precorneal drug retention	[23]
Solid lipid nanoparticle	Cyclosporine A	Topical	Sheep	Prolonged drug release	[29]
	Tobramycin	Topical	Rabbit	Increased drug retention	[30]
Polymeric NPs (Eudragit®)	Amikacin	Topical	Rabbit	Improved ocular penetration Controlled release	[36]
Polymeric NPs (chitosan)	Cerium oxide		Mice	Improved corneal penetration	[35]
Dendrimers	Acetazolamide	Topical	Rabbit	Enhanced drug residence time	[44]
	Puerarin		Rabbit	Improved corneal permeation	[45]
	Anti-VEGF			Reduced CNV progression	[46]
Nanosuspension	IBU sodium salt	Topical	Rabbit	Increased penetration Prolonged drug release	[50]
	Glucocorticoid			Prolonged drug absorption	[51]
Nanoemulsion	Terbinafine hydrochloride			Improved drug residence time Increased bioavailability	[58]
	Acyclovir			Increased corneal permeation	[59]

4. INORGANIC NANOPARTICLES

Inorganic nanoparticles comprise mainly metallic nanoparticles (Figure 1) and quantum dots. Metallic nanoparticles have gained much attention over the past decade since Faraday proved that they could exist in solution [35]. They can be categorized into four different groups: metal nanoparticles, metal oxide nanoparticles, doped metal–metal/oxide– metal nanoparticles, and metal sulfide, and metal–organic frameworks. Metal NPs such as silver (Ag), gold (Au), copper (Cu), magnesium (Mg), titanium (Ti), platinum (Pt), zinc (Zn), and iron (Fe) nanoparticle have been investigated in different research areas and have succeeded as efficient, stable drug delivery platforms with fewer side effects and powerful imaging probes. Doped metallic nanoparticles, like ZnO doped with Co, were reported to have lower toxicities than single metals with better properties [60,61]. In the ocular drug and gene delivery field, metals such as cerium oxide (CeO2) NPs and gold NPs have demonstrated their potential antioxidant capabilities with a high safety profile after delivery to the eye and high chemical stability.

4.1 Gold NPs (AuNPs)

AuNPs are one of the noble metals that are known for their unique optical properties induced by the popular p henomenon of localized surface plasmon resonance (LSPR). This phenomenon is strongly influenced by the AuNP shape and is the main reason for its penetrative ability to the biological tissues [62]. AuNPs have other advantages that make them popular in the nanomedicine field, like their chemical stability, surface functionalization, biocompatibility, and unique surface and size characteristics [63]. The biocompatibility and internalization capability of different AuNP morphologies (rods, spheres, and cubes) have been assessed in retinal pigment epithelium (RPE) cells. Spheres and cubes of 50 and 100 nm showed no cytotoxicity and good internalization properties; however, rod-shaped particles were less biocompatible [64]. Dong et al. [65] showed that resveratrol coated AuNPs with a median size of 20 nm and doses of 200–300 mg/kg injected in streptozotocin-induced diabetic rats for 3 months could provide a protective effect against diabetic retinopathy. This protective effect of AuNPs could help to regain the balance of the stimulators and inhibitors of the angiogenesis process through the inhibitory effects of the ERK1/2 pathway and nuclear factor kappa B (NF-kb) expression, which could reduce inflammation and permeability of the blood-retinal barrier in the diabetic rats. Furthermore, there was a significant reduction of all the retinal mRNA expressions of vascular endothelial growth factor (VEGF-1), tumor necrosis factor-alpha (TNFα), and interleukin 6 (IL-6) [65]. In another study of the topical application of AuNPs, Pereira et al. [66] showed that AuNPs could decrease the intraocular oxidative damage and inflammation in endotoxin-induced uveitis rat models.

The treatment of choroid-retina endothelial (RF/6A) cells with AuNPs of 3–5 nm successfully inhibited the VEGF-induced RF/6A cell migration through the Akt/eNOS pathways. In their cell viability and cell adhesion studies, they found neither cytotoxic effects on AuNPS on RF/6A nor detrimental effects on the normal physiological adhesion of cells to fibronectin [62]. Hayashi et al. [67] examined the feasibility, biodistribution, and effects of the subretinal injections of immunoglobulin G (IgG) adsorbed on gold nanoparticles (GNPs) in rabbits' eyes for three months and in ARPE-19 cultured cells, and they found that GNPs was successfully delivered into photoreceptor cells and RPE in rabbits and reported no cytotoxicity in ARPE-19 cells. Moreover, in inherited retinal dystrophies gene delivery therapies, AuNPs have been examined as a safer and more efficient alternative to the viral-based approaches in RPE cells. The preliminary results of Trigueros et al. [68] have shown that plasmid DNA-wrapped AuNPs with a size of 40 nm could be successfully internalized in differentiated ARPE-19 cells with good transfection efficiency. These results were supported by the early expression of a reporter gene that was noted at 16 h post-transfection. Their results show a potentially successful gene delivery route to RPE cells using AuNPs; nevertheless, they propose that the interaction mechanism of cell to AuNPs must be better understood to increase the transfection efficiency of these particles and escape the autophagic pathways of the particles to ensure sustained gene expression effect of their system [68]. Generally, AuNPs are considered as a core-shell system, at which gold is the inert inner core, and the drug can be conjugated to either its core or outer active layer through thiol linkages. As mentioned earlier, gold is biocompatible, and it is approved for internal medicine; however, the main challenge for AuNPs to be used as drug delivery vehicles in the clinical practice is their non-degradable gold core that elongates their excretion from the body [69]. Interestingly, the successful passage of AuNPs of the size 20 nm through the blood-retinal barrier (BRB) and their distribution in all retinal layers after their intravenous injection into C57B1/6 mice without showing any structural damage or increased cell death compared to cells without NPs by Kim et al. [70] raise their potential existence in future clinical trials.

4.2 Silver Nanoparticles (AgNPs)

Silver nanoparticles (AgNPs) have been used widely in the nanomedicine field due to their advantages of the unique chemical and physical characteristics, large surface area to volume ratio, low production cost, and biocompatibility that make them good candidates as drug delivery carriers [71]. Various physical and chemical methods are being used to synthesize and stabilize the AgNPs. Out of these, the chemical method is most common, which includes chemical reduction, electrochemical techniques, reducing agents, physiochemical reduction, and radiolysis [72]. AgNPs synthesis using these chemical methods is experiencing problems such as stability and agglomeration of NPs, shape, and size of particles.

The synthesis of AgNPs requires (a) silver precursors, (b) reducing agents that include organic (e.g., ethylene glycol, ascorbic acid, glucose) and inorganic (e.g., hydrazine, sodium borohydride), (c) capping agents or stabilizing agents such as polyvinylpyrrolidone (PVP) [72]. Capping agents offer electrostatic and steric stabilization effects for AgNPs dispersion in suspensions [73].

AgNPs have been reported as effective nanoparticles in the treatment of cancer and infections due to their powerful tumor-killing and bactericidal effects. In the ocular field, AgNPs have shown successful delivery of therapeutic agents to eyes and have been used as a coating agent for contact lenses. In addition, AgNPs alone or in conjugation with natural plant extracts demonstrated significant anti-angiogenic, antioxidant, anti-glycation end products, and anti-cataractogenic effects in various cell culture systems and animal models for ocular diseases [74]. Anbukkarasi et al. [75] investigated the antioxidant and anti- cataractogenic effect of AgNPs with a size range of 15–50 nm formulated with the ethanolic extract of *T. divaricate* leaves, which reportedly have antioxidant and anti-cataractogenic pharmacological effects. The study designed an in vitro selenite-induced cataractogenic model to evaluate the effects of AgNPs against the dense opacification of Wistar rat lenses. The results suggested that the formulated AgNPs acted as a potent antioxidant by dose- dependent scavenging of DPPH and H2O2 free radicals, which are the main triggers of cataract formation [75]. Kalishwaralal et al. [76] tested the anti-angiogenic properties of formulated AgNPs in a size range of 40–50 nm, acting as an inhibitor of VEGF, which induces cell proliferation and migration in bovine retinal endothelial cells (BRECs). The findings of Gurunathan et al. [77] suggested that AgNPs could inhibit angiogenesis via the inhibition of the PI3K/Akt signaling pathways. Sheikpranbabu et al. [78] studied the effects of AgNPs with a size of 50 nm on advanced glycation end-products (AGEs)-induced endothelial cell permeability in porcine retinal endothelial cells (PRECs), and the studies demonstrated that AGE-induced permeability was inhibited by AgNPs through the Src kinase pathway.

Since the nineteenth century, silver nitrate has been known for its potential antimicro- bial effects as it was widely used against ophthalmia neonatorum. Xu et al. [79] investigated the antifungal effects of polymeric AgNPs of sizes ranging from 20 to 30 nm against three different types of filamentous fungi isolates that cause fungal keratitis. The antifungal effects of AgNPs in the fungal cultures were compared to those of the antifungal natamycin, and the AgNPs demonstrated a significantly superior antifungal activity that was 4–32 times higher than natamycin.

Despite the advantages that AgNPs offers as antifungal, antioxidant, anti-angiogenic, and anti-inflammatory agents, their limited advancement in the ocular drug delivery field goes back to their reported toxicity in several studies.

The widely reported toxicity is related to the release of silver ions from the nanoparticle surface that induces irritation to different ocular parts. Jun et al. [80] investigated the toxicity of silver NPs at a dose of 0.4 mg/L and reported down regulation of several lens crystalline genes by the possible cell death or the nuclear DNA or RNA export blockage. Kim et al. [81] also observed edema and conjunctival redness after a one-hour exposure of New Zealand white rabbits' eyes to AgNPs. In a study, AgNPs with sizes of 22.4 nm and 42.5 nm, respectively, induced cytotoxic effects and activation of the reactive oxygen species (ROS) system in bovine retinal endothelial cells (BRECs) [82].

4.3 Cerium Oxide Nanoparticles (Nanoceria-CeO2-NPs)

Cerium is a rare earth element in the lanthanide series of the periodic table and is considered a potent antioxidant. The mechanism by which CeO2-NPs of 3–5 nm exhibits its catalytic antioxidant reaction involves two oxidation states, a reduction of Ce^{4+} to Ce^{3+} and a loss of an oxygen atom [83]. This reaction creates vacancies for reactive oxygen species (ROS) in cells to replace the loss of oxygen, which mimics the antioxidant enzymatic activities of superoxide dismutase (SOD) and catalase [84]. Nanoceria has proved their efficacy as antioxidants to provide protection in many diseases and act as neuroprotective, radioprotective, cardioprotective, as well as anti-inflammatory agents [85].

In the ophthalmic field, CeO2-NPs have been investigated in several animal models of age-related macular degeneration (AMD) and showed promising antioxidant, anti- inflammatory, and anti-angiogenic effects. Moreover, they were reported to provide long-term neuroprotection for photoreceptors in retinitis pigmentosa (RP) models without collateral adverse effects. They were also used as an effective anti-cataractogenic material and were reported to be safe on lens cells [86]. Nanoceria is mainly fabricated by simple wet chemistry methods, as explained by Karakoti et al. [87], and their PEGylation and liposomal encapsulation enable them to permeate the cornea [86]. Chen et al. [88] reported that fabricated CeO2-NPs of 20 nm particle size could prevent the increased concentration of free radicals in primary cell cultures of rat retinas. These findings prompted them to test the NPs in vivo, where a group of albino rats was exposed to 2700 lux of light for 6 h. The light exposure normally results in damage to 50–60% of photoreceptors, but their electroretinogram (ERG) data showed that a significant number of photoreceptor cells were rescued after posttreatment with CeO2-NPs [88]. Kong et al. [85] testing of nanoceria on P34 mice including ROS reduction, increase in the expression of neuroprotection associated genes, and inhibition of the apoptosis signaling pathways, all of which delayed the degenerative condition of photoreceptor cells.

In a study by Fiorani et al. [89], pretreatment of albino Sprague–Dawley rats with intravenous and intravitreal injections of CeO2-NPs for 3 weeks could prevent neuronal death induced by 1000 lux light for 24 h and helped preserve the integrity of the outer nuclear layer. The study also demonstrated that nanoceria remained stable in the outer layer and could inhibit the microglial activation and migration toward the outer nuclear layer for an extended time [89]. Intravitreal injections of nanoceria were also tested by Zhou et al. [90], who found inhibition of neovascularization in Vldlr knockout mice through their inhibitory effects of the VEGF cascade and their initiation of antioxidant genes expression pathways; they reported these results as promising findings that could make nanoceria good candidates in the treatment of AMD and diabetic retinopathy diseases. In a study by Bhargava et al. [91], nanoceria of 5–10 nm could increase the survival of rod and cone photoreceptors in culture by helping the maintenance of the cell line, which allows for improved drug screening experiments on these cell lines.

4.4 Mesoporous Silica Nanoparticles (MSNs)

Mesoporous silica nanoparticles (MSNs), one of the most well-studied inorganic nanoparticles for drug delivery. These are mesoporous systems made up of silica with particle sizes of 30–300 nm. They can absorb exceptionally large quantities of drugs in their pores. Unique properties of MSNs include large specific surface area and pore volume, controllable particle size and better biocompatibility make them excellent nanoplatform for biomedical applications and drug delivery [92]. Park et al. [93] conducted a study to evaluate the cytotoxicity of three different sizes (50 nm, 100 nm and 150 nm) of silica nanoparticles (SiNPs) on ocular surface cells such as human corneal epithelial cells (HCECs). Experimental results suggested that cellular autophagy and mammalian target of rapamycin (mTOR) pathways activated with the addition of SiNPs without inducing any significant cytotoxicity in cultured HCECs.

An eye drop, brimonidine showed limited effectiveness in the treatment of glaucoma because of its rapid clearance from preocular space. To resolve this issue, a study was conducted by Kim et al. [94] to deliver brimonidine using amino-functionalized mesoporous silica (AMS) particles so that AMS particles form adhesion with the mucous layer and allow high preocular retention time. Because of the presence of mesopores, AMS encapsulated brimonidine released the drug in a sustained manner over 8 h. BMD-AMS stayed in preocular space for up to 12 h when topically administered into the eyes of a rabbit. To measure the in vivo efficacy, the variance in IOP and brimonidine concentration was examined after administering BMD-AMS and compared it with that induced by the marketed brimonidine eye drops, i.e., Alphagan P. The results showed that in BMD-AMS, the duration in the decrease in IOP and the area under the drug concentration in the aqueous humor-time curve was found to be 12 h and 2.68 μg/mL, respectively, which is twice as compared to Alphagan P. These finding indicated enhancement in ocular bioavailability of brimonidine with BMD-AMS.

4.5 Magnetic Nanoparticles (MNPs)

Magnetic nanoparticles (MNPs) belong to the group of nanotechnology-based materials consisting of magnetic elements such as iron, cobalt, chromium, manganese. It can be used for various biomedical applications such as drug delivery, magnetic resonance imaging, tissue repair, transfection and tissue targeting [95]. It forms a strong drug delivery system as their reactive surface can be functionalized with biocompatible coatings or bioactive molecules, which increase their specificity toward cellular targets and avoid their interaction with healthy cells [96]. In addition to this, MNPs also provides two other advantages; first, it can be controlled by noncontact forces, second, it has potential to be used in MRI tracking and could be used in targeted nano-drug delivery system [96].

Most used MNPs include iron-oxide magnetic particles. The coating of biocompatible material on iron-oxide magnetic particles avoids aggregation, biodegradation, and alterations of particles from their original state and allows entrapping of the bioactive agent on the particle via covalent attachment or adsorption [97]. It has a magnetite or maghemite core surrounded by a shell that contains a layer of polymer or functional group (such as antibodies, biotin, amines and streptavidin) [69]. The drug molecules are attached to the shell of magnetic nanoparticles. The main advantage of using magnetic iron oxide nanoparticles is that they can be visualized by magnetic resonance imaging, and the drug-loaded nanoparticles can be held in place by using a magnetic field. In a study done by Yanai et al. [98], super paramagnetic iron oxide nanoparticles (SPIONs) were used to magnetized rat mesenchymal stem cells (MSCs) in order to deliver cells to the diseased area in the dystrophic retina. Mesenchymal stem cells (MSCs) labeled with fluidMAG-D were injected intravitreally in a retinal degenerative transgenic rat. In vitro studies revealed that cells remain viable and retained differentiation ability even after magnetizing them using fluidMAG-D. The results showed that magnetic MSC delivery to the retina increased tenfold as compared to the normal intravitreal injected cells. Cryosection imaging revealed that magnetic MSC cells had migrated into the inner as well as outer retina. Moreover, magnetic nanoparticle treatment along with orbital magnet resulted in significantly higher concentrations of anti-inflammatory molecule IL-10 and hepatocyte growth factor in the retina. These findings suggested that this approach may provide optimal benefit in outer retinal diseases like AMD where controlled delivery to focal cells is required because it can deliver a higher drug load to the site of interest and resulted in therapeutically useful biochemical changes in the dystrophic retina.

In a study on 44 Sprague–Dawley rats, polymer-coated 50 nm or 4 μm MNPs were injected in the left eye, and the same amount of PBS into the right eye and their effect can be evaluated on IOP, corneal endothelial cell count, the morphology of retina and functioning of photoreceptors at different times [99]. Measurement of IOP, ERG and histology showed no toxicity after injecting magnetic nano- and microparticles. However, microparticles showed small toxicity in corneal endothelial cell counts and iron deposition in tissues. In another study, polymer-coated MNPs were found to be nontoxic to the photoreceptors at histologic as well as electrophysiologic levels [100]. Table 2. Summarizes the applications of inorganic nanoparticles for drug delivery to ocular surfaces.

Table 2. Applications of inorganic nanoparticles for drug delivery to ocular surfaces.

AuNPs	Resveratrol	Injection		Reduced retinal inflammation	[65]
		Topical	Rat	Decrease the intraocular oxidative damage	[66]
	Immunoglobulin-G	Subretinal injection	Rabbit	Improved biodistribution	[67]
	DNA			Treat retinal dystrophies	[68]
AgNPs			Rat	Produced antioxidants Exhibit anticataractogenic	[75]
				Anti-vasopermeability	[76]
				Antifungal activity	[79]
Cerium oxide NPs			Rat	Reduced photoreceptor damaged	[88]
			Mice	Decreased reactive oxygen species	[85]
		Intravenous Injection	Rat	Decrease neurodegenerative	[89]
		Intravitreal injection	Mice	Inhibit neovascularization	[90]
MSNs	Brimonidine	Topical	Rabbit	Enhancement in ocular bioavailability	[91]

MNPs	Mesenchymal stem cells (MSCs)	Intravitreal injection	Rat	Increase tenfold in delivery of drug load to the site of interest	[98]
	Polymer-coated MNPs	Intravitreal injection	Rat	polymer-coated MNPs found to be nontoxic to the photoreceptors	[99]

4.6 Implant Devices

Implant devices made from polymeric materials can be used as a mode of drug delivery. They showed controllable release of the drug over a long period of time [69]. Based on the material, implants can be classified into two types; biodegradable and non- biodegradable devices. Non-biodegradable implants have shown more accurate control of drug release and longer release periods as compared to biodegradable implants. However, surgical removal of non-biodegradable devices may cause problems for a patient. Based on the places of implantation, the release of drugs in the retina may range from five weeks to six months. Currently, there are several types of commercial implants that are available for ocular disease treatment, such as Ozurdex (dexamethasone biodegradable implant; Allergan, Inc., Irvine, CA, USA), Trivaris (triamcinolone acetonide suspension; Allergan, Inc.), Kenalog (triamcinolone acetonide suspension; Bristol-Myers Squibb, Princeton, NJ, USA), Triesence (triamcinolone acetonide suspension; Alcon, Fort Worth, TX, USA), Retisert (fluocinolone acetonide non-biodegradable implant; Bausch and Lomb, Inc., Rochester, NY, USA), Iluvien (fluocinolone acetonide non-biodegradable implant; Alimera Sciences, Inc., Alpharetta, GA, USA) [69].

5. POTENTIAL OCULAR NANOMEDICINE IN CLINICAL TRIALS AND ON THE MARKET

The astounding potential of nanomedicine in the pharmaceutical field to improve health care has captured scientists' attention, with extensive research globally to acquire a market position [101,102]. The first product of nanomedicine was placed on the market almost two decades ago. As of 2012, 33 nanotherapies were available on the market with more than 100 therapeutics in development, and around 50% of nanomedicine companies were based in the United States [103,104]. Numerous nanomedicine products have been under investigation in clinical trials, and some have been approved by the US Food and Drug Administration (FDA) and are available to use in clinical settings for the treatment of conditions such as cancer, autoimmune disease, infectious disease, age-related macular degeneration and others [105]. The great evolution of nanotechnology has dominated the drug delivery field, leading to the progression of formulations for ocular drug delivery, with many in clinical trials and some already introduced in the market.

Table 3. Nanomedicine for ocular diseases under clinical trial and approved by FDA.

Restasis®	Nanoemulsion	Dry eye	Approved	[7]
Durezol®	Nanoemulsion	Eye inflammation	Approved	[7]
Ozurdex	Dexamethasone biodegradable implant	Macular edema, Non-infectious uveitis	Approved	
Trivaris	Triamcinolone acetonide suspension	Uveitis	Approved	
Kenalog	Triamcinolone acetonide suspension	Macular edema	Approved	
Retisert	Fluocinolone acetonide non-biodegradable implant	Non-infectious uveitis	Approved	[69]
Iluvien	Fluocinolone acetonide non-biodegradable implant	Diabetic macular edema	Approved	[106]
Triesence	Triamcinolone acetonide suspension	Macular edema	Approved	[106]
Visudyne®	Liposome	AMD	Approved	[107]
Macugen®	Aptamer–polymer nanoparticle	Wet AMD	Approved	[107]
TLC399 (ProDex)	Lipid-based nanoparticle	Macular edema	Phase II	[108]
POLAT-001	Liposome	Glaucoma	Phase II	
SYSTANE®	Propylene glycol-based nanoemulsion	Dry eye	Phase IV	

Many nanoformulations for ocular therapies have been developed and commercialized in the market, but some formulations such as drug-loading emulsions are still slow in progressing for therapy even though drug-free nanoemulsion has been approved to treat dry eye. Restasis® is the first product on the market developed as a nanoemulsion containing cyclosporin A for chronic dry eye treatment, and Durezol®, a nanoemulsion containing the drug difluprednate, is approved to treat inflammation of the eye [7]. An- other ocular nanomedicine is Visudyne®, a liposomal formulation containing verteporfin marketed by Novartis Pharma Ag that was approved by the FDA in 2000 for intravenous administration in the treatment of choroidal neovascularization due to age-related macular generation, pathological myopia, and ocular histoplasmosis syndrome. Macugen® is a PEG anti-vascular endothelial growth factor aptamer administered by intravitreal injection that was approved by the FDA in 2004 for wet age-related macular degeneration treatment [106]. According to Grumezescu, many ophthalmic nanoformulations are under development in clinical evaluation, including TLC399 (ProDex), which contains dexamethasone sodium phosphate and is currently in phase II for treatment of macular edema due to retinal vein occlusion. In addition, latanoprost coated liposome (POLAT-001) has completed phase II clinical trials for ocular hypertension and primary open-angle glaucoma treatment [107]. Based on clinicaltrials.gov, a website developed by the U.S. National Institutes of Health, Department of Health and Human Services, one current ocular nanomedicine under clinical trial in 2020 is SYSTANE®, a propylene glycol-based eye drops nanoemulsion, which has completed phase IV for dry eye disease treatment. The development of nanotechnology in ophthalmic therapies seems to be a promising strategy with an increasing number of nano- based formulations in clinical trials and on the market, although more research and studies are necessary regarding nanostructure delivery to the eye [108]. Table 3. Nanomedicine for ocular diseases under clinical trial and approved by FDA.

6. CONCLUSIONS

The field of nanotechnology has boomed in recent years, accounting for half of the patent filings and one-third of all publications in the US alone. The nanotechnology market has a forecasted compound annual growth rate (CGAR) of 22%, making it a sector with significant investment potential. Around 200 companies and several potential startups have been identified as active in nanomedicine development worldwide. Nanomedicines accounts for approximately 5% of research publications in the field of nanotechnology worldwide. Several nanoparticles, either organic or inorganic, have provided tools that can overcome the problems associated with the conventional drug delivery of ophthalmic formulations. The nanomedicine market is in the early stages of development, and several drug delivery devices are in the clinical testing stage. Various Intravitreal, polymeric drug delivery implants for ocular diseases have been approved by the FDA.

The application of nanotech has shown great potential in ocular nanomedicine re- search. The nanomaterials could also be used in increasing the bioavailability of various therapeutic agents. Examples of such nanomaterials include liposomes, dendrimers, niosomes, metal NPs. Nanomedicine also has started to showcase its applied potential in improving the pharmacokinetic properties of drug delivery systems; many studies are focusing on integrating hybrid systems with hydrogel using micelles, dendrimers, and cyclo-dextrins [109]. Other aspects that are associated with ocular nanomedicine development have made uniform NPs with reproducible features on an enormous scale. To stimulate clinical translations of nanomedicine, it is critical to investigate the safety and toxicity profiles of such nanomedicine formulations. There could be many other factors that can influence toxicity in nanomedicine, such as administration dose, shape, and size of the particle, surface charge, and functional groups.

Advanced nanofabrication technologies like particle replication in non-wetting tem- plates (PRINT) [110] and the hydrogel template method [111] have been introduced to create ocular nanomedicine. PRINT technology can be used to create monodispersed NPs and microparticles with controlled shape, size, and surface modification at a large- scale. Another applied example that demonstrated the use of PRINT technology is the AR13503 implant, which was manufactured to provide sustained release of API for more than two months in vitro [112]. This implant was prepared using biodegradable polymers PLGA/PDLA/poly(ester amide) (PEA) with rod-shaped and size suitable for injection through a 27 gauge needle. Clinical trials for AR 13,503 are underway and are expected to be used for treating diabetic macular edema (DME) and wet AMD. The hydrogel template method is another advanced nanofabrication technology that can be used to fabricate large amounts of homogenous nanoparticles and microparticles [111]. For example, nanowafer, an ultra-thin transparent lens prepared by filling drug solutions into the dissoluble PVA template. The amount of the drug can be controlled by adjusting the size of the particle.

With several advantages like drug targeting, sustainability, and increased bioavailability, nanomedicine promises to revolutionize the medical markets. Considering the numerous benefits of nanomedicine in ocular drug delivery, aggregation, toxicity, and clearance of nanoparticles remain a concern. Further study of nanodrug carrier development is necessary to provide a more inside understanding of the safety issue and progress in ocular drug delivery.

7. CONCISE SUMMARY

Ocular diseases can deteriorate vision to the point of blindness and thus can have a major impact on the daily life of an individual. Conventional therapies are unable to provide absolute therapy for all ocular diseases due to the several limitations during drug delivery across the blood-retinal barrier, making it a major clinical challenge. With recent developments, the vast number of publications undergird the need for nanotechnology-based drug delivery systems in treating ocular diseases. The tool of nanotechnology provides several essential advantages, including sustained drug release and specific tissue targeting. Additionally, comprehensive in vitro and in vivo studies have suggested a better uptake of nanoparticles across ocular barriers. Nanoparticles can overcome the blood-retinal barrier and consequently increase ocular penetration and improve the bioavailability of the drug. In this review, we aim to summarize the development of organic and inorganic nanoparticles for ophthalmic applications. We highlight the potential nanoformulations in clinical trials as well as the products that have become a commercial reality.

REFERENCES

1. Bourne, R.R.A.; Flaxman, S.R.; Braithwaite, T.; Cicinelli, M.V.; Das, A.; Jonas, J.B.; Keeffe, J.; Kempen, J.H.; Leasher, J.; Limburg, H.; et al. Vision Loss Expert Group. Magnitude, temporal trends, and projections of the global prevalence of blindness and distance and near vision impairment: A systematic review and meta-analysis. Lancet Glob. Health 2017, 5, e888–e897. [CrossRef]
2. Kim, Y.C.; Chiang, B.; Wu, X.; Prausnitz, M.R. Ocular delivery of macromolecules. J. Control. Release 2014, 190, 172–181. [CrossRef] [PubMed]
3. Agrahari, V.; Mandal, A.; Agrahari, V.; Trinh, H.M.; Joseph, M.; Ray, A.; Hadji, H.; Mitra, R.; Pal, D.; Mitra, A.K. A comprehensive insight on ocular pharmacokinetics. Drug Deliv. Transl. Res. 2016, 6, 735–754. [CrossRef]
4. Urtti, A. Challenges and obstacles of ocular pharmacokinetics and drug delivery. Adv. Drug Deliv. Rev. 2006, 58, 1131–1135. [CrossRef] [PubMed]
5. Weng, Y.; Liu, J.; Jin, S.; Guo, W.; Liang, X.; Hu, Z. Nanotechnology-based strategies for treatment of ocular disease. Acta Pharm. Sin. 2017, 7, 281–291. [CrossRef] [PubMed]
6. Nasimi, P.; Haidari, M. Medical use of nanoparticles: Drug delivery and diagnosis diseases. Int. J. Green Nanotechnol. 2013, 1. [CrossRef]
7. Reimondez-Troitiño, S.; Csaba, N.; Alonso, M.J.; Fuente, M. Nanotherapies for the treatment of ocular diseases. Eur. J. Pharm. Biopharm. 2015, 95, 279–293. [CrossRef]
8. Das, S.; Suresh, P.K. Drug delivery to eye: Special reference to nanoparticles. Int. J. drug Deliv. 2010, 2, 12–21. [CrossRef]
9. Rehman, A.U.; Akram, S.; Seralin, A.; Vandamme, T.; Anton, N. Lipid nanocarriers: Formulation, properties, and applications. In Smart Nanocontainers; Elsevier: Amsterdam, The Netherlands, 2020; pp. 355–382.
10. Asasutjarit, R.; Managit, C.; Phanaksri, T.; Treesuppharat, W.; Fuongfuchat, A. Formulation development and in vitro evaluation of transferrin-conjugated liposomes as a carrier of ganciclovir targeting the retina. Int. J. Pharm. 2020, 577, 119084. [CrossRef]
11. Law, S.L.; Huang, K.J.; Chiang, C.H. Acyclovir-containing liposomes for potential ocular delivery. Corneal penetration and absorption. J. Control. Release 2000, 63, 135–140. [CrossRef]
12. Shimazaki, H.; Hironaka, K.; Fujisawa, T.; Tsuruma, K.; Tozuka, Y.; Shimazawa, M.; Takeuchi, H.; Hara, H. Edaravone-loaded liposome eyedrops protect against light-induced retinal damage in mice. Investig. Ophthalmol. Vis. Sci. 2011, 52, 7289–7297. [CrossRef] [PubMed]

13. Natarajan, J.V.; Chattopaddyay, S.; Ang, M.; Darwitan, A.; Foo, S.; Zhen, M.; Koo, M.; Wong, T.T.; Venkatramam, S.S. Sustained release of an anti-glaucoma drug: Demonstration of efficacy of a liposomal formulation in the rabbit eye. PLoS ONE 2011, 6, e24513. [CrossRef] [PubMed]

14. Bochot, A.; Fattal, E.; Gulik, A.; Couarraze, G.; Couvreur, P. Liposomes dispersed within a thermosensitive gel: A new dosage form for ocular delivery of oligonucleotides. Pharm. Res. 1998, 15, 1364–1369. [CrossRef] [PubMed]

15. Moghassemi, S.; Hadjizadeh, A. Nano-niosomes as nanoscale drug delivery systems: An illustrated review. J. Control. Release 2014, 185, 22–36. [CrossRef]

16. Pardakhty, A.; Moazeni, E. Nano-niosomes in drug, vaccine and gene delivery: A rapid overview. Nanomed. J. 2013, 1, 1–12.

17. Herrero-Vanrell, R.; Torre, V.D.L.; Andres-Guerrero, V.; Barbosa-Alfaro, D.; Molina-Martinez, I.T.; Bravo-Osuna, I. Nano and microtechnologies for ophthalmic administration, an overview. J. Drug Deliv. Sci. Technol. 2013, 23, 75–102. [CrossRef]

18. Ge, X.; Wei, M.; He, S.; Yuan, W.E. Advances of Non-Ionic Surfactant Vesicles (Niosomes) and Their Application in Drug Delivery. Pharmaceutics 2019, 11, 55. [CrossRef]

19. Kaur, I.P.; Garg, A.; Singla, A.K.; Aggarwal, D. Vesicular systems in ocular drug delivery: An overview. Int. J. Pharm. 2004, 269, 1–14. [CrossRef]

20. Chen, S.; Hanning, S.; Falconer, J.; Locke, M.; Wen, J. Recent advances in non-ionic surfactant vesicles (niosomes): Fabrication, characterization, pharmaceutical and cosmetic applications. Eur. J. Pharm. Biopharm. 2019, 144, 18–39. [CrossRef]

21. Gugleva, V.; Titeva, S.; Rangelov, S.; Momekova, D. Design and in vitro evaluation of doxycycline hyclate niosomes as a potential ocular delivery system. Int. J. Pharm. 2019, 567, 118431. [CrossRef]

22. Abdelbary, G.; El-Gendy, N. Niosome-encapsulated gentamicin for ophthalmic controlled delivery. AAPS PharmSciTech 2008, 9, 740–747. [CrossRef] [PubMed]

23. Zeng, W.; Li, Q.; Wan, T.; Liu, C.; Pan, W.; Wu, Z.; Zhang, G.; Pan, J.; Qin, M.; Lin, Y.; et al. Hyaluronic acid-coated niosomes facilitate tacrolimus ocular delivery: Mucoadhesion, precorneal retention, aqueous humor pharmacokinetics, and transcorneal permeability. Colloids Surf B Biointerfaces 2016, 141, 28–35. [CrossRef] [PubMed]

24. Ekambaram, P.; Sathali, A.H.; Priyanka, K. Solid lipid nanoparticles: A review. Sci. Rev. Chem. Commun. 2012, 2, 80–102.

25. Wilczewska, A.Z.; Niemirowicz, K.; Markiewicz, K.H.; Car, H. Nanoparticles as drug delivery systems. Pharmacol. Rep. 2012, 64, 1020–1037. [CrossRef]

26. Mukherjee, S.; Ray, S.; Thakur, R.S. Solid lipid nanoparticles: A modern formulation approach in drug delivery system. Indian J. Pharm. Sci. 2009, 71, 349–358. [CrossRef] [PubMed]

27. Seyfoddin, A.; Shaw, J.; Al-Kassas, R. Solid lipid nanoparticles for ocular drug delivery. Drug Deliv. 2010, 17, 467–489. [CrossRef]

28. Mohanty, B.; Majumdar, D.K.; Mishra, S.K.; Panda, A.K.; Patnaik, S. Development and characterization of itraconazole-loaded solid lipid nanoparticles for ocular delivery. Pharm. Dev. Technol. 2015, 20, 458–464. [CrossRef]

29. Basaran, E.; Demirel, M.; Sirmagul, B.; Yazan, Y. Cyclosporine—A incorporated cationic solid lipid nanoparticles for ocular delivery. J. Microencapsul. 2010, 27, 37–47. [CrossRef]

30. Cavalli, R.; Gasco, M.R.; Chetoni, P.; Burgalassi, S.; Saettone, M.F. Solid lipid nanoparticles (SLN) as ocular delivery system for tobramycin. Int. J. Pharm. 2002, 238, 241–245. [CrossRef]

31. Attama, A.A.; Reichl, S.; Müller-Goymann, C.C. Diclofenac sodium delivery to the eye: In vitro evaluation of novel solid lipid nanoparticle formulation using human cornea construct. Int. J. Pharm. 2008, 355, 307–313. [CrossRef]

32. Yadav, H.K.; Almokdad, A.A.; Shaluf, S.I.M.; Debe, M.S. Polymer-based nanomaterials for drug-delivery carriers. In Nanocarriers for Drug Delivery; Elsevier: Amsterdam, The Netherlands, 2019; pp. 531–556.

33. Mishra, D.K.; Shandilya, R.R.; Mishra, P.K. Lipid based nanocarriers: A translational perspective. Nanomedicine 2018, 14, 2023–2050. [CrossRef] [PubMed]

34. Agarwal, P.; Huang, D.; Thakur, S.S.; Rupenthal, I.D. Nanotechnology for ocular drug delivery. In Design of Nanostructures for Versatile Therapeutic Applications; Elsevier: Amsterdam, The Netherlands, 2018; pp. 137–188.

35. Yu, F.; Zheng, M.; Zhang, A.Y.; Han, Z. A cerium oxide loaded glycol chitosan nano-system for the treatment of dry eye disease. J. Control. Release 2019, 315, 40–54. [CrossRef] [PubMed]

36. Sharma, U.K.; Verma, A.; Prajapati, S.K.; Panday, H.; Panday, A.C. In vitro, in vivo and pharmacokinetic assessment of amikacin sulphate laden

polymeric nanoparticles meant for controlled ocular drug delivery. Appl. Nanosci. 2015, 5, 143–155. [CrossRef]

37. Li, F.; Wen, Y.; Zhang, Y.; Zheng, K.; Ban, J.; Xie, Q.; Wen, Y.; Liu, Q.; Chen, F.; Mo, Z.; et al. Characterisation of 2-HP-β-cyclodextrin-PLGA nanoparticle complexes for potential use as ocular drug delivery vehicles. Artif. Cells Nanomed. Biotechnol. 2019, 47, 4097–4108. [CrossRef] [PubMed]
38. Kalomiraki, M.; Thermos, K.; Chaniotakis, N.A. Dendrimers as tunable vectors of drug delivery systems and biomedical and ocular applications. Int. J. Nanomed. 2016, 11, 1–12.
39. Chaplot, S.P.; Rupenthal, I.D. Dendrimers for gene delivery—A potential approach for ocular therapy? J. Pharm. Pharmacol. 2014, 66, 542–556. [CrossRef] [PubMed]
40. Villanueva, J.R.; Navarro, M.G.; Villanueva, L.R. Dendrimers as a promising tool in ocular therapeutics: Latest advances and perspectives. Int. J. Pharm. 2016, 511, 359–366. [CrossRef]
41. Yavuz, B.; Pehlivan, B.S.; Ünlü, N. Dendrimeric systems and their applications in ocular drug delivery. Sci. World J. 2013, 732340. [CrossRef]
42. Nance, E.; Zang, F.; Mishra, M.K.; Zhang, Z.; Kambhampati, S.P.; Kannan, R.M.; Kannan, S. Nanoscale effects in dendrimer- mediated targeting of neuroinflammatio. Biomaterials 2016, 101, 96–107. [CrossRef]
43. Jevprasesphant, R.; Penny, J.; Jalal, R.; Attwood, D.; Mckeown, N.B.; D'Emanuele, A. The influence of surface modification on the cytotoxicity of PAMAM dendrimers. Int. J. Pharm. 2003, 252, 263–266. [CrossRef]
44. Mishra, V.; Jain, N.K. Acetazolamide encapsulated dendritic nano-architectures for effective glaucoma management in rabbits. Int. J. Pharm. 2014, 461, 380–390. [CrossRef] [PubMed]
45. Yao, W.J.; Sun, K.X.; Liu, Y.; Liang, N.; Mu, H.J.; Yao, C.; Liang, R.C.; Wang, A.P. Effect of poly(amidoamine) dendrimers on corneal penetration of puerarin. Biol. Pharm. Bull. 2010, 33, 1371–1377. [CrossRef] [PubMed]
46. Marano, R.; Toth, I.; Wimmer, N.; Brankov, M.; Rakoczy, P.E. Dendrimer delivery of an anti-VEGF oligonucleotide into the eye: A long-term study into inhibition of laser-induced CNV, distribution, uptake and toxicity. Gene Ther. 2005, 12, 1544–1550. [CrossRef] [PubMed]
47. Patravale, V.B.; Date, A.A.; Kulkarni, R.M. Nanosuspensions: A promising drug delivery strategy. J. Pharm. Pharmacol. 2004, 56, 827–840. [CrossRef] [PubMed]

48. Jiraporn, C.; Walailak, J. Nanosuspension technology for drug delivery. Walailak J. Sci. Technol. 2007, 4, 139–153.

49. Jacob, S.; Nair, A.B.; Shah, J. Emerging role of nanosuspensions in drug delivery systems. Biomater. Res. 2020, 24, 3. [CrossRef]

50. Pignatello, R.; Bucolo, C.; Ferrara, P.; Maltese, A.; Puleo, A.; Puglisi, G. Eudragit RS100®nanosuspensions for the ophthalmic controlled delivery of ibuprofen. Eur. J. Pharm. Sci. 2002, 16, 53–61. [CrossRef]

51. Kassem, M.; Rahman, A.A.; Ghorab, M.M.; Ahmed, M.B.; Khalil, R.M. Nanosuspension as an ophthalmic delivery system for certain glucocorticoid drugs. Int. J. Pharm. 2007, 340, 126–133. [CrossRef]

52. Kamaleddin, M.A. Nano-ophthalmology: Applications and considerations. Nanomedicine 2017, 13, 1459–1472. [CrossRef]

53. Gupta, A. Nanoemulsions. In Nanoparticles for Biomedical Applications; Elsevier: Amsterdam, The Netherlands, 2020; pp. 371–384.

54. Gupta, A.; Eral, H.B.; Hatton, T.A.; Doyle, P.S. Nanoemulsions: Formation, properties and applications. Soft Matter 2016, 12, 826–841. [CrossRef]

55. Ismail, A.; Nasr, M.; Sammour, O. Nanoemulsion as a feasible and biocompatible carrier for ocular delivery of travoprost: Improved pharmacokinetic/pharmacodynamic properties. Int. J. Pharm. 2020, 583, 119402. [CrossRef] [PubMed]

56. Morsi, N.; Ibrahim, M.; Refai, H.; Sorogy, H.E. Nanoemulsion-based electrolyte triggered in situ gel for ocular delivery of acetazolamide. Eur. J. Pharm. Sci. 2017, 104, 302–314. [CrossRef] [PubMed]

57. Ammar, H.O.; Salama, H.A.; Ghorab, M.; Mahmoud, A.A. Nanoemulsion as a potential ophthalmic delivery system for dorzolamide hydrochloride. AAPS PharmSciTech 2009, 10, 808. [CrossRef] [PubMed]

58. Tayel, S.A.; El-Nabarawi, M.A.; Tadros, M.I. Promising ion-sensitive in situ ocular nanoemulsion gels of terbinafine hydrochloride: Design, in vitro characterization and in vivo estimation of the ocular irritation and drug pharmacokinetics in the aqueous humor of rabbits. Int. J. Pharm. 2013, 443, 293–305. [CrossRef] [PubMed]

59. Mahboobian, M.M.; Mohammadi, M.; Mansouri, Z. Development of thermosensitive in situ gel nanoemulsions for ocular delivery of acyclovir. J. Drug Deliv. Sci. Technol. 2020, 55, 101400. [CrossRef]

60. Yaqoob, A.A.; Ahmad, H.; Praveen, T.; Ahmad, A.; Oves, M.; Ismail, I.M.; Qari, H.A.; Umar, K.; Ibrahim, M.N.M. Recent Advances in Metal Decorated

Nanomaterials and Their Various Biological Applications: A Review. Front. Chem. 2020, 8, 341. [CrossRef]

61. Guo, B.L.; Han, P.; Guo, L.C.; Cao, Y.Q.; Li, A.D.; Kong, J.Z.; Zhai, H.F.; Wu, D. The Antibacterial Activity of Ta-doped ZnO Nanoparticles. Nanoscale Res. Lett. 2015, 10, 1047. [CrossRef]
62. Boken, J.; Khurana, P.; Thatai, S.; Kumar, D.; Prasad, S. Plasmonic nanoparticles and their analytical applications: A review. Appl. Spectrosc. Rev. 2017, 52, 74–820. [CrossRef]
63. De Matteis, V.; Rizzello, L. Noble Metals and Soft Bio-Inspired Nanoparticles in Retinal Diseases Treatment: A Perspective. Cells 2020, 9, 679. [CrossRef]
64. Karakocak, B.B.; Raliya, R.; Davis, J.T.; Chavalmane, S.; Wang, W.N.; Ravi, N.; Biswas, P. Biocompatibility of gold nanoparticles in retinal pigment epithelial cell line. Toxicol. In Vitro 2016, 37, 61–69. [CrossRef]
65. Dong, Y.; Wan, G.; Yan, P.; Qian, C.; Li, F.; Peng, G. Fabrication of resveratrol coated gold nanoparticles and investigation of their effect on diabetic retinopathy in streptozotocin induced diabetic rats. J. Photochem. Photobiol. B 2019, 195, 51–57. [CrossRef] [PubMed]
66. Pereira, D.V.; Petronilho, F.; Pereira, H.R.; Vuolo, F.; Mina, F.; Possato, J.C.; Vitto, M.F.; de Souza, D.R.; da Silva, L.; da Silva Paula, M.M.; et al. Effects of gold nanoparticles on endotoxin-induced uveitis in rats. Investig. Ophthalmol. Vis. Sci. 2012, 53, 8036–8041. [CrossRef] [PubMed]
67. Hayashi, A.; Naseri, A.; Pennesi, M.E.; de Juan, E., Jr. Subretinal delivery of immunoglobulin G with gold nanoparticles in the rabbit eye. Jpn. J. Ophthalmol. 2009, 53, 249–256. [CrossRef] [PubMed]
68. Trigueros, S.; Domenech, E.B.; Toulis, V.; Marfany, G. In Vitro Gene Delivery in Retinal Pigment Epithelium Cells by Plasmid DNA-Wrapped Gold Nanoparticles. Genes 2019, 10, 289. [CrossRef] [PubMed]
69. Shen, H.H.; Chan, E.C.; Lee, J.H.; Bee, Y.S.; Lin, T.W.; Dusting, G.J.; Liu, G.S. Nanocarriers for treatment of ocular neovasculariza- tion in the back of the eye: New vehicles for ophthalmic drug delivery. Nanomedicine 2015, 10, 2093–2107. [CrossRef]
70. Kim, J.H.; Kim, J.H.; Kim, K.W.; Kim, M.H.; Yu, Y.S. Intravenously administered gold nanoparticles pass through the blood-retinal barrier depending on the particle size, and induce no retinal toxicity. Nanotechnology 2009, 20, 505101. [CrossRef]

71. Mathew, T.V.; Kuriakose, S. Photochemical and antimicrobial properties of silver nanoparticle-encapsulated chitosan functionalized with photoactive groups. Mater. Sci. Eng. C Mater. Biol. Appl. 2013, 33, 4409–4415. [CrossRef]

72. Iravani, S.; Korbekandi, H.; Mirmohammadi, S.V.; Zolfaghari, B. Syntheisi of silver nanoparticles: Chemial, physical, and biological methods. Res. Pharm. Sci. 2014, 9, 385–406.

73. Ajitha, B.; Reddy, Y.A.K.; Reddy, P.S.; Jeon, H.J.; Ahn, C.W. Role of capping agents in controlling silver nanoparticles size, antibacterial activity and potential application as optical hydrogen peroxide sensor. RSC Adv. 2016, 6, 36171–36179. [CrossRef]

74. Ge, L.; Li, Q.; Wang, M.; Ouyang, J.; Li, X.; Xing, M.M.Q. Nanosilver particles in medical applications: Synthesis, performance, and toxicity. Int. J. Nanomed. 2014, 9, 2399–2407.

75. Anbukkarasi, M.; Thomas, P.A.; Sheu, J.R.; Gerladine, P. In vitro antioxidant and anticataractogenic potential of silver nanoparticles biosynthesized using an ethanolic extract of Tabernaemontana divaricata leaves. Biomed. Pharmacother. 2017, 91, 467–475. [CrossRef] [PubMed]

76. Kalishwaralal, K.; Banumathi, E.; Pandian, S.R.K.; Deepak, V.; Muniyandi, J.; Eom, S.H.; Gurunathan, S. Silver nanoparticles inhibit VEGF induced cell proliferation and migration in bovine retinal endothelial cells. Colloids Surf. B Biointerfaces 2009, 73, 51–57. [CrossRef] [PubMed]

77. Gurunathan, S.; Lee, K.J.; Kalishwaralal, K.; Sheikpranbabu, S.; Vaidyanathan, R.; Eom, S.H. Antiangiogenic properties of silver nanoparticles. Biomaterials 2009, 30, 6341–6350. [CrossRef] [PubMed]

78. Sheikpranbabu, S.; Kalishwaralal, K.; Lee, K.J.; Vaidyanathan, R.; Eom, S.H.; Gurunathan, S. The inhibition of advanced glycation end-products-induced retinal vascular permeability by silver nanoparticles. Biomaterials 2010, 31, 2260–2271. [CrossRef] [PubMed]

79. Xu, Y.; Gao, C.; Li, X.; He, Y.; Zhou, L.; Pang, G.; Sun, S. In vitro antifungal activity of silver nanoparticles against ocular pathogenic filamentous fungi. J. Ocul. Pharmacol. Ther. 2013, 29, 270–274. [CrossRef]

80. Jun, Y.; Zi, Z.; Wanga, Y.; Zhaoa, G.; Liu, J.X. Silver nanoparticles affect lens rather than retina development in zebrafish embryos. Ecotoxicol Environ Saf 2018, 163, 279–288.

81. Kim, J.S.; Song, K.S.; Sung, J.H.; Ryu, H.R.; Choi, B.G.; Cho, H.S.; Lee, J.K.; Yu, I.J. Genotoxicity, acute oral and dermal toxicity, eye and dermal irritation and corrosion and skin sensitisation evaluation of silver nanoparticles. Nanotoxicology 2013, 7, 953–960. [CrossRef]

82. Maneewattanapinyo, P.; Banlunara, W.; Thammacharoen, C.; Ekgasit, S.; Kaewamatawong, T. An evaluation of acute toxicity of colloidal silver nanoparticles. J. Vet. Med. Sci. 2011, 73, 1417–1423. [CrossRef]

83. Patil, S.; Seal, S. Role of trivalent La and Nd dopants in lattice distortion and oxygen vacancy generation in cerium oxide nanoparticles. Appl. Phys. Lett. 2006, 88, 243110. [CrossRef]

84. Korsvik, C.; Patil, S.; Seal, S.; Self, W.T. Superoxide dismutase mimetic properties exhibited by vacancy engineered ceria nanoparticles. Chem. Commun. 2007, 10, 1056–1058. [CrossRef]

85. Kong, L.; Cai, X.; Zhou, X.; Wong, L.L.; Karakoti, A.S.; Seal, S.; McGinnis, J.F. Nanoceria extend photoreceptor cell lifespan in tubby mice by modulation of apoptosis/survival signaling pathways. Neurobiol. Dis. 2011, 42, 514–523. [CrossRef] [PubMed]

86. Maccarone, R.; Tisi, A.; Passacantando, M.; Ciancaglini, M. Ophthalmic Applications of Cerium Oxide Nanoparticles. J. Ocul. Pharmacol. Ther. 2019, 36, 376–383. [CrossRef] [PubMed]

87. Karakoti, A.; Monteiro-Riviere, N.A.; Aggarwal, R.; Davis, J.P.; Narayan, R.J.; Self, W.T.; McGinnis, J.; Seal, S. Nanoceria as antioxidant: Synthesis and biomedical applications. Jom 2008, 60, 33–37. [CrossRef] [PubMed]

88. Chen, J.; Patil, S.; Seal, S.; McGinnis, J.F. Rare earth nanoparticles prevent retinal degeneration induced by intracellular peroxides. Nat. Nanotechnol. 2006, 1, 142–150. [CrossRef]

89. Fiorani, L.; Passacantando, M.; Santucci, S.; Marco, S.D.; Bisti, S.; Maccarone, R. Cerium Oxide Nanoparticles Reduce Microglial Activation and Neurodegenerative Events in Light Damaged Retina. PLoS ONE 2015, 10, e0140387. [CrossRef]

90. Zhou, X.; Wong, L.L.; Karakoti, A.S.; Seal, S.; McGinnis, J.F. Nanoceria inhibit the development and promote the regression of pathologic retinal neovascularization in the Vldlr knockout mouse. PLoS ONE 2011, 6, e16733. [CrossRef]

91. Bhargava, N.; Shanmugaiah, V.; Saxena, M.; Sharma, M.; Sethy, N.K.; Singh, S.K.; Balakrishnan, K.; Bhargava, K.; Das, M. Nanocerium oxide increases the survival of adult rod and cone photoreceptor in culture by abrogating hydrogen peroxide- induced oxidative stress. Biointerphases 2016, 11, 031016. [CrossRef]

92. Pednekar, P.P.; Godiyal, S.C.; Jadhav, K.R.; Kadam, V.J. Mesoporous silica nanoparticles: A promising multifunctional drug delivery system.

In Nanostructures for Cancer Therapy; Elsevier: Amsterdam, The Netherlands, 2017; pp. 593–621.

93. Park, J.H.; Jeong, H.; Hong, J.; Chang, M.; Kim, M.; Chuck, R.S.; Lee, J.K.; Park, C.Y. The effect of silica nanoparticles on human corneal epithelial cells. Sci. Rep. 2016, 6, 37762. [CrossRef]

94. Kim, S.N.; Ko, S.A.; Park, C.G.; Lee, S.H.; Huh, B.K.; Park, Y.H.; Kim, Y.K.; Ha, A.; Park, K.H.; Choy, Y.B. Amino-functionalized mesoporous silica particles for ocular delivery of brimonidine. Mol. Pharm. 2018, 15, 3143–3152. [CrossRef]

95. Gupta, A.K.; Naregalkar, R.R.; Vaidya, V.D.; Gupta, M. Recent advances on surface engineering of magnetic iron oxide nanoparticles and their biomedical applications. Future Med. 2007. [CrossRef]

96. Giannaccini, M.; Giannini, M.; Calatayud, M.P.; Goya, G.F.; Cuschieri, A.; Dente, L.; Raffa, V. Magnetic nanoparticles as intraocular drug delivery system to target retinal pigmented epithelium (RPE). Int. J. Mol. Sci. 2014, 15, 1590–1605. [CrossRef] [PubMed]

97. Bruschi, M.L.; de Toledo, L.D.A.S. Pharmaceutical applications of iron-oxide magnetic nanoparticles. Magnetochemistry 2019, 5, 50. [CrossRef]

98. Yanai, A.; Häfeli, U.O.; Metcalfe, A.L.; Soema, P.; Addo, L.; Gregory-Evans, C.Y.; Po, K.; Shan, X.; Moritz, O.L.; Gregory-Evans, K. Focused magnetic stem cell targeting to the retina using superparamagnetic iron oxide nanoparticles. Cell Transplant. 2012, 21, 1137–1148. [CrossRef] [PubMed]

99. Raju, H.B.; Hu, Y.; Vedula, A.; Dubovy, S.R.; Goldberg, J.L. Evaluation of magnetic micro-and nanoparticle toxicity to ocular tissues. PLoS ONE 2011, 6, e17452. [CrossRef] [PubMed]

100. Häfeli, U.O.; Riffle, J.S.; Harris-Shekhawat, L.; Carmichael-Baranauskas, A.; Mark, F.; Dailey, J.P.; Bardenstein, D. Cell uptake and in vitro toxicity of magnetic nanoparticles suitable for drug delivery. Mol. Pharm. 2009, 6, 1417–1428. [CrossRef] [PubMed]

101. Morigi, V.; Tocchio, A.; Pellegrini, C.B.; Sakamoto, J.H.; Arnone, M.; Tasciotti, E. Nanotechnology in medicine: From inception to market domination. J. Drug Deliv. 2012, 2012, 389485. [CrossRef] [PubMed]

102. Nijhara, R.; Balakrishnan, K. Bringing nanomedicines to market: Regulatory challenges, opportunities, and uncertainties. Nanomedicine 2006, 2, 127–136. [CrossRef] [PubMed]

103. Weissig, V.; Pettinger, T.K.; Murdock, N. Nanopharmaceuticals (part 1): Products on the market. Int. J. Nanomed. 2014, 9, 4357. [CrossRef]

104. Wagner, V.; Dullaart, A.; Bock, A.K.; Zweck, A. The emerging nanomedicine landscape. Nat. Biotechnol. 2006, 24, 1211–1217. [CrossRef]

105. Ventola, C.L. The nanomedicine revolution: Part 2: Current and future clinical applications. Pharm. Ther. 2012, 37, 582–591.

106. Pooja, D.; Kadari, A.; Kulhari, H.; Sistla, R. Lipid-based nanomedicines: Current clinical status and future perspectives. In Lipid Nanocarriers for Drug Targeting, 1st ed.; William Andrew Publishing: Cambridge, MA, USA, 2018; pp. 509–528.

107. Grumezescu, A.M. Design of Nanostructures for Versatile Therapeutic Applications, 1st ed.; William Andrew Publishing: Cambridge, MA, USA, 2018.

Medicine USNLo. Study of Efficacy and Tolerability of SYSTANE Complete in Patients with Dry Eye Disease. Available online: https://clinicaltrials.gov/ct2/show/NCT03492541? term=nano&cond=ocular&cntry=US&draw=2&rank=1 (accessed on 27 June 2020).

108. Liu, S.; Jones, L.W.; Gu, F. Nanotechnology and naomaterials in ophthalmic drug delivery. In Nano-Biomaterials for Ophthalmic Drug Delivery; Springer: Cham, Switzerland, 2016.

109. Perry, J.L.; Herlihy, K.P.; Napier, M.E.; DeSimone, J.M. Print: A novel platform toward shape and size specific nanoparticle theranostics. Acc. Chem. Res. 2011, 44, 990–998. [CrossRef] [PubMed]

110. Acharya, G.; Shin, C.S.; McDermott, M.; Mishra, H.; Park, H.; Kwon, I.C.; Park, K. The hydrogel template method for fabrication of homogeneous nano/microparticles. J. Control. Release 2010, 141, 314–319. [CrossRef] [PubMed]

111. Glendenning, A.; Crews, K.; Sturdivant, J.; DeLong, M.A.; Kopczynski, C.; Lin, C.W. Sustained release, biodegradable PEA implants for intravitreal delivery of the ROCK/PKC in hibitor AR-13503. ARVO Ann. Meet. Abstr. 2018, 59, 5672.

10 PRODUCTION OF NANOPARTICLES OF CERIUM OXIDE UTILIZING DIFFERENT TECHNIQUES:

IMPLICATIONS FOR BIOMEDICAL APPLICATIONS

1. INTRODUCTION: AN OVERVIEW

The synthesis of cerium oxide nanoparticles has garnered considerable attention in the realm of nanotechnology, owing to its diverse applications, particularly in the biomedical field. This burgeoning interest stems from the unique properties exhibited by cerium oxide nanoparticles, such as their potent antioxidant capabilities and versatile physicochemical characteristics. Various methods have been employed to synthesize these nanoparticles, each offering distinct advantages and implications for their biomedical use. This exploration into the synthesis techniques and their implications opens avenues for understanding how cerium oxide nanoparticles can be harnessed for innovative biomedical applications, holding promise for advancements in diagnostics and therapeutics. In this context, this review delves into the synthesis methodologies, shedding light on their relevance and potential contributions to the ever-evolving landscape of biomedical research. During the latter years of the 1960s, scientists dedicated to miniaturized delivery systems introduced nanoparticle-based drug delivery systems and vaccines [1]. Since that application of nanoparticulates into medicine, the use of nanomaterials has yielded great advances in the diagnosis and treatment of numerous pathologies [2,3]. Cerium belongs to a class of lanthanide metals in the periodic table [4]. In oxide form, cerium has a fluorite structure. The nanoscale form, cerium oxide nanoparticles retain the fluorite structure with oxygen deficiencies. This yields cerium oxide nanoparticles with (CeO^{2-x}) vacancies which provide sites for reduction-oxidation reactions. The arrangement of the surfaces of the fluorite structure determines the catalytic performance of the nanoparticles. The (100), (110), and (111) are the best possible surfaces on a cerium oxide nanocrystal (see Figure 1 below) [5]. These represent the lattice arrangement of ions (Ce^{3+} and O^{2-}) in the fluorite structure. The (111) and the (100) possess the o-terminal endings, while the (110) arrangement exposes the Ce center and the O ions [5,6]. These properties enable these nanoparticles to be very useful in industrial applications such as the removal of carbon monoxide, hydrocarbons, and nitric oxide species from the exhaust gas. This property is facilitated by the ability of cerium to occur in trivalent (Ce^{+3}) and tetravalent (Ce^{+4}) states [7,8]. Cerium oxide nanoparticles have been shown to mitigate oxidative stress damage, which has been linked to the development of neurodegenerative diseases such as Parkinson's and Alzheimer's disease. The ability of cerium oxide nanoparticles to switch between valency states enables them to mimic specific enzyme functions such as superoxide dismutase (SOD), Catalase (CAT) and phosphatase, oxidase peroxidase, and phosphotriesterase [9,10].

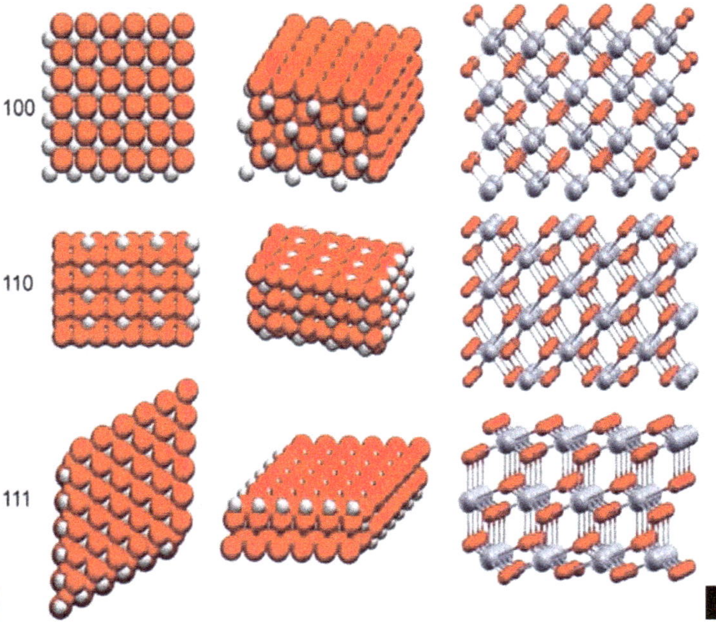

Figure 1. Crystal lattice and (100), (110), and (111) surfaces of cerium oxide nanoparticles [11], used with no changes under the terms of the Creative Commons Attribution 3.0 International License (Creative Commons Attribution 3.0 License).

Cerium oxide nanoparticles are cost-effective and maintain their catalytic properties under harsh environments (Figure 2). Although they are not naturally occurring enzymes, they possess powerful SOD-like activities. Their ability to scavenge reactive oxygen species in an enzyme-like fashion renders them a suitable biocompatible alternative for natural bio-scavengers such as superoxide dismutase [12]. The surface properties of nanoparticles impact on their enzyme mimetic activities. The formation of nanostructures of cerium oxide alters the oxygen non-stoichiometry ratio of Ce^{3+}/Ce^{4+} on the surface of nanostructures.

The activity of cerium oxide nanoparticles correlates with the oxygen non-stoichiometry of nanoparticles. The SOD-like enzyme-mimetic activities of cerium oxide nanoparticles have been reported to be dependent on the Ce^{3+} fraction [10,13]. Another report evaluates the surface properties of cerium oxide nanoparticles using x-ray photoelectron spectroscopy (XPS) and UV-vis. The study illustrates that treatment of the nanoparticles with hydrogen peroxide decreases the Ce^{3+}/Ce^{4+} ratio which correlates directly with the loss of SOD-mimetic activity [13,14].

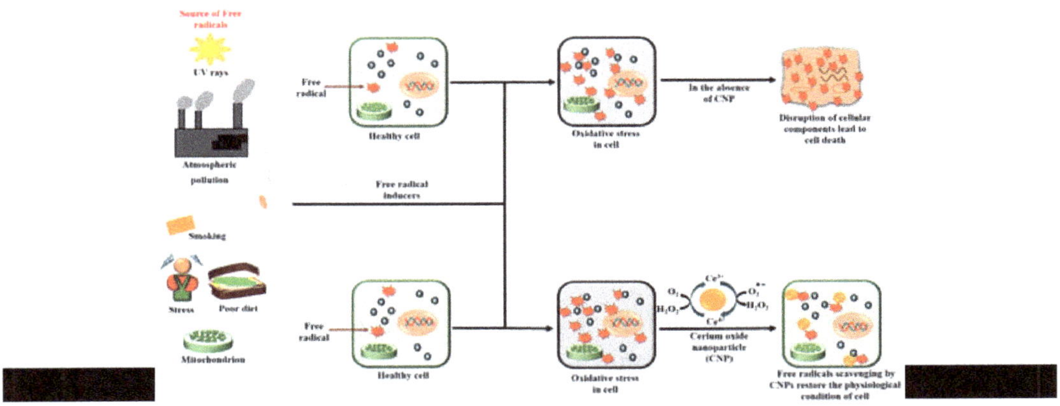

Figure 2. A depiction of some of the causes of oxidative stress and the redox action of cerium oxide nanoparticles (CNPs) [15], used with no changes under the terms of the Creative Commons Attribution 4.0 International License (http://creativecommons.org/publicdomain/zero/1.0/).

Cerium oxide nanoparticles have also shown catalase-mimicking activities. A recent study exploring the mimetic activities of these nanoparticles employed the pulsed electron evaporation as the method of synthesis. The catalase-mimicking activities were measured by the addition of hydrogen peroxide to the cerium oxide nanoparticle (CNP) suspension and was measured at the 380 nm absorption. Vazirov and co-workers suggested that surface doping of metals onto the CNPs increases the Ce^{3+}/Ce^{4+} ratio and the number of vacancies on the nanoparticle surface [16]. This consequently increases the points of interaction and reaction on the surfaces of the nanoparticles. The catalase-like activities are dependent on the Ce^{4+} fraction. A study assessed the catalase-like activities of CNPs using the Amplex Red assay to detect the physiological levels/concentration of peroxide. The test demonstrated catalase-like activity in the suspension with high levels of Ce^{4+} compared to a suspension with high Ce^{3+} [14,17].

In general, phosphorylation and dephosphorylation mechanisms are very valuable in the regulation of critical physiological conditions. The process involves the addition and removal of phosphate groups for energy maintenance. This, in turn, is the fundamental process in the synthesis of adenosine triphosphate

(ATP) molecule. Adenosine diphosphate (ADP) is the hydrolyzed form of ATP. Upon hydrolysis, energy (H^+) is released and the phosphate is used in multiple cellular applications such as cellular physiological and metabolism. Cerium oxide nanoparticles and HMT-CNPs (Hexamethylenetetramine-cerium oxide nanoparticles) were tested for phosphate-like activity. HMT-CNPs decrease the levels of ATPs, thus exhibiting substantial ATPase (phosphatase) activity. This causes a decrease in cell viability which leads to an increase ADP, resulting in toxicity [16].

Moreover, pro and antioxidant properties are dependent on the method of synthesis, the ratio of +3/+4 and their ability to switch between the two species, which most studies have found to be the increase of this ratio that provides (+3) antioxidant properties after internalization of the nanoceria [9,18].

2. SYNTHESIS OF CERIUM OXIDE NANOPARTICLES AND THEIR PROPERTIES

Studies have demonstrated different methods for the synthesis of cerium oxide nanoparticles for different applications. The differences in the conditions of synthesis influence the end product. This then implies that the resulting nanostructures will possess different physical/morphological as well as chemical properties, thus affecting the behaviour of each. It is very important to consider the intended application of the nanostructures in order to determine the method of synthesis. The final features of the nanostructures are very important, especially in the medical application, as their properties affect the interaction at the biological interface [19].

While numerous studies have documented the therapeutic and beneficial effects of cerium oxide nanoparticles, some studies have suggested that cerium oxide nanoparticles may exhibit toxic and harmful effects on cells [20,21]. However, there are studies that show that the resultant properties such as pro/antioxidant and toxicity of cerium oxide are mainly dependent on synthesis conditions such as pH, temperature, and method of synthesis which confer behaviour-altering physicochemical properties such as size/agglomeration, morphology, surface chemistry, and zeta potential [22,23,24,25]. These properties then dictate the behaviour and therapeutic effectiveness upon application.

2.1. Size of Nanoparticles

In general, the size of nanoparticles enables the delivery of therapeutics across biological membranes without compromising the integrity of the membrane. Changes in nanoparticle sizes directly affect their characteristics and behaviour.

These changes influence biological parameters such as the biological half-life, diffusivity, and immunogenicity [26]. Decreases in the sizes of the particles increases the surface area to volume ratio. The large surface to volume ratio influences the catalytic properties of cerium oxide nanoparticles. The large surface area to volume ratio of minute nanoparticles enables particles to entrap free radicals on the surface [27]. Studies suggest that as the size of the cerium oxide nanoparticles decreases it promotes the reducibility and high oxygen storage capacity [28]. The increase in the oxygen vacancies enhances the movement of oxygen through the crystal lattice, thus assisting in the ability of ceria nanoparticles to oxidize or reduce molecules. It has been reported that a decrease in the size of cerium oxide nanoparticles correlates with an expansion of the lattice [15]. This arrangement increases the oxygen vacancies on the surfaces of the nanoparticles [29].

A study by Morones and co-workers demonstrated that the smaller the size of the nanoparticles the more potent the bacterial properties of silver nanoparticles [30]. Eriksson and co-workers demonstrated that cerium oxide nanoparticles between 3–5 nm offer exceptional antioxidative properties when doped into gadolinium for MRI imaging [26]. Studies have also identified size as one of the causes of nanoparticle toxicity. For these reasons, the size of nanoparticles is important in engineering the desired effects of nanotherapeutics. Zhang et al. were able to control the sizes of cerium oxide nanoparticles by adjusting a couple of parameters. In order to obtain particle sizes of between 3–12 nm, they adjusted the reaction times. However, in order to obtain larger nanoparticles, they adjusted temperature was adjusted between 400–800 °C for 30 min [29]. Dhall and Self suggested that careful consideration of the interaction of cerium oxide nanoparticles with the biological milieu is needed when synthesizing nanoparticles for therapeutic intervention [31]. It has also been suggested that the size of the nanoparticles contributes in the vascular clearance of cerium oxide nanoparticles by the immune system. “Bare” or un-coated cerium oxide nanoparticles tend to aggregate, which increases the size. However, applying a stabilizer prevents aggregation and increases the retention time in the body. This increase in size due to aggregation causes rapid clearance by the reticuloendothelial system [32]. Xu and Qu suggested that the body does not possess mechanisms for clearance of cerium oxide as it is not naturally produced in the human body [33]. However, when cerium oxide nanoparticles are ingested, they are not quickly absorbed and they are excreted via feces [32]. Negatively charged cerium oxide nanoparticles under the size of 6 nm have been also found to be excreted via the renal system [34].

2.2. Aggregation and Agglomeration of the Particle System

The use and meaning of the words ‘aggregation’ and ‘agglomeration’ are discipline-specific. Here we define ‘aggregation’ as strongly bonded or fused collection of particles while 'agglomeration' involves loosely clumped nanoparticles (or aggregates) held by weaker van der Waals forces. The biological environment and nanoparticle physicochemical characteristics (e.g., surface properties, particle size) are believed to influence the agglomeration and aggregation of the nanoparticles [35,36]. The Brownian motion is a major factor in this physical behaviour as it causes continuous collisions between the particles. As a result of this agglomeration occurs as the energy of attraction exceeds the energy of repulsion [37]. The forces involved in the collisions include Born repulsion, diffuse double layer potential, and van der Waals attraction. It is also known that aggregate size in the solution depends on properties such as initial particle size and concentration [38]. Also, the aggregate size may vary among different particle types, zinc oxide nanoparticles (NPs) dispersed in aqueous solutions aggregate in a wide range of sizes [39,40,41] whereas TiO2 NPs showed a uniform distribution and agglomeration [42].

2.3. Particle Morphology

The particle morphology is one of the vital physical properties that are to be considered for efficacious nano-therapeutics in biological systems (Figure 3) [19]. Interaction between nanoparticles and biological components can be affected by the morphology of the nanoparticles. Studies have suggested that the method of synthesis may affect particle morphology. Nanoparticles exist in various shapes including spherical, polygonal, cube, or rod shapes. Nanoparticles with sharp edges are less biocompatible as they may inflict mechanical damage on cell membranes and organelles. Particle shape is known to play an important role in the fate and behaviour of manufactured NPs into their environment. This could be as a result of the differences in diffusion rates of the material change with the aspect ratio of the material (e.g., higher drag on a tubular structure compared to a perfect sphere) or because of steric hindrance in the collisions as the morphology limits inter-particular interactions [43]. Several reports have addressed the role of shape and size on cellular internalization [44,45,46].

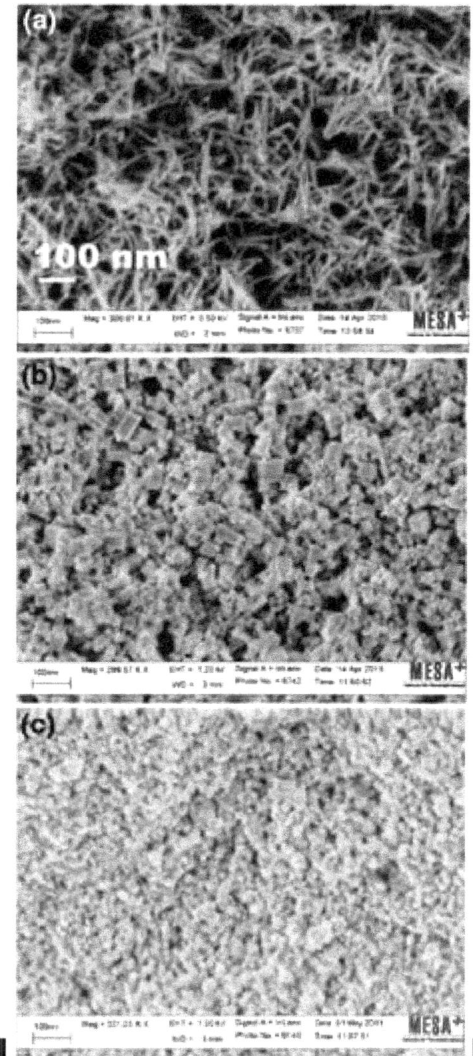

Figure 3. HRSEM images of cerium oxide nanoparticles in different morphologies; (a) nanorods, (b) nanocubes, and (c) nanospheres [19], used with no changes under the terms of the Creative Commons Attribution 4.0 International License (http://creativecommons.org/licenses/by/4.0/).

2.4. Chemical Composition

The chemical composition of the medium is known to have a crucial impact on the electrostatic surface charge of the nanoparticles, thereby affecting the rate at which these nanoparticles agglomerate/aggregate and thus affecting nanoparticle stability. Most of the manufactured nanoparticles nowadays are coated with surfactants to increase the stability of the suspension. The presence of a surface coating on manufactured NPs may significantly modify their surface chemistry, compared with the uncoated equivalents [18,47]. Similar kinds of changes may happen when pristine particles are treated with complex media such as humic acids. For example, adsorption of humic acids on the surface of small aggregates of silver NPs is known to result in the disaggregation of the NPs [48]. Similarly, using stabilizers on the surfaces of cerium oxide nanoparticle limits inter-particle interactions which may result in aggregation [21]. The authors also suggest that the application of stabilizers prevents agglomeration-induced toxicity. Andreescu and colleagues suggested that the environment to which cerium oxide nanoparticles are exposed to may affect their surface chemistry. The interaction of the hydrogen ions in the solution as well as the oxygen atoms in the facets of the lattice alters the surface chemistry of the nanoparticles and consequently alters their catalytic abilities (see Figure 1). The authors also suggest that the nanoparticle surface chemistry is highly sensitized to components of an environment such as pH, proteins, oxidizing, and/reducing agents [49]. Szymanski and company showed that in a biological environment, cerium oxide nanoparticles showed different oxidation states in different parts of the cells or organelles. Nanoparticles in the intracellular environment were shown to have a high Ce^{3+}/Ce^{4+} ratio compared to those in the extracellular environment. This suggests that a noticeable reduction of the cerium oxide nanoparticles occurs in the cells. The authors postulate that differences in pH of the endocytic vesicles (pH = 7.4) versus lysosome (pH = 4.5) may also be the reason for these differences in surface chemistry [50]. Qi and company synthesized PPEG (Phosphonated-Polyethylene glycol oligomers) surface modified nano-powder for re-dispersion in aqueous solutions. The authors showed that the coated cerium oxide nanoparticles were successfully re-dispersed in various organic solvents such as acetone, chloroform, and ethanol. The complexing of the oligomer increased the stability of these nanoparticles when compared to their uncoated/bare counterparts [51]. These examples suggest that the chemical composition of the environment to which cerium oxide nanoparticles are exposed, impacts the surface chemistry and ultimately the behaviour of the nanoparticles.

2.5. Surface Chemistry and Physical Properties

Redox state: Redox reactions (oxidation and reduction processes) occurring on the surface of nanoparticles result in an altered crystalline nature. For example, cerium oxide NPs. Cerium occurs in both trivalent (III) as well as tetravalent (IV) states and has the unique ability to switch readily between these two states [18,52]. This low energy change confers unique catalytic properties on the cerium oxide nanoparticles [53]. However, it is known that the oxidation state is spatially variable within an individual particle and is dependent on size [54], so understanding of the redox mechanism on the particle surface is crucial.

Zeta potential: 'zeta potential' is the potential difference between the dispersion medium and the layer of fluid attached to the dispersed particle and is often used as an analog for colloidal stability, although this is only relevant where NPs are charge stabilized.

Solubility/Dissolution: Dissolution is a dynamic process in which the contents of the dissolving substance migrate from the surface to the bulk solution through a diffusion layer [55]. The thermodynamic parameter that controls this process is described as solubility [56]. Metal-based NPs such as zinc oxide are known to dissolve quickly and release ions that are themselves known to be toxic. Thus, the extent of dissolution and the relative toxicities of both the nanoparticulate and dissolved forms need to be considered to better understand the potential NP effects on organisms over time [57].

Surface functionalization of nanostructures has been used to synthesize non-toxic, stable, and biocompatible nano-therapeutics for applications in various conditions. Polymer chemistry in nano-therapeutics is a means of improving bioavailability, biocompatibility, solubility, as well as bio-retention of water-soluble and insoluble therapeutics for biomedical applications [58,59]. A number of these biodegradable polymers have been applied in the synthesis of multiple stable cerium oxide nanoparticle-systems [60,61]. Cerium oxide nanoparticles can be covered by hydroxyl groups. Therefore, biocompatible and biodegradable polymers that intrinsically possess hydroxyl moieties are capable of stabilizing the CNPs.

Applying the polymers as capping/stabilizing agents can assist in logically controlling the diameter of NPs [62]. The major challenge of suspending nanoparticles in solution is the ability to maintain their physicochemical properties. The surface charge of the nanoparticles, the low volume to surface area ratio of the nanoparticles causes them agglomerate, thus increasing the size of the particles from nano-range to micro-range [63,64]. This then negatively impacts on their therapeutic properties. In order to mitigate this, the surface coating is often employed to prevent the particles from agglomerating. This application of biocompatible coating enhances the solubility of the nanoparticles. Applying these coatings onto the surface of particles increases the stability and mitigates the toxicity of nanoparticles for efficient therapy.

2.6. Precipitation Method

Numerous studies have looked at various precursors to synthesize cerium oxide nanoparticles using the precipitation method, Table 1. Changes in any of the factors such as precursor, pH, and temperature alter the physicochemical properties of the cerium oxide nanoparticles. Researchers have investigated and reported on the precipitation method for the synthesis of cerium oxide nanoparticles using various chemical ingredients. Some of the reagents that are used in the precipitation method include precursors such as cerium nitrate hexahydrate and sodium hydroxide [26,65,66].

Table 1. Table showing the synthesis of cerium oxide nanoparticles using the precipitation method.

-	Cerium nitrate hexahydrate	9–18	Cubic Hexagonal	[67]
PVP	Cerium nitrate hexahydrate	27	Spherical	[67]
MEEETES Samarium	Cerium nitrate hexahydrate	10	Polyhedral	[68]
Ethylene glycol Samarium	Cerium nitrate hexahydrate	5–10	Square	[69]
Dextran	Cerium nitrate hexahydrate	3–5	Spherical	[70]
Sarium	Cerium nitrare hexahydrate	10–13	Spherical	[22,71,72]

Metal oxide powders are acquired by the addition of a ligand such as ammonia into solution of metal ions. In order to obtain a precipitate from this solution or reaction, the point of solubility must be surpassed. In a study by Chen and Chang, the effect of the atmosphere, as well as the reaction temperature in the synthesis of cerium oxide nanoparticles, were studied. The precursor aqueous cerium nitrate and ammonia water were used. The temperature, stirring rate, the atmosphere in the reaction chamber, and the pH of 9.5 was set. The results suggest that increases in temperature change the morphology of the nanoparticles from cubic to hexagonal. Increases in oxygen content or/and decrease in temperature of the reaction atmosphere decreases the size of the nanoparticles [68].

The application of surfactants, capping, or surface doping agents has also been demonstrated in controlling the physicochemical and even catalytic properties of nanoparticles. In a study by Ramachandran and co-workers sought to synthesize cerium oxide nanoparticles with a preferred morphology. It is to be noted that controlling the resulting morphology of nanostructures can be a challenging task. In order to synthesize the nanoparticles, the pH was varied between one and 12 in order to investigate its effect on nanoparticle properties. The synthesis was carried out using cerium nitrate, polyvinyl/pyrrilidone as a surfactant. The synthesis was conducted at different pH ranges of between nine and 12. The XPS results suggest that at the pH of 12, the Ce^{4+} ion was abundant on the surfaces of the particles in the sample. The results also showed that the size of nanoparticles with spherical morphology decreases as the pH increases towards the pH 12 [69].

Cerium oxide nanoparticles possess catalytic, optical, and luminescence properties that have been assessed in imaging for therapeutics. A number of studies have proven that doping metals onto the surfaces of nanoparticles may enhance the properties of the nanoparticles. In order to improve these imaging qualities while maintaining the oxygen vacancies that provide them with their catalytic properties, Nanda sought to surface modify cerium oxide nanoparticles. In their study, they used ammonia-induced ethylene glycol-assisted precipitation to synthesize Samarium-doped cerium oxide nanoparticles (Sm-doped cerium oxide nanoparticles). The dynamic light scattering results from this study revealed that the nanoparticles had a hydrodynamic size of 236 nm for the Sm-doped nanoparticles. This was larger compared to that of the (6-{2-[2-(2-methoxy-ethoxy)-ethoxy]-ethoxy}-hexyl) triethoxysilane (MEEETES)-Sm-doped nanoparticles (MEEETES-Sm-doped nanoparticles), which has a hydrodynamic size of 116.4 nm [70]. Along with the doped cerium oxide nanoparticles, a hydrophilic MEEETES was used as a functional moiety on the surface. The application of biocompatible surface modifying moiety, MEEETES inhibits aggregation and enables dispersion in media. Their results comparing the zeta potential of the nanoparticles at day 0 and at day 180 showed that the zeta potential changed from between -11 and -12 mV at day 0 to between -14 and -13 mV on day 180 [71]. These insignificant changes over a period of 180 days suggests that the zeta potential of nanoparticle can be stabilized over time by the application of stabilizing/doping agents. These studies also suggest that using hydrophilic moieties such as MEEETES on the surface of the nanoparticles may influence not only the hydrodynamic size but also the stability of the zeta potential.

The pH is a vital factor to consider when synthesizing cerium oxide nanoparticles for therapeutic applications. The pH of environment to which the nanoparticles are applied can affect the catalytic properties of the nanoparticles. Therapeutic properties of dextran-coated nanoparticles were investigated in an in vitro study using osteosarcoma cells. The synthesis was carried out using cerium nitrate solution, 0.1 M dextran and ammonium hydroxide. The resulting product consisted of spherical particles of sizes between 3–5 nm. The investigation was carried out in an acidic pH of 3, 5, and 6 to replicate the acidic cancerous conditions, neutral pH of 7, and in an alkaline pH of 9. Characterization of the dextran-coated nanoparticles using XPS showed that the nanoparticles with the Ce^{3+} oxidation states on there were synthesized using this method. However, the effect of the pH was observed when the surfaces of the nanoparticles had Ce^{4+} in abundance at pH 6 and pH 7. The pH was also shown to have an effect on the zeta potential. The recorded zeta potential at pH 7 was 0.58 mV compared to that of 16.68 mV at the pH of 6 and 6.32

mV at pH 9. A study by Yan and company suggests that altering the synthesis pH can alter the morphology and increases the surface area of the nanoparticles. This shows that applying a capping agent such as dextran and controlling the synthesis conditions such as pH may yield cerium oxide nanoparticles with predetermined properties and may enhance the therapeutic properties [72,73,74].

The abovementioned studies of cerium oxide nanoparticle synthesis using the precipitation method show that a number of studies use cerium nitrate hexahydrate as a precursor for synthesis. The morphology was predominantly spherical. Employing the precipitation method produced nanoparticles with sizes 3–27 nm. The use of capping agents/solvents/doping agents narrowed the polydispersity of the nanoparticles. The studies also suggest that the synthesis conditions also play a vital role in the resulting nanoparticles. Therefore, careful control of the synthesis conditions such as temperature, concentration of the reactants, pH, atmosphere, as well as stabilizing agents influence the properties of nanoparticles when employing the simple and economic precipitation method.

2.7. Microemulsification Method

Micro-emulsion is a method that utilizes a polar aqueous medium (water) and a non-polar aqueous medium (oil) in the presence of a surfactant. This method produces nanoparticles of controlled size and structure. The application of this method has yielded a number of insoluble products such as nanoparticles, Table 2, quantum dots of metal oxides, metals, silica, as well as lanthanide fluorides [75,76,77,78].

Table 2. Table showing the synthesis of cerium oxide nanoparticles using the microemulsion method.

AOT DDAB DTAB Brij35	Cerium nitrate hexahydrate/Cerous chloride	6–13 (surfactant) 21 (No surfactant)	Cubic	[75]
OP-10	Cerium nitrate hexahydrate	2–6	-	[77]
Hexamethyl tetraamine	Cerium nitrate hexahydrate	7–10	Spherical	[79]
Ethylene glycol Aβ	Cerium nitrate hexahydrate	3–5	Spherical	[80]

Properties of nanoparticles are crucial in their function. Some of those properties are surface area and oxygen storage capacity. Nanoparticles with high surface area and high oxygen storage capacity enhance the catalytic properties of cerium oxide nanoparticles. However, increased temperature treatment of between

600–800 °C increases particle size thus decreases the surface area. Bumajdad and co-workers sought to synthesize cerium oxide nanoparticles with exposed surfaces and high thermal stability. The microemulsion method included sodium bis(2-ethylhexyl) sulfosuccinate (AOT), di-n-didodecyldimethylammonium bromide (DDAB), and DTAB+Brij 35 as surfactants. The results suggested that the application of surfactants in the synthesis process yielded cerium oxide nanoparticles with high surface area compared to the surfactant-free synthesis. Moreover, application of surfactant such as DDAB increased the surface area and produced cerium oxide nanoparticles which are stable at high temperatures. The use of cerium nitrate hexahydrate and cerium chloride heptahydrate as precursors in the synthesis of nanoparticles, solvents, and surfactants were used at the temperature of 50 °C. After calcination at 400 °C, the XPS results suggest that the surface of the nanoparticles consists of both Ce^{4+} and Ce^{3+}. Upon calcination, the Ce^{4+}/Ce^{3+} ratio increased [75].

In a study by Kockrick and co-workers, the team employed the microemulsion method to synthesize cerium oxide nanoparticles. The nanoparticles were calcined at a temperature of between 100–600 °C and the results suggest a size-dependent catalytic activity. The increase in temperature also aligns with an increase in size of the nanoparticles and a decrease in catalytic activity. The size was narrowly controlled using surfactants to between 6–16 nm [76]. In another study, the pH treatment of the nanoparticles was conducted and the zeta potential was measured. The zeta potential of the nanoparticles synthesized using this method was negative compared to a positive zeta potential of nanoparticles obtained from hydrothermal synthesis. The use of NH_4OH resulted in the zeta potential of -16.26 mV. The study also found that positive zeta potential assists in protein adsorption as opposed to a negative zeta potential [78].

Arya and co-workers used cerium oxide nanoparticles in primary cortical cells as a means of protection against apoptosis via mitochondrial membrane depolarization. The authors also looked at the NADH/NAD+ and ATP concentration post cerium oxide nanoparticle treatment. Precursors such as cerium nitrate hexahydrate and hexamethyltetraamine were used in the microemulsion method of synthesis. The results reveal that spherical nanostructure with a diameter of between 7–10 nm was synthesized. The cortical primary cultures were treated with cerium oxide nanoparticles and the results suggest that cerium oxide nanoparticles prevented apoptosis by stabilizing the mitochondrial membrane potential [79].

Masui and co-workers prepared ultrafine cerium oxide nanoparticles with narrow size distribution and increased oxidative activity using the microemulsion method. The narrow size distribution was achieved by controlling the concentration of starting reagents. The ultra-fine nanoparticles synthesized by this microemulsion were not found to have the quantum-size effect. This was seen when the particle size had no direct and indirect optical energy gaps [80].

Cimini and co-workers investigated the neuro-regenerative properties of anti-Aβ-conjugated polyethylene glycol (PEG)-coated cerium oxide nanoparticles in a model of oxidative stress-induced Alzheimer's disease. The XPS results showed that the amine-functionalized nanoparticles had mixed oxidation state (Ce^{4+}/Ce^{3+}) on the surface. The zeta potential of amine-functionalized was +16 mV for bare cerium oxide nanoparticles and -37 mV for PEG-functionalized nanoparticles [81].

These studies suggest that the temperature, capping agents, and pH in the synthesis of cerium oxide nanoparticles using the microemulsion method of synthesis can control and narrow the particle size, morphology, and catalytic activity of cerium oxide nanoparticles. The various studies mentioned above show that the particle size was limited to between 2–21 nm. The predominant shape of the nanoparticles is seen to be spherical.

2.8. Hydrothermal Method

Particle sizes of the cube-shaped nanoparticles increases as the concentration of the solvent increased from 10–15 M. The hydrothermal method used produced rod-shaped nanoparticles at temperatures between 70–100 degrees Celsius as shown in Table 3. At a constant temperature of 100 degrees Celsius, the size of the nanosized particles decreased. Upon characterization of the nanoparticles using the temperature program reduction (TPR), the results showed an increase in the catalytic activity of nanorods versus nanocubes. This was due to the high concentration of surface oxygen on the surfaces of the nanoparticles. The high Ce^{3+} on the surfaces of the nanoparticles was confirmed by XPS analysis [82].

Table 3. Table showing the synthesis of cerium oxide nanoparticles using the hydrothermal method.

-	Cerium nitrate hexahydrate	8–16	Cubes; rods	[79]
-	Cerium hydroxide/Cerium acetate	5–54	Cubes; amorphous	[81]
Citric acid	Cerium chloride	<5	Spherical	[82]
Trisodium phosphate dodecahydrate	Cerium nitrate hexahydrate	5–60	Rods; Cubes; octahedral	[83]
Dithio-polydopamine	Cerium nitrate hexahydrate	L = 60 d = 5.8	Rods	[80]
-	Cerium nitrare hexahydrate	3–5	Spherical; Cubes; Rods	[84]

In a study that applied the hydrothermal method, Tok and co-workers sought to investigate the use of cerium hydroxide and/or cerium acetate as precursors.

Thermal degradation and the duration of synthesis on the shape and crystallinity were investigated. The authors explored the effects of acidic (pH 4) and basic (pH 10) pH on the synthesis using either precursor prior to hydrothermal treatment. The results suggest that the crystalline morphology of the cerium oxide nanoparticles synthesized from both precursors yielded cubic structured nanostructures across different time-points (0 h, 6 h, 12 h, 18 h, and 24 h). At the 24 h time-point, the results suggest that the cerium acetate-derived/synthesized nanostructures were larger compared to the cerium hydroxide-synthesized nanostructures. Thermal treatments of cerium hydroxide-synthesized nanostructures at 500 °C and 1000 °C for 2 h grew the crystalline size to 8.8 and 47.4 nm, respectively from 5–6 nm. Thermal treatment of the cerium acetate-synthesized grew the crystalline size to 17.7 and 53.6 nm, respectively from 10–15 nm. The result also showed that in both precursor-derived nanostructures, agglomeration was observed [83].

Masui and co-workers investigated the application of citric acid as a protectant against particle growth. The surface of the particles was homogeneously covered with citric acid. Large particles formed prior to hydrothermal treatment when using Ce (III) salts were smaller in comparison with synthesis involving citric acid. This is because of the adsorption of citric acid onto the surface of the nanoparticles, thus controlling the nucleation process. The particle sizes of 5 nm were obtained from this method, with less agglomeration observed [80].

Cerium oxide nanoparticles with different morphologies were synthesized using hydrothermal synthesis. Different synthesis parameters such as temperature and concentrations were applied in order to manipulate the morphology of the resulting nanoparticles. The resulting nanostructures were nanocubes, nanorods, nano-octahedral, and submicronic cerium oxide nanoparticles. Thus, suggesting that altering synthesis parameters does alter the morphology and size of the nanoparticles. An analysis of the nanoparticles using XPS showed an oxidation state of $Ce3+$ on the surfaces of less than 1% [84].

In a different study, Zhang and co-workers synthesized degradable and stimuli-responsive polymer-coated cerium oxide nanorods for cancer therapy. The authors used dithio-polydopamine for the coating of the nanorods. The hydrothermal method of synthesis using cerium nitrate in mili-Q water (18.2 MΩ). In order to target cancerous cells, a lactose derivative moiety was attached to the surface of the nanoparticles. The resulting nanoparticles (spherical, cube, and rods) had hydrodynamic radii of 370, 194, and 192 nm and a set potential of -9.98, -10.86, and -5.9 mV [85].

Singh and co-workers evaluated the morphology and size-dependent cellular internalization at the nano-bio interface. The synthesis method used was hydrothermal synthesis. Varying concentrations of cerium nitrate hexahydrate

(0.45–0.6 M) in deionized water were used. The results suggest that the resulting nanoparticles had a diameter of 3–5 nm. The results also suggest that the concentration altered the shape of the nanoparticles from spherical to nanorod to nanocube [86].

A myriad of morphologies can result from employing the hydrothermal method. Depending on the type of solvents, stabilizing agents, and synthesis conditions, resultant nanoparticles possess different properties.

2.9. Green Synthesis Routes

The green synthesis method is an environmentally innocuous technique that employs naturally occurring substances such as plant extracts, sugars, biodegradable polymers, and microorganisms in the synthesis of nanoparticles, Table 4. These naturally occurring components function as reducing as well as capping agents [87,88].

Table 4. Table showing the synthesis of cerium oxide nanoparticles using the green synthesis method.

Acalypha indica	Cerium chloride heptahydrate	8–54	Eliptical spherical	[87]
Fructose; Glucose; Lactose	Ammonnium Cerium nitrate	2–6	Spherical/Agglomerate	[86]
Hibiscus Sabdariffa	Cerium nitrate hexahydrate	3.9	Amorphous	[89]
Egg White	Cerium nitrate hexahydrate	25	Spherical	[90]

In a study by Thovhogi and co-workers, physicochemical properties of cerium oxide nanoparticles synthesis from Hibiscus Sabdariffa flower extract. The green synthesis method yielded nano-spheres with a diameter of 3.9 nm. An XPS investigation of the oxidation states on the surface of nanoparticles post-functionalization was undertaken. The results revealed a low 3+/4+ ratio. This shows the ability of environmentally friendly-green synthesis to alter the physical properties of nanoparticles [89].

Cerium oxide nanopowder with spherical particle sizes of 25 nm was synthesized using fresh egg white as a capping agent. The fresh egg white was used to control the particle distribution of the nanopowder. The eco-friendly capping and stabilizing agent is bio-degradable and possesses numerous amino acids. The ovalbumin and lysosome are the two major proteins that are available on the egg white. The interaction of the egg white with water and its ability to associate well with metal ions make it suitable for application as a shape-controlling stabilizing agent. The cell-viability of periodontal fibroblast cells treated with a range of CNP concentration of between 12.5–800 μg/mL was done. The results showed no toxicity of CNP concentrations below 800 μg/mL [90].

2.10. Solvothermal Method

Solvothermal synthesis employs the use of organic solvents in a chamber under high pressure and temperature to produce nanomaterials of varying sizes, Table 5. Teflon-lined autoclaves are used as they can withstand high temperature and pressure. Properties such as size were regulated by manipulating the temperature, reaction time, the concentration of the reagents, as well as the type of solvent. The 1,4-butanediol prohibited particle growth because of its viscous nature compared to water. This suggests that the viscous nature of a capping agent/solvent influences particle growth upon nucleation [91,92].

Table 5. Table showing the synthesis of cerium oxide nanoparticles using the solvothermal method.

[illegible]	[illegible]	[illegible]	[illegible]	[illegible]
1,4-butanediol	Ceric ammonium nitrate	5–10	-	[92]
Ethylenediamine	Cerium nitrate hexahydrate	2.5–8	-	[91]
Ethylene glycol	Cerium nitrate hexahydrate	-	Plate; Spherical	[93]

Kar and co-workers sought to synthesize ultra-small, water-soluble cerium oxide nanoparticles with the application of ethylenediamine as a surface capping/ stabilizing agent at room temperature. The relatively monodispersed nanoparticles of 2.5 nm diameter showed mixed valence on the nanoparticle surface. The ultra-small particle exhibited a mixed oxidation state/valence on the surface of the nanoparticles [94]. The use of a solvent controls the physical properties of the cerium oxide nanoparticles such as morphology, particle size, and crystal growth. In this study, solvents/media such as water, ethanol, and ethylene glycol at different compositions were used. These were used to control the morphology, oxygen storage capacity, and electrical properties of cerium oxide nanoparticles. Water produced plate-shaped nanoparticles, spherical-shaped cerium oxide nanoparticles with a decreased size of the nanoparticles were produced with a 70:30 (water: ethanol) and the 70:30 (water: ethylene glycol) which resulted in porous nanoparticles [93].

2.11. Sol-Gel Method

Precursors such as cerium nitrate hexahydrate $Ce(NO)_3$.6H2O undergo express hydrolysis, resulting in the production of a cerium hydroxide. The metal hydroxide then undergoes condensation thus forming a gel, Table 6. The gel is then exposed to a drying process to yield the final product [95].

Table 5. Table showing the synthesis of cerium oxide nanoparticles using the solvothermal method.

1,4-butanediol	Ceric ammonium nitrate	5–10	-	[92]
Ethylenediamine	Cerium nitrate hexahydrate	2.5–8	-	[91]
Ethylene glycol	Cerium nitrate hexahydrate	-	Plate; Spherical	[93]

Yu and co-workers used cerium (III) nitrate in a reaction with diphenyl ether as well as surfactants such as oleylamine to produce nanoparticles with spherical, wire, and tadpole morphology. The synthesis of nanostructures with different morphologies was achieved by using cerium nitrate and diphenyl ether in the presence of other surfactants. To obtain spherical morphologies oleylamine was used at the temperature of 320 °C. Nanowires were obtained from the application of the co-surfactants, oleylamine, and oleic acid at the temperature of 320 °C. Longer nanowires were synthesized by the addition of more oleic acid. Tadpole-like nanostructures were synthesized by increasing the amount of aleylamine in the co-surfactant mixture. The morphology and size of these nanostructures were synthesized/controlled by altering experimental parameters such as surfactant molar ratio as well as the duration of the reaction [95]. Another study investigated the effect of methanol as a solvent and the temperature of 400 °C on the size of nanoparticle morphology and optical properties. The results showed that the high ratio of UV to visible emission suggested good nanocrystal quality, i.e., good surface density. Nanoparticles with uniform spherical morphology with a size of 8 nm were synthesized [96].

A solution of cerium nitrate hexahydrate in a mixture of distilled water and Lu seeds and calcined at between 400–600 °C. The biosynthesis of these nanoparticles using the Lu seeds provided nanoparticles with altered physicochemical properties. Elahi and co-workers attached the 99Tc radio-labeller onto the nanoparticles in order to track the biological mobility/bioavailability of the nanoparticles in vivo. The results show on agglomerated morphology of the nanoparticles with an average size of 21–32 nm aligning with the calcination temperature of between 400–600 °C. These suggest that calcination, as well as the calcination temperature, has an effect on particle size. The smallest size showed no significant cytotoxicity on A549 cells. A bio-distribution to the kidneys and poor uptake by the stomach and thyroid were observed on an in vivo study using Wistar rats [97].

Another in vivo study on Wistar rats investigated the toxic effects of sol-gel prepared cerium oxide nanoparticles. The oxidative stress status, biochemical as well as hematological parameters were investigated. Pullulan was used as a capping agent to the cerium oxide nanoparticles. The resulting cerium oxide nanoparticles had a diameter of 12 nm. The results showed that pullulan-mediated cerium oxide nanoparticles possessed no cytotoxic effects upon acute administration on Neuro2A cells. There were also no toxic effects observed in the liver, spleen, and other organs. There was no damaged reported on the relevant antioxidant enzymes. The cerium oxide nanoparticles showed oxidative stress-protective properties and no hemotoxic effects or tissue damage upon histological analysis [98].

The green synthesis of cerium oxide nanoparticle-embedded hydrogel was conducted. The hydrogel was synthesized from carrageen as a stabilizing agent for the sol-gel synthesis of cerium oxide nanoparticles. Nourmohammadi and co-workers investigated the effectivity of carrageen hydrogel as capping agents on cerium oxide nanoparticles. They also investigated the cytotoxicity of these hydrogel-embedded nanoparticles on WEHI 164 cancer cells. The synthesis was conducted using cerium nitrate hexahydrate in double-distilled water and carrageen powder. The resulting nanoparticles possessed a spherical morphology with an average size of 34 nm. The results also showed no significant signs of toxicity in vitro even at high concentrations. However, the cellular metabolic activity was decreased with an increase in dose [99].

In a different study, starch was used as a capping agent on CNPs. The 6 nm spherical CNPs obtained were shown to have no cytotoxic effects in an in vitro study using the Neuro2A cell line [96]. The van der Waals forces between the particles increase the particle size during storage and upon contact with media. This causes the particles to be toxic to cells. Therefore, the application of starch as a stabilizing agent limited the inter-particular interactions which often result in aggregation and an increase in particle size. The cell-viability studies demonstrated no toxic effects on CNPs at a concentration of 5 μg/mL.

2.12. Ball Milling

A laboratory aluminar-zirconia ceramic mill and zirconia balls (ZrO2wt % 95%, diameter 0.5–0.3 mm, Mohs hardness 9) are used to create nanosized cerium oxide powder of 50 nm from cerium oxide powder with average particle sizes of 7.4 μm [100].

In a study by Yadav and co-workers, nanoparticles were produced from micron-sized cerium powder following 30 h of high energy ball milling, Table 7. The process yielded spherical nanoparticles with sizes between 8–12 nm with no defects. The conditions that yielded optimum results were noted to be 30 h of high energy ball milling at ice-cold temperatures [100]. Another study applied poly acrylic acid (PAA) to use as a dispersant of the cerium oxide nanoparticles. Normally, ball milling has been shown to produce nanoparticles with a diameter of 210 nm [101]. Li and co-workers used hydrated cerium carbonate and sodium hydroxide during high energy ball milling. In a different study, an organic base was used to synthesize cerium oxide nanoparticles with a nearly-spherical shape as opposed to an inorganic base during ball milling [102]. In this study, by He and co-workers, the use of PAA reduced the size of the nanoparticles from 128 nm to 50 nm and provided more stable cerium oxide nanoparticle slurry [103].

Table 7. Table showing the synthesis of cerium oxide nanoparticles using the various synthesis methods.

[illegible]	[illegible]	[illegible]	[illegible]	[illegible]	[illegible]
Ball Milling	-	Cerium powder (5µm)	10	Spherical	[104]
Ball Milling	Tetrabutyl ammonium hydroxide	Cerium ammonium nitrate	5–56	Spherical	[100]
Flame spray pyrolysis	Folic acid	Cerium acetate hydrate	7	-	[102]
Reverse phase evaporation	Lipid anionic mixture	Cerium nitrate hexahydrate	12–230	Spherical	[103]

2.13. Flame Spray Pyrolysis

In an interesting study, Vassie and co-workers synthesized folic acid (FA)-coated cerium oxide nanoparticles by flame spray pyrolysis, Table 7. The aim was to investigate whether FA-coated cerium oxide nanoparticles can modulate intracellular activities and control ROS levels when compared to bare/uncoated cerium oxide nanoparticles. The liquid precursor, cerium 2-ethylhexanoate in xylene was used for the experiment. Using a syringe, the mixture was fed through the flame at a determined rate. Folic acid was also conjugated onto the nanoparticles to form FA-cerium oxide nanoparticles. The results show cytoplasmic localization of both the FA-cerium oxide nanoparticles and bare nanoparticles. There was also an increase in the uptake of the FA-coated nanoparticles in ovarian cancer cells that led to the induction of cell death via an increased concentration of ROS [105].

2.14. Reverse-Phase Evaporation

Liposome-entrapped cerium oxide nanoparticles were synthesized to enhance their therapeutic properties, Table 7. Cerium oxide nanoparticles with a size of 12 nm were loaded into liposomes. The sizes of nanoparticle-loaded liposomes were 230 nm. The liposomes enhanced the lipophilicity of the cerium oxide nanoparticles. The in vitro toxicity test was conducted on normal human dermal fibroblast (NHDF) cells. The study findings showed that the cerium oxide nanoparticle-loaded liposomes displayed colloidal stability and retained their antioxidant properties [106].

2.15. Reverse Micelle

An in vivo model of cardio-toxicity using rats was created to elucidate the ameliorative effects of cerium oxide nanoparticles, Table 7. The rat model was created by subcutaneous injection of isoproterenol hydrochloride (30 mg/kg) in saline. This study compared the ameliorative effects of Captopril and cerium oxide nanoparticles in oxidative-stress-induced cardio-toxicity. The cerium oxide loaded reverse micelles with sizes of 5 nm were synthesized. The results showed that cerium oxide-loaded micelles showed better prophylactic and ameliorative effects of cardio cytotoxicity compared with Captopril [107].

3. TOXICITY AND CYTOTOXICITY OF CERIUM OXIDE NANOPARTICLES

The toxicity and cytotoxicity of cerium oxide has long been a point of contention among researchers exploring the application of cerium oxide nanoparticles in biomedicine. The lanthanide-derived nanoparticles have been applied immensely in industrial or non-medical applications. These applications include environmental chemistry, cosmetics, gas-emissions, fuel cells, oxygen sensors, and corrosion protectors [31,33].

Studies have suggested that the successful application of these metal oxides in the abovementioned fields may not infer positive application in biomedical systems. This is may be due to the complex nature of physiological systems. The properties of these nanoparticles and the manner in which interact with the biological milieu needs to be considered in order to synthesize biocompatible nanoparticles for biomedical application. This is because the properties and the manner in which these nanoparticles interact with the biological system may have toxic effects [108,109]. A study by Asati and co-workers investigated the surface charge-dependent internalization of polymer-coated cerium oxide nanoparticles into normal versus malignant cells. Transformed lung cells A549 and MCF-7 breast cells as well as untransformed HEK293 and H9c2 were chosen. Poly-(acrylic acid), aminated poly (acrylic acid), and/or dextran were applied as coating on the nanoparticles. These polymers furnished the nanoparticles with different charge characteristics such as negative, positive and neutral, respectively.

The internalization of the nanoparticles was observed in every cell type except the MCF-7 cell line. The negatively charged nanoparticles were internalized in the A549 cell line only and were not taken up by the MCF-7, thus the toxicity to the A549 lung carcinoma cells. Neutral nanoparticles were localized in the cytoplasm of all cell types in this experiment and were non-toxic. The study thus found that factors such as internalization, localization, and surface charge of the cerium oxide nanoparticles play an important role in the toxicity of cerium oxide nanoparticles [110]. In another study, alkaline-precipitation as well as the inverse micro-emulsion methods were employed to synthesize ultrafine cerium oxide nanoparticles. Li and co-workers synthesized water-insoluble (CeO2-P), water-soluble, dextran coated, PAA coated, and EDA coated nanoparticles were synthesized. The polymers provided different functional groups on nanoparticle surfaces such as -OH, -COOH, and -NH_2. The results suggest a concentration-dependent cytotoxicity. The results also suggest that the "bare" cerium oxide nanoparticles (without functional groups) induced acute cytotoxicity on BGC-803 cell lines. The dextran, PAA, and EDA coated nanoparticles showed some viability as high concentrations (20 mg/mL) [14].

The toxicity of cerium oxide nanoparticles has also been attributed to their size. The smaller the size the more toxic they are. This has been suggested to be due to the large surface area to volume ratio. This means they possess more catalytic activity on the surface, thus making them very reactive and consequently toxic. Their size may also result in increased internalization into the cells which may result in toxicity. The tendency of the cerium oxide nanoparticles to agglomerate has also been suggested to be one of the causes of their toxicity [111]. The interaction of these metal oxide nanoparticles with their biological milieu may induce toxic effects. The synthesis method and the storage conditions have also been implicated in their toxicity [112]. These studies suggest that the physico-chemical properties of cerium oxide nanoparticles as well as the manner they interact with the biological environment may cause them to be toxic.

4. CONCLUSIONS

These studies clearly demonstrate the therapeutic effects and the behaviour of cerium oxide nanoparticles in a bio-relevant environment. They also demonstrate the effects of synthesis and surface modification on the therapeutic effects of cerium oxide nanoparticles in the biological interface. This highlights the importance of the synthesis method or post-synthesis modifications of cerium oxide nanoparticles. The application of biocompatible and biodegradable capping and stabilizing agents clearly prove necessary in the synthesis of nanoparticles that are able to exert therapeutic effects. These capping agents have to be considered when synthesizing biologically-relevant nanoparticles in order to avoid unwanted interaction between the nanoparticles as well as between the nanoparticles and their environment. More studies looking into the therapeutic effects of cerium oxide nanoparticles in systemic conditions caused inter alia by oxidative stress, inflammation, and bacteria. Therapeutic effects of these nanoparticles in diseases that require tissue regeneration (scaffolds) need to be further explored.

5. CONCISE SUMMARY

Cerium oxide nanoparticles have been used in a number of non-medical products over the years. The therapeutic application of these nanoparticles has mainly been due to their oxidative stress ameliorating abilities. Their enzyme-mimetic catalytic ability to change between the Ce^{3+} and Ce^{4+} species makes them ideal for a role as free-radical scavengers for systemic diseases as well as neurodegenerative diseases. In this review, we look at various methods of synthesis (including the use of stabilizing/capping agents and precursors), and how the synthesis method affects the physicochemical properties, their behaviour in biological environments, their catalytic abilities as well as their reported toxicity.

REFERENCES

1. Kreuter, J. Nanoparticles—A Hist. *Perspect. IntJ. Pharm* 2007, *331*, 1–10. [Google Scholar] [CrossRef]
2. Attia, M.S.; Al-Radadi, N.S. Nano optical sensor binuclear Pt-2-pyrazinecarboxylic acid -bipyridine for enhancement of the efficiency of 3-nitrotyrosine biomarker for early diagnosis of liver cirrhosis with minimal hepatic encephalopathy. *Biosens. Bioelectron.* 2016, *86*, 406–412. [Google Scholar] [CrossRef]
3. Panja, S. Biological application of green silver nanoparticle synthesized from leaf extract of Rauvolfia serpentina Benth. *Asian Pac. J. Trop. Dis.* 2016, *6*, 549–556. [Google Scholar] [CrossRef]
4. Weeks, M.E. The discovery of the elements. *Xvi Rare Earth Elem. J. Chem. Educ.* 1751, *1932*, 9. [Google Scholar] [CrossRef]
5. Yao, S.Y.; Xu, W.Q.; Johnston-Peck, A.C.; Zhao, F.Z.; Liu, Z.Y.; Luo, S.; Senanayake, S.D.; Martínez-Arias, A.; Liu, W.J.; Rodriguez, J.A. Morphological effects of nanostructured ceria supports on the activity and stability of CuO/ CeO2 catalysts for the water-gas shift reaction. *Phys. Chem. Chem. Phys.* 2014, *16*, 17183–17195. [Google Scholar] [CrossRef] [PubMed]
6. Huang, W.; Gao, Y. Morphology-dependent surface chemistry and catalysis of CeO_2 nanocrystals. *Catal. Sci. Technol.* 2014, *4*, 3772–3784. [Google Scholar] [CrossRef]
7. Medalia, A. Byrne Spectrophotometric, B. of Cerium (IV). *Anal. Chem.* 1951, *23*, 453–456. [Google Scholar] [CrossRef]
8. Dowding, J.M.; Das, S.; Kumar, A.; Dosani, T.; McCormack, R.; Gupta, A.; Sayle, T.X.; Sayle, D.C.; von Kalm, L.; Seal, S.; et al. Cellular interaction and toxicity depend on physicochemical properties and surface modification of

redox-active nanomaterials. *Acs Nano* 2013, *7*, 4855–4868. [Google Scholar] [CrossRef][Green Version]

9. Yang, Y.; Mao, Z.; Huang, W.; Liu, L.; Li, J.; Li, J.; Wu, Q. Redox enzyme-mimicking activities of CeO_2 nanostructures: Intrinsic influence of exposed facets. *Sci. Rep.* 2016, *6*, 35344. [Google Scholar] [CrossRef]

10. Baldim, V.; Bedioui, F.; Mignet, N.; Margaill, I.; Berret, J.F. The enzyme-like catalytic activity of cerium oxide nanoparticles and its dependency on Ce3+ surface area concentration. *Nanoscale* 2018, *10*, 6971–6980. [Google Scholar] [CrossRef][Green Version]

11. Naaz, F.; Farooq, U.; Ahmad, T. Ceria as an Efficient Nanocatalyst for Organic Transformations. In *Nanocatalysts*; IntechOpen: London, UK, 2019. [Google Scholar]

12. Charbgoo, F.; Ahmad, M.B.; Darroudi, M. Cerium oxide nanoparticles: Green synthesis and biological applications. *Int. J. Nanomed.* 2017, *12*, 1401–1413. [Google Scholar] [CrossRef] [PubMed][Green Version]

13. Heckert, E.G.; Karakoti, A.S.; Seal, S.; Self, W.T. The role of cerium redox state in the SOD mimetic activity of nanoceria. *Biomaterials* 2008, *29*, 2705–2709. [Google Scholar] [CrossRef] [PubMed][Green Version]

14. Li, C.; Shi, X.; Shen, Q.; Guo, C.; Hou, Z.; Zhang, J. Hot Topics and Challenges of Regenerative Nanoceria in Application of Antioxidant Therapy. *J. Nanomater.* 2018, *2018*, 1–12. [Google Scholar] [CrossRef][Green Version]

15. Thakur, N.; Manna, P.; Das, J. Synthesis and biomedical applications of nanoceria, a redox active nanoparticle. *J. Nanobiotechnol.* 2019, *17*, 84. [Google Scholar] [CrossRef] [PubMed]

16. Vazirov, R.A. Physicochemical characterization and antioxidant properties of cerium oxide nanoparticles. In *Journal of Physics Conference Series*; IOP Publishing: Bristol, UK, 2018. [Google Scholar]

17. Pirmohamed, T.; Dowding, J.M.; Singh, S.; Wasserman, B.; Heckert, E.; Karakoti, A.S.; King, J.E.; Seal, S.; Self, W.T. Nanoceria exhibit redox state-dependent catalase mimetic activity. *Chem. Commun. (Camb. Engl.)* 2010, *46*, 2736–2738. [Google Scholar] [CrossRef][Green Version]

18. Pulido-Reyes, G.; Rodea-Palomares, I.; Das, S.; Sakthivel, T.S.; Leganes, F.; Rosal, R.; Seal, S.; Fernández-Piñas, F. Untangling the biological effects of cerium oxide nanoparticles: The role of surface valence states. *Sci. Rep.* 2015, *5*, 15613. [Google Scholar] [CrossRef][Green Version]

19. Kovacevic, M.; Mojet, B.L.; van Ommen, J.G.; Lefferts, L. Effects of Morphology of Cerium Oxide Catalysts for Reverse Water Gas Shift Reaction. *Catal. Lett.* 2016, *146*, 770–777. [Google Scholar] [CrossRef][Green Version]

20. Kumari, M.; Singh, S.P.; Chinde, S.; Rahman, M.F.; Mahboob, M.; Grover, P. Toxicity study of cerium oxide nanoparticles in human neuroblastoma cells. *Int. J. Toxicol.* 2014, *33*, 86–97. [Google Scholar] [CrossRef]

21. Berret, J.-F. In vitro toxicity of nanoceria: Effect of coating and stability in biofluids. *Nanotoxicology* 2014, *8*, 799–811. [Google Scholar]

22. Alpaslan, E.; Geilich Benjamin, M.; Yazici Hilal Webster Thomas, J. pH-Controlled Cerium Oxide Nanoparticle Inhibition of Both Gram-Positive and Gram-Negative Bacteria Growth. *Sci. Rep.* 2017, *7*, 45859. [Google Scholar] [CrossRef]

23. Kalyanaraman,V.;Naveen,S.V.;Mohana,N.;Balaje,R.M.;Navaneethakrishnan, K.R.; Brabu, B.; Murugan, S.S.; Kumaravel, T.S. Biocompatibility studies on cerium oxide nanoparticles-combined study for local effects, systemic toxicity and genotoxicity via implantation route. *Toxicol. Res. (Camb.)* 2019, *8*, 25–37. [Google Scholar] [CrossRef] [PubMed][Green Version]

24. Karakoti, A.S.; Singh, S.; Kumar, A.; Malinska, M.; Kuchibhatla, S.V.; Wozniak, K.; Self, W.T.; Seal, S. PEGylated nanoceria as radical scavenger with tunable redox chemistry. *J. Am. Chem. Soc.* 2009, *131*, 14144–14145. [Google Scholar] [CrossRef] [PubMed][Green Version]

25. Yang, X.; Pan, H.; Wang, P.; Zhao, F.J. Particle-specific toxicity and bioavailability of cerium oxide (CeO_2) nanoparticles to Arabidopsis thaliana. *J. Hazard. Mater.* 2017, *322*, 292–300. [Google Scholar] [CrossRef] [PubMed]

26. Eriksson, P.; Tal, A.A.; Skallberg, A.; Brommesson, C.; Hu, Z.; Boyd, R.D.; Olovsson, W.; Fairley, N.; Abrikosov, I.A.; Zhang, X.; et al. Cerium oxide nanoparticles with antioxidant capabilities and gadolinium integration for MRI contrast enhancement. *Sci. Rep.* 2018, *8*, 6999. [Google Scholar] [CrossRef]

27. Das, S.; Dowding, J.M.; Klump, K.E.; McGinnis, J.F.; Self, W.; Seal, S. Cerium oxide nanoparticles: Applications and prospects in nanomedicine. *Nanomedicine* 2013, *8*, 1483–1508. [Google Scholar] [CrossRef]

28. Arul, N. Sabari, Strong quantum confinement effect in nanocrystalline cerium oxide. *Mater. Lett.* 2011, *65*, 2635–2638. [Google Scholar] [CrossRef]

29. Zhang, F.; Chan, S.W.; Spanier, J.E.; Apak, E.; Jin, Q.; Robinson, R.D.; Herman, I.P. Cerium Oxide Nanoparticles: Size-Selective Formation and Structure Analysis. *Appl. Phys. Lett.* 2002, *80*, 127–129. [Google Scholar] [CrossRef] [Green Version]

30. Morones, J.R.; Elechiguerra, J.L.; Camacho, A.; Holt, K.; Kouri, J.B.; Ramirez, J.T.; Yacaman, M.J. The bactericidal effect of silver nanoparticles. *Ssrn Electron. J.* 2005, *16*, 2346–2353. [Google Scholar] [CrossRef][Green Version]

31. Dhall, A.; Self, W. Cerium Oxide Nanoparticles: A Brief Review of Their Synthesis Methods and Biomedical Applications. *Antioxidants* 2018, *7*, 97. [Google Scholar] [CrossRef][Green Version]

32. Molina, R.M.; Konduru, N.V.; Jimenez, R.J.; Pyrgiotakis, G.; Demokritou, P.; Wohlleben, W.; Brain, J.D. Bioavailability, distribution and clearance of tracheally instilled, gavaged or injected cerium dioxide nanoparticles and ionic cerium. *Environ. Sci. Nano* 2014, *1*, 561–573. [Google Scholar] [CrossRef]

33. Xu, C.; Qu, X. Cerium oxide nanoparticle: A remarkably versatile rare earth nanomaterial for biological applications. *Npg Asia Mater.* 2014, *6*, e90. [Google Scholar] [CrossRef]

34. Walkey, C.; Das, S.; Seal, S.; Erlichman, J.; Heckman, K.; Ghibelli, L.; Traversa, E.; McGinnis, J.F.; Self, W.T. Catalytic Properties and Biomedical Applications of Cerium Oxide Nanoparticles. *Environ. Sci. Nano* 2015, *2*, 33–53. [Google Scholar] [CrossRef] [PubMed]

35. Rasmussen, K.; Rauscher, H.; Mech, A.; Riego Sintes, J.; Gilliland, D.; Gonzalez, M.; Kearns, P.; Moss, K.; Visser, M.; Groenewold, M.; et al. Physico-chemical properties of manufactured nanomaterials-Characterisation and relevant methods. An outlook based on the OECD Testing Programme. *Regul. Toxicol. Pharm.* 2018, *92*, 8–28. [Google Scholar] [CrossRef] [PubMed]

36. Farré, M.; Sanchís, J.; Barceló, D. Analysis and assessment of the occurrence the fate and the behavior of nanomaterials in the environment. *Trac. Trends Anal. Chem.* 2011, *30*, 517–527. [Google Scholar] [CrossRef]

37. Azimi, S.S.; Kalbasi, M. A molecular dynamics simulation of Brownian motion of a nanoparticle in a nanofluid. *Nanoscale Microscale Thermophys. Eng.* 2017, *21*, 263–277. [Google Scholar] [CrossRef]

38. Phenrat, T.; Saleh, N.; Sirk, K.; Tilton, R.D.; Lowry, G.V. Aggregation and Sedimentation of Aqueous Nanoscale Zerovalent Iron Dispersions. *Environ. Sci. Technol.* 2007, *41*, 284–290. [Google Scholar] [CrossRef]

39. Pipan-Tkalec, Ž. Zinc bioaccumulation in a terrestrial invertebrate fed a diet treated with particulate ZnO or ZnCl2 solution. *Toxicology* 2010, *269*, 198–203. [Google Scholar] [CrossRef]

40. Wang, H.; Wick, R.L.; Xing, B. Toxicity of nanoparticulate and bulk ZnO, Al_2O_3 and TiO_2 to the nematode Caenorhabditis elegans. *Env. Pollut.* 2009, *157*, 1171–1177. [Google Scholar] [CrossRef]

41. Lin, D.; Xing, B. Root uptake and phytotoxicity of ZnO nanoparticles. *Env. Sci. Technol.* 2008, *42*, 5580–5585. [Google Scholar] [CrossRef]

42. Jemec, A.; Drobne, D.; Remškar, M.; Sepčić, K.; Tišler, T. Effects of ingested nano-sized titanium dioxide on terrestrial isopods (Porcellio scaber). *Environ. Toxicol. Chem. Setac* 2008, *27*, 1904–1914. [Google Scholar] [CrossRef]

43. Handy, R.D.; Von der Kammer, F.; Lead, J.R.; Hassellöv, M.; Owen, R.; Crane, M. The ecotoxicology and chemistry of manufactured nanoparticles. *Ecotoxicology* 2008, *17*, 287–314. [Google Scholar] [CrossRef]

44. Barreneche, C. Influence of nanoparticle morphology and its dispersion ability regarding thermal properties of water used as phase change material. *Appl. Therm. Eng.* 2018, *128*, 121–126. [Google Scholar] [CrossRef] [Green Version]

45. Geng, Y.A.N.; Dalhaimer, P.; Cai, S.; Tsai, R.; Tewari, M.; Minko, T.; Discher, D.E. Shape effects of filaments versus spherical particles in flow and drug delivery. *Nat. Nanotechnol.* 2007, *2*, 249–255. [Google Scholar] [CrossRef] [PubMed]

46. Champion, J.A.; Mitragotri, S. Role of target geometry in phagocytosis. *Proc. Natl. Acad. Sci. USA* 2006, *103*, 4930–4934. [Google Scholar] [CrossRef] [PubMed] [Green Version]

47. Badawy, A.M.E.; Luxton, T.P.; Silva, R.G.; Scheckel, K.G.; Suidan, M.T.; Tolaymat, T.M. Impact of environmental conditions (pH, ionic strength, and electrolyte type) on the surface charge and aggregation of silver nanoparticles suspensions. *Env. Sci. Technol.* 2010, *44*, 1260–1266. [Google Scholar] [CrossRef] [PubMed]

48. Fabrega, J.; Fawcett, S.R.; Renshaw, J.C.; Lead, J.R. Silver Nanoparticle Impact on Bacterial Growth: Effect of pH, Concentration, and Organic Matter. *Environ. Sci. Technol.* 2009, *43*, 7285–7290. [Google Scholar] [CrossRef]

49. Andreescu, D.; Bulbul, G.; Özel, R.E.; Hayat, A.; Sardesai, N.; Andreescu, S. Applications and implications of nanoceria reactivity: Measurement tools and environmental impact. *Environ. Sci. Nano* 2014, *1*, 445–458. [Google Scholar] [CrossRef]

50. Szymanski, C.J.; Munusamy, P.; Mihai, C.; Xie, Y.; Hu, D.; Gilles, M.K.; Tyliszczak, T.; Thevuthasan, S.; Baer, D.R.; Orr, G. Shifts in oxidation states of cerium oxide nanoparticles detected inside intact hydrated cells and organelles. *Biomaterials* 2015, *62*, 147–154. [Google Scholar] [CrossRef][Green Version]

51. Qi, L.; Sehgal, A.; Castaing, J.C.; Chapel, J.P.; Fresnais, J.; Berret, J.F.; Cousin, F. Redispersible Hybrid Nanopowders: Cerium Oxide Nanoparticle Complexes with Phosphonated-PEG Oligomers. *ACS Nano* 2008, *2*, 879–888. [Google Scholar] [CrossRef][Green Version]

52. Suzuki, T.; Kosacki, I.; Anderson, H.U.; Colomban, P. Electrical Conductivity and Lattice Defects in Nanocrystalline CeO_2 Thin Films. *J. Am. Ceram. Soc.* 2004, *84*, 2007–2014. [Google Scholar] [CrossRef]

53. Turner, S.; Lazar, S.; Freitag, B.; Egoavil, R.; Verbeeck, J.; Put, S.; Strauven, Y.; Van Tendeloo, G. High resolution mapping of surface reduction in ceria nanoparticles. *Nanoscale* 2011, *3*, 3385–3390. [Google Scholar] [CrossRef] [PubMed]

54. Paier, J.; Penschke, C.; Sauer, J. Oxygen defects and surface chemistry of ceria: Quantum chemical studies compared to experiment. *Chem Rev.* 2013, *113*, 3949–3985. [Google Scholar] [CrossRef]

55. Borm, P.; Klaessig, F.C.; Landry, T.D.; Moudgil, B.; Pauluhn, J.; Thomas, K.; Trottier, R.; Wood, S. Research strategies for safety evaluation of nanomaterials, part V: Role of dissolution in biological fate and effects of nanoscale particles. *Toxicol. Sci.* 2006, *90*, 23–32. [Google Scholar] [CrossRef] [PubMed][Green Version]

56. Misra, S.K.; Dybowska, A.; Berhanu, D.; Luoma, S.N.; Valsami-Jones, E. The complexity of nanoparticle dissolution and its importance in nanotoxicological studies. *Sci. Total Env.* 2012, *438*, 225–232. [Google Scholar] [CrossRef] [PubMed]

57. Avramescu, M.L.; Rasmussen, P.E.; Chenier, M.; Gardner, H.D. Influence of pH, particle size and crystal form on dissolution behaviour of engineered nanomaterials. *Env. Sci. Pollut. Res. Int.* 2017, *24*, 1553–1564. [Google Scholar] [CrossRef] [PubMed][Green Version]

58. Li, Y.; Liao, C.; Tjong, S.C. Synthetic Biodegradable Aliphatic Polyester Nanocomposites Reinforced with Nanohydroxyapatite and/or Graphene Oxide for Bone Tissue Engineering Applications. *Nanomaterials* 2019, *9*, 590. [Google Scholar] [CrossRef] [PubMed][Green Version]

59. Chen, C.-L.; Zuckermann, R.N.; DeYoreo, J.J. Surface-Directed Assembly of Sequence-Defined Synthetic Polymers into Networks of Hexagonally Patterned Nanoribbons with Controlled Functionalities. *ACS Nano* 2016, *10*, 5314–5320. [Google Scholar] [CrossRef]

60. Soppimath, K.S.; Aminabhavi, T.M.; Kulkarni, A.R.; Rudzinski, W.E. Biodegradable polymeric nanoparticles as drug delivery devices. *J. Control. Release* 2001, *70*, 1–20. [Google Scholar] [CrossRef]

61. Fathi, M.; Barar, J. Perspective highlights on biodegradable polymeric nanosystems for targeted therapy of solid tumors. *Bioimpacts* 2017, *7*, 49–57. [Google Scholar] [CrossRef][Green Version]

62. Wang, J.; Xia, H.; Zhang, Y.; Lu, H.; Kamat, R.; Dobrynin, A.V.; Cheng, J.; Lin, Y. Nucleation-Controlled Polymerization of Nanoparticles into Supramolecular Structures. *J. Am. Chem. Soc.* 2013, *135*, 11417–11420. [Google Scholar] [CrossRef]

63. Khan, S.; Ansari, A.A.; Rolfo, C.; Coelho, A.; Abdulla, M.; Al-Khayal, K.; Ahmad, R. Evaluation of in vitro cytotoxicity, biocompatibility, and changes in the expression of apoptosis regulatory proteins induced by cerium oxide nanocrystals. *Sci. Technol. Adv. Mater.* 2017, *18*, 364–373. [Google Scholar] [CrossRef] [PubMed][Green Version]

64. Gatoo, M.A.; Naseem, S.; Arfat, M.Y.; Mahmood Dar, A.; Qasim, K.; Zubair, S. Physicochemical properties of nanomaterials: Implication in associated toxic manifestations. *Biomed. Res. Int.* 2014, *2014*, 498420. [Google Scholar] [CrossRef] [PubMed]

65. Pujar, M.S.; Hunagund, S.M.; Desai, V.R.; Patil, S.; Sidarai, A.H. One-step synthesis and characterizations of cerium oxide nanoparticles in an ambient temperature via Co-precipitation method. In *AIP Conference Proceedings*; AIP Publishing: Melville, NY, USA, 2018; p. 050026. [Google Scholar]

66. Farahmandjou, M.; Zarinkamar, M.; Firoozabadi, T.P. Synthesis of Cerium Oxide (CeO_2) nanoparticles using simple CO-precipitation method. *Rev. Mex. De Física* 2016, *62*, 496–499. [Google Scholar]

67. Chen, P.-L.; Chen, I.W. Reactive Cerium(IV) Oxide Powders by the Homogeneous Precipitation Method. *J. Am. Ceram. Soc.* 1993, *76*, 1577–1583. [Google Scholar]

68. Chen, H.-I.; Chang, H.-Y. Synthesis of nanocrystalline cerium oxide particles by the precipitation method. *Ceram. Int.* 2005, *31*, 795–802. [Google Scholar] [CrossRef]

69. Ramachandran, M.; Subadevi, R.; Sivakumar, M. Role of pH on synthesis and characterization of cerium oxide (CeO_2) nano particles by modified co-precipitation method. *Vacuum* 2019, *161*, 220–224. [Google Scholar] [CrossRef]

70. Nanda, H.S. Preparation and Biocompatible Surface Modification of Redox Altered Cerium Oxide Nanoparticle Promising for Nanobiology and Medicine. *Bioengineering* 2016, *3*, 28. [Google Scholar] [CrossRef][Green Version]

71. Nethi, S.K.; Nanda, H.S.; Steele, T.W.; Patra, C.R. Functionalized nanoceria exhibit improved angiogenic properties. *J. Mater. Chem. B* 2017, *5*, 9371–9383. [Google Scholar] [CrossRef]

72. Corsi, F.; Caputo, F.; Traversa, E.; Ghibelli, L. Not Only Redox: The Multifaceted Activity of Cerium Oxide Nanoparticles in Cancer Prevention and Therapy. *Front. Oncol.* 2018, *8*, 309. [Google Scholar] [CrossRef][Green Version]

73. Gliga, A.R.; Edoff, K.; Caputo, F.; Källman, T.; Blom, H.; Karlsson, H.L.; Ghibelli, L.; Traversa, E.; Ceccatelli, S.; Fadeel, B. Cerium oxide nanoparticles inhibit differentiation of neural stem cells. *Sci. Rep.* 2017, *7*, 9284. [Google Scholar] [CrossRef]

74. Celardo, I.; De Nicola, M.; Mandoli, C.; Pedersen, J.Z.; Traversa, E.; Ghibelli, L. Ce(3)+ ions determine redox-dependent anti-apoptotic effect of cerium oxide nanoparticles. *ACS Nano* 2011, *5*, 4537–4549. [Google Scholar] [CrossRef]

75. Bumajdad, A.; Zaki, M.I.; Eastoe, J.; Pasupulety, L. Microemulsion-Based Synthesis of CeO2 Powders with High Surface Area and High-Temperature Stabilities. *Langmuir ACS J. Surf. Colloids* 2005, *20*, 11223–11233. [Google Scholar] [CrossRef] [PubMed]

76. Kockrick, E. Synthesis and catalytic properties of microemulsion-derived cerium oxide nanoparticles. *J. Solid State Chem.* 2008, *181*, 1614–1620. [Google Scholar] [CrossRef]

77. Richard, B.; Lemyre, J.L.; Ritcey, A.M. Nanoparticle Size Control in Microemulsion Synthesis. *Langmuir* 2017, *33*, 4748–4757. [Google Scholar] [CrossRef] [PubMed]

78. Patil, S.; Sandberg, A.; Heckert, E.; Self, W.; Seal, S. Protein adsorption and cellular uptake of cerium oxide nanoparticles as a function of zeta potential. *Biomaterials* 2007, *28*, 4600–4607. [Google Scholar] [CrossRef] [PubMed][Green Version]

79. Arya, A.; Sethy, N.K.; Das, M.; Singh, S.K.; Das, A.; Ujjain, S.K.; Sharma, R.K.; Sharma, M.; Bhargava, K. Cerium oxide nanoparticles prevent apoptosis in primary cortical culture by stabilizing mitochondrial membrane potential. *Free Radic Res.* 2014, *48*, 784–793. [Google Scholar] [CrossRef]

80. Masui, T. Characterization of Cerium(IV) Oxide Ultrafine Particles Prepared Using Reversed Micelles. *Chem. Mater.* 1997, *9*, 2197–2204. [Google Scholar] [CrossRef]

81. Cimini, A.; D'Angelo, B.; Das, S.; Gentile, R.; Benedetti, E.; Singh, V.; Monaco, A.M.; Santucci, S.; Seal, S. Antibody-conjugated PEGylated cerium oxide nanoparticles for specific targeting of Aβ aggregates modulate neuronal survival pathway. *Acta Biomater* 2012, *8*, 2056–2067. [Google Scholar] [CrossRef]

82. López, J.M.; Gilbank, A.L.; García, T.; Solsona, B.; Agouram, S.; Torrente-Murciano, L. The prevalence of surface oxygen vacancies over the mobility of bulk oxygen in nanostructured ceria for the total toluene oxidation. *Appl. Catal. B Environ.* 2015, *174*, 403–412. [Google Scholar] [CrossRef][Green Version]

83. Tok, A.I.Y. Hydrothermal synthesis of CeO_2 nano-particles. *J. Mater. Process. Technol.* 2007, *190*, 217–222. [Google Scholar] [CrossRef]

84. Trenque, I.; Magnano, G.C.; Bolzinger, M.A.; Roiban, L.; Chaput, F.; Pitault, I.; Briançon, S.; Devers, T.; Masenelli-Varlot, K.; Bugnet, M.; et al. Shape-selective synthesis of nanoceria for degradation of paraoxon as a chemical warfare simulant. *Phys. Chem. Chem. Phys.* 2019, *21*, 5455–5465. [Google Scholar] [CrossRef]

85. Zhang, Y.; Wu, X.; Hou, C.; Shang, K.; Yang, K.; Tian, Z.; Pei, Z.; Qu, Y.; Pei, Y. Dual-responsive dithio-polydopamine coated porous CeO_2 nanorods for targeted and synergistic drug delivery. *Int. J. Nanomed.* 2018, *13*, 2161–2173. [Google Scholar] [CrossRef][Green Version]

86. Singh, S.; Ly, A.; Das, S.; Sakthivel, T.S.; Barkam, S.; Seal, S. Cerium oxide nanoparticles at the nano-bio interface: Size-dependent cellular uptake. *Artif. Cells Nanomed. Biotechnol.* 2018, *43* (Suppl. 3), S956–S963. [Google Scholar] [CrossRef] [PubMed][Green Version]

87. Kannan, S.K.; Sundrarajan, A.M. Approach for the Synthesis of a Cerium Oxide Nanoparticle: Characterization and Antibacterial Activity. *Int. J. Nanosci.* 2014, *13*, 1450018. [Google Scholar] [CrossRef]

88. Sangsefidi, F.S.; Nejati, M.; Verdi, J.; Salavati-Niasari, M. Green synthesis and characterization of cerium oxide nanostructures in the presence carbohydrate sugars as a capping agent and investigation of their cytotoxicity on the mesenchymal stem cell. *J. Clean. Prod.* 2017, *156*, 741–749. [Google Scholar] [CrossRef][Green Version]

89. Thovhogi, N.; Diallo, A.; Gurib-Fakim, A.; Maaza, M. Nanoparticles green synthesis by Hibiscus Sabdariffa flower extract: Main physical properties. *J. Alloy. Compd.* 2015, *647*, 392–396. [Google Scholar] [CrossRef]

90. Kargar, H.; Ghazavi, H.; Darroudi, M. Size-controlled and bio-directed synthesis of ceria nanopowders and their in vitro cytotoxicity effects. *Ceram. Int.* 2015, *41*, 4123–4128. [Google Scholar] [CrossRef]

91. Zhang, H.; He, X.; Zhang, Z.; Zhang, P.; Li, Y.; Ma, Y.; Kuang, Y.; Zhao, Y.; Chai, Z. Nano-CeO_2 exhibits adverse effects at environmental relevant concentrations. *Env. Sci. Technol.* 2011, *45*, 3725–3730. [Google Scholar] [CrossRef]

92. Soren, S.; Jena, S.R.; Samanta, L.; Parhi, P. Antioxidant Potential and Toxicity Study of the Cerium Oxide Nanoparticles Synthesized by Microwave-Mediated Synthesis. *Appl. Biochem. Biotechnol.* 2015, *177*, 148–161. [Google Scholar] [CrossRef]

93. Machmudah, S.; Winardi, S.; Kanda, H.; Goto, M. Synthesis of Ceria Zirconia Oxides using Solvothermal Treatment. In *MATEC Web of Conferences*; EDP Sciences: Les Ulis, France, 2018. [Google Scholar]

94. Kar, S.; Direct, S.; Chemistry, C.; Patel, C.; Temperature, S. Synthesis of Valence State Engineered UltraSmall Ceria Nanoparticles Investigation on the Role of Ethylenediamine as a Capping Agent. *J. Phys.* 2009, *113*, 4862–4867. [Google Scholar]

95. Yu, T.; Joo, J.; Park, Y.I.; Hyeon, T. Large-scale nonhydrolytic sol-gel synthesis of uniform-sized ceria nanocrystals with spherical, wire, and tadpole shapes. *Angew. Chem. Int. Ed. Engl.* 2005, *44*, 7411–7414. [Google Scholar] [CrossRef]

96. Darroudi, M. Green synthesis and evaluation of metabolic activity of starch mediated nanoceria. *Ceram. Int.* 2014, *40*, 2041–2045. [Google Scholar] [CrossRef]

97. Elahi, B.; Mirzaee, M.; Darroudi, M.; Sadri, K.; Oskuee, R.K. Bio-based synthesis of Nano-Ceria and evaluation of its bio-distribution and biological properties. *Colloids Surf. B Biointerfaces* 2019, *181*, 830–836. [Google Scholar] [CrossRef] [PubMed]

98. Gnanam, S.; Rajendran, V. Synthesis of CeO_2 or α–Mn_2O_3 nanoparticles via sol-gel process and their optical properties. *J. Solgel Sci. Technol.* 2011, *58*, 62–69. [Google Scholar] [CrossRef]

99. Nourmohammadi, E. Cytotoxic activity of greener synthesis of cerium oxide nanoparticles using carrageenan towards a WEHI 164 cancer cell line. *Ceram. Int.* 2018, *44*, 19570–19575. [Google Scholar] [CrossRef]

100. Yadav, T.P.; Srivastava, O.N. Synthesis of nanocrystalline cerium oxide by high energy ball milling. *Ceram. Int.* 2012, *38*, 5783–5789. [Google Scholar] [CrossRef]

101. Li, X. Nanoscale Structural and Mechanical Characterization of a Natural Nanocomposite Material: The Shell of Red Abalone. *Nano Lett.* 2004, *4*, 613–617. [Google Scholar] [CrossRef]

102. Lu, J.; Fang, Z.Z. Synthesis and Characterization of Nanoscaled Cerium (IV) Oxide via a Solid-State Mechanochemical Method. *J. Am. Ceram. Soc.* 2006, *89*, 842–847. [Google Scholar] [CrossRef]

103. He, H. Preparation and Dispersion of Nanosize Ceria in High Electrolyte Slurry by Ball-Milling. *Integr. Ferroelectr.* 2015, *161*, 36–44. [Google Scholar] [CrossRef]

104. Khorrami, M.B.; Sadeghnia, H.R.; Pasdar, A.; Ghayour-Mobarhan, M.; Riahi-Zanjani, B.; Hashemzadeh, A.; Zare, M.; Darroudi, M. Antioxidant and toxicity studies of biosynthesized cerium oxide nanoparticles in rats. *Int. J. Nanomed.* 2019, *14*, 2915–2926. [Google Scholar] [CrossRef][Green Version]

105. Vassie, J.A.; Whitelock John, M.; Lord Megan, S. Targeted Delivery and Redox Activity of Folic Acid-Functionalized Nanoceria in Tumor Cells. *Mol. Pharm.* 2018, *15*, 994–1004. [Google Scholar] [CrossRef]

106. Grillone, A.; Li, T.; Battaglini, M.; Scarpellini, A.; Prato, M.; Takeoka, S.; Ciofani, G. Preparation, Characterization, and Preliminary In Vitro Testing of Nanoceria-Loaded Liposomes. *Nanomaterials* 2017, *7*, 276. [Google Scholar] [CrossRef] [PubMed][Green Version]

107. El Shaer, S.S.; Salaheldin, T.A.; Saied, N.M.; Abdelazim, S.M. In vivo ameliorative effect of cerium oxide nanoparticles in isoproterenol-induced cardiac toxicity. *Exp. Toxicol. Pathol.* 2017, *69*, 435–441. [Google Scholar] [CrossRef] [PubMed]

108. Shanmugam, R.; Naik, P. Synthesis and biomedical applications of Cerium oxide nanoparticles – A Review. *Biotechnol. Rep.* 2017, *17*. [Google Scholar] [CrossRef]

109. Reed, K. Exploring the properties and applications of nanoceria: Is there still plenty of room at the bottom? *Environ. Sci. Nano* 2014, *1*, 390–405. [Google Scholar] [CrossRef][Green Version]

110. Asati, A.; Santra, S.; Kaittanis, C.; Perez, J.M. Surface-charge-dependent cell localization and cytotoxicity of cerium oxide nanoparticles. *ACS Nano* 2010, *4*, 5321–5331. [Google Scholar] [CrossRef][Green Version]

111. Gagnon, J.; Fromm, K. Toxicity and Protective Effects of Cerium Oxide Nanoparticles (Nanoceria) Depending on Their Preparation Method, Particle Size, Cell Type, and Exposure Route. *Eur. J. Inorg. Chem.* 2015, *2015*, 4510–4517. [Google Scholar] [CrossRef][Green Version]

112. Kumar, A.; Das, S.; Munusamy, P.; Self, W.; Baer, D.R.; Sayle, D.C.; Seal, S. Behavior of nanoceria in biologically-relevant environments. *Environ. Sci. Nano* 2014, *1*, 516–532. [Google Scholar] [CrossRef]

11 POLYCAPROLACTONE COMPOSITE NANOFIBROUS MATERIAL: AN ALTERNATIVE TO RESTRICTED ACCESS MEDIA FOR DIRECT EXTRACTION AND SEPARATION OF NON-STEROIDAL ANTI-INFLAMMATORY DRUGS

1. INTRODUCTION: AN OVERVIEW

In recent years, the development of innovative materials for efficient extraction and separation of pharmaceutical compounds has gained substantial attention in analytical chemistry. One such promising avenue involves the utilization of polycaprolactone (PCL) composite micro/nanofibrous materials as an alternative to traditional restricted access media. This research explores the potential of employing PCL-based materials for direct extraction and separation of non-steroidal anti-inflammatory drugs (NSAIDs) from human serum, employing a column-switching chromatography approach.

Non-steroidal anti-inflammatory drugs (NSAID) are widely used in pain, osteoarthritis, and rheumatoid arthritis treatment. A mechanism of their action lies in the inhibition of cyclo-oxygenase enzyme, an enzyme serving prostaglandin biosynthesis. These drugs play a crucial role in the suppression of the inflammatory response of an organism. The reduction in prostaglandin levels results in antipyretics, analgesics, and anti-inflammatory activity of NSAID. The most widely used NSAIDs in clinical practice are acetylsalicylic acid, paracetamol, indomethacin, ibuprofen, ketoprofen, sodium diclofenac, flurbiprofen, mefenamic acid, piroxicam, and nabumeton. Monitoring of NSAID in body fluid is essential for toxicological and pharmacokinetics studies. Their determination as a part of rational pharmacotherapy that aims to reduce risks and achieve disease treatment goals is less common. The levels are specifically monitored either in patients with an impaired renal function when the insufficient excretion occurs or when non-compliance is suspected [1,2]. The pharmaceutical industry is another sector where NSAID are determined in pharmaceutical formulations to ensure safety and effectiveness of drugs [3].

Numerous methods for NSAID determination in biological fluids including gas chromatography [4,5,6], capillary electrophoresis [7,8,9], and liquid chromatography [10,11] were developed over the past years. Although the analytical instruments were designed to provide increasingly faster and more sensitive analysis, the treatment of the biological matrixes prior to the instrumental analysis remains the bottleneck in the drug level monitoring [12]. The NSAID are usually determined in plasma, where the direct analysis of untreated matrix can lead to lower sensitivity, unprecise results, and in the worst case to irreversible damage of the chromatography column caused by protein precipitation. Thus, the sample pretreatment step cannot be omitted. This step is mainly based on time-consuming, laborious, and error-prone methods, such as protein precipitation, solid-phase extraction, and liquid–liquid extraction that can be potentially hazardous for laboratory staff [13]. Several approaches have been developed to overcome these shortcomings. The emphasis has been primarily placed on higher speed, automation, reduced consumption of organic solvents, smaller sample volumes, and increased selectivity [13,14]. Restricted access media (RAM) represent one of the applicable methods [15].

RAM sorbents separate the low molecular analytes from macromolecular interferences mostly via size exclusion effect. This analyte extraction and macromolecules removal occurs simultaneously after the injection of the untreated biological fluids [15]. The sample preparation is simplified and the analysis time is shortened while maintaining the desired extraction efficiency and avoiding column clogging [16]. The extraction mechanism of the original RAM is primarily based on their two surfaces. The outer surface is covered by hydrophilic groups avoiding access of macromolecules in the material, and the hydrophobic functional groups on the inner pore surface that are responsible for analyte retention. Additionally, the pores in the sorbent material contribute to the matrix removal by the size-exclusion mechanism [17].

The potential of RAM can be adequately utilized in on-line liquid chromatography extractions. The RAM cartridges are most often coupled with a column-switching chromatography system [16]. This setup was reported several times for NSAID analysis [18,19]. The main component of this system is a double position six-port selection valve that redirects the mobile phases between both extraction cartridge and analytical column. The extraction begins after the sample is injected in the cartridge. The undesired matrix components are removed from the analytes using the washing mobile phase. The column-switching valve then changes its position to complete the extraction step, and the analytes are eluted in the analytical column where they are separated using the separation mobile phase.

The washing mobile is meanwhile transferred in the waste container. This process can be carried out either in a straight-flush or in a back-flush mode differing in the way the analytes are eluted from the extraction cartridge. They are injected and eluted from the extraction cartridge in the same direction as washing mobile phase flow in the former approach while the mobile phase elutes the analytes in the opposite direction to the sample injection in the latter [20].

Advanced approaches combine the RAM concept with the molecularly imprinted polymers, magnetism, or supramolecular solvents to enhance selectivity and specificity of the analyte extraction [21,22,23]. The use of nanomaterials is also included in these developments. For example, application of the restricted access nanotubes was reported [24]. We demonstrated first nanofibers as an alternative to RAM elsewhere [25]. We confirmed that the composite material produced from poly-ε-caprolactone exhibited satisfactory efficiency in extraction of parabens from human serum and bovine milk. The idea of using nanofibers as an alternative to RAM emerged as a result of their characteristics. The nanofibers have a large surface area to volume ratio enabling them to capture considerable quantities of low molecular weight analytes. In contrast, we speculated that the macromolecules are not retained because of the curvature of the fibers. The macromolecules appeared not to be flexible enough to be attached at multiple points to the highly curved nanofiber for their sufficient retention. This is not an issue with the low-molecular weight analytes [25]. Additionally, the large spaces between the fibers support the macromolecule passage through the sorbent to the waste.

We described the advantages of material combining poly-ε-caprolactone micro- and nanofibers (micro/nano PCL) for the on-line extraction of analytes from milk and serum matrixes in our previous study [25]. We confirmed the protein removal capability, good mechanical stability in the extraction/HPLC system, and re-usability of this sorbent for more than 300 analyses. The current study extends exploration of the potential of micro/nano PCL material in therapeutic drug monitoring and a method for NSAID determination in human serum samples was developed. This method was validated according to the International Council for Harmonization guideline and used for the determination of diclofenac content in a real patient serum.

2. MATERIALS AND METHODS

2.1. Reagents and Materials

Ketoprofen (≥98.0%, KET), naproxen (≥98.0%, NAP), sodium diclofenac (≥98.5%, DCF), and indomethacin (≥98.5%, IND) were purchased from Sigma-Aldrich (Darmstadt, Germany) and used as model analytes. Phosphoric acid (Honeywell, Morris Plains, NJ, USA), acetonitrile (ACN), and methanol (MeOH) (VWR, Paris, France) served as mobile phase modifiers. The standard solutions were acidified by acetic acid from Penta (Prague, Czech Republic). The proteinaceous matrix was prepared by dissolution of lyophilized human serum Lyo hum N (Erba Lachema, Brno, Czech Republic). Water was purified by Millipore Milli-Q Direct Water Purification System from Merck (Darmstadt, Germany). The composite poly-ε-caprolactone sorbent was fabricated at the Technical University of Liberec from poly-ε-caprolactone (Mw 43.000) provided by Polysciences (Heddesheim, Germany), and chloroform (≥97%) and ethanol (99.97%) bought from Penta Chemicals (Pardubice, Czech Republic).

2.2. Instrumentation

2.2.1. Column-Switching Chromatography System

A Shimadzu Prominence instrument (Shimadzu Corporation, Kyoto, Japan) equipped with three LC-20AD Prominence pumps, a DGU-20A Prominence on-line degasser, a SIL 20AC Prominence auto-sampler, a CTO-20AC Prominence column oven, an SPD-M20A Prominence UV/VIS photodiode array detector, and a CBM-20A Prominence system controller was used for the separation and analysis. Switching between the analytical column and the extraction cartridge was provided via the six-port high-pressure flow line switching valve FCV-12AH. LC Solution software (version 5.97, Shimadzu Corporation, Kyoto, Japan) controlled the instrument. The separation of NSAID extracted from the human serum was carried out on the Ascentis Express RP-Amide (100 × 4.6 mm, 5 μm) analytical column combined with the RP-Amide (5 × 4.6 mm, 2.7 μm) guard column (Sigma-Aldrich, Darmstadt, Germany). The system configuration in extraction and separation mode is depicted in the Figure 1.

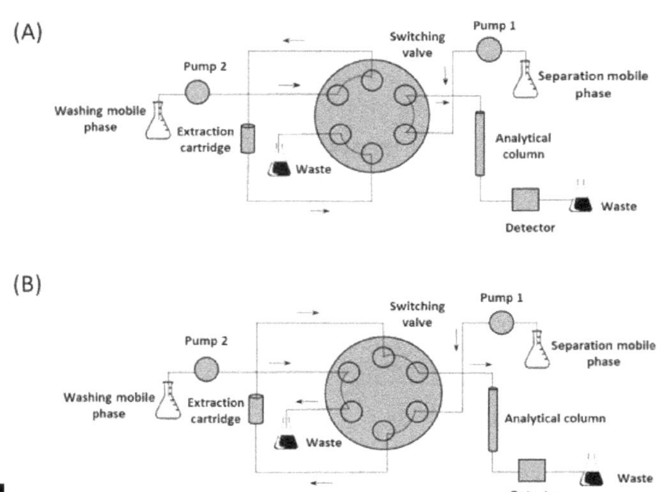

Figure 1. Column-switching chromatographic system in (A) on-line extraction and (B) on-line extraction-separation mode.

2.2.2. Meltblown and Electrospun Fibers

Poly-ε-caprolactone nanofibers and microfibers were fabricated using a process described in Supporting Information and was firstly reported elsewhere [26,27]. The meltblown equipment (Laboratory equipment J&M Laboratories, Ashland, OH, USA) in combination with DC electro-spinning lab-made system based on a multi-needle spinner was used for the preparation of the fibers. The production equipment consisting of both electro-spinning and meltblown systems is depicted in Supporting Information section. A scanning electron microscope VEGA 3 (Tescan, Brno, Czech Republic) was used for the imaging of the fibrous sorbent.

2.3. Preparation of Extraction Cartridge

A commercially available PEEK cartridge 4.6 × 10 mm (Merck, Darmstadt, Germany) was manually filled with 44 mg of the composite micro/nanofibrous PCL. The cartridge in a plastic holder was then placed in the chromatographic system and washed with the mobile phase containing 10% ACN in 0.085% (*v*/*v*) aqueous phosphoric acid for 15 min and neat ACN was then injected six times. This washing was carried out to ensure complete removal of impurities that could remain in fibers after their production.

2.4. Preparation of Standard and Matrix Solutions

The stock KET, NAP, DCF, and IND solutions were prepared at a concentration of 1 mg mL^{-1} by dissolving the individual standard substances in MeOH. These solutions were stored in the dark at °4C. The mix stock solution containing 0.2 mg mL^{-1} NSAID was prepared by mixing the stock solutions. An appropriate volume of the mix stock solution was diluted with a water–acetonitrile–acetic acid solution (2/48/50, *v*/*v*/*v*) to obtain standards solutions at seven concentration levels of 50; 100; 500; 1000; 5000; 10,000; and 20,000 ng mL^{-1}. These solutions were used for the plotting the calibration curve. Commercially available lyophilized human serum was reconstituted according to manufacturer instructions and then diluted ten times with 20% aqueous ACN. This serum solution was spiked with the mix stock solution to obtain matrix solutions at concentrations of 50; 100; 500; 1000; 5000; 10,000; and 20,000 ng mL^{-1} and these solutions were centrifuged at 14,000× *g* rpm (21,578× *g*) for 15 min. The supernatant was then injected in the chromatographic system and the matrix calibration curve plotted. Matrix solutions were prepared fresh daily.

2.5. Real Sample

A real human serum was obtained from a patient administered with 250 mL continual intravenous infusion containing 75 mg of sodium diclofenac. Blood sampling was carried out in the Department of Clinical Biochemistry and Diagnostics of the

University Hospital Hradec Králové. This real life serum was handled the same way as the matrix working solutions. First, it was diluted 10 times with 20% aqueous ACN and then centrifuged at 14,000× *g* rpm (21,578× *g*) for 15 min.

2.6. Analytical Method

The analytical run started with the extraction step when 100 μL sample, standard or matrix solution, was injected in the extraction cartridge filled with micro/nano PCL. The NSAID were captured by the sorbent while the proteins and other potentially interfering macromolecular substances were removed by the washing with the mobile phase composed of 10% ACN in 0.085% aqueous H_3PO_4. Simultaneously, the analytical column was conditioned using the 30% ACN in 0.085% aqueous H_3PO_4. The column-switching valve redirected this separation mobile phase after 1.5 min in the extraction cartridge. Hereby, the analytes were eluted in the analytical column and separated using the ACN gradient increasing the ACN percentage to 45% in 2.5 min and this mobile phase was pumped through the column for another 1.5 min. Then, the percentage was ramped to 55% in 4 min and to 75% in 0.5 min. Finally, the ACN percentage decreased to the initial concentration in 1 min at which the analytical column was re-equilibrated for 3 min. The flow rate of both mobile phases was held on 1 mL min^{-1}. The total analysis time was 15 min. The separated analytes were detected using the diode array detector. KET, DCF, and IND were monitored at 270 nm while NAP at 232 nm. The column was held at 20 °C since the micro/nano PCL dissolve in ACN at higher temperatures [26].

3. RESULTS AND DISCUSSION

3.1. Optimization of Chromatographic Separation

We tested ACN and MeOH as organic components of the mobile phase with ACN produced peaks with a better symmetry. The NSAID drugs are weak acids. Therefore, 0.085% aqueous H_3PO_4 solution pH 2.6 was used as the aqueous part of the mobile phase to increase the retention in the column. We tested two analytical columns Ascentis Express RP-Amide (4.6 × 100 mm, 5 μm) and Ascentis Express F4.6 × 100) 5 mm, 5 μm) for the separations. Both columns use the core-shell particles technology improving the efficiency and reducing the flow resistance in column switching system. However, they are more susceptible to impurities from proteinaceous samples resulting in an increase in back pressure and eventually column clogging. Therefore, the Ascentis Express (4.6 × 5 mm, 2.7 μm) guard columns packed with the same stationary phase were inserted in the system to protect the analytical column. The Ascentis Express RP-Amide analytical column enabled a better separation of the analytes. Ascentis Express F5 stationary phase was not well suited because the extensive peak broadening and coelution of ketoprofen-naproxen and sodium diclofenac-indomethacin pairs as presented in Figure 2.

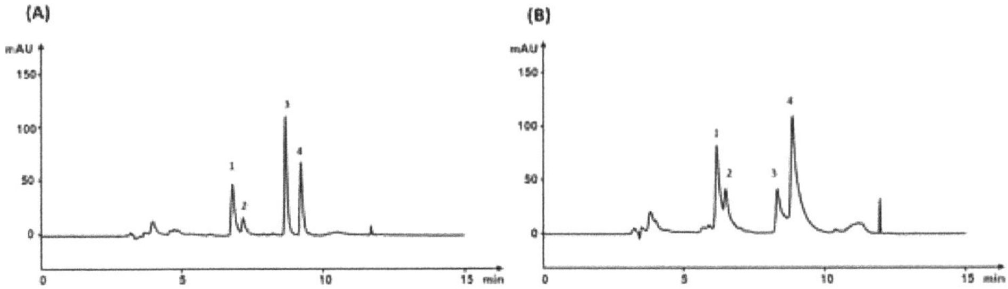

Figure 2. Chromatograms of standard solution spiked with 5000 ng mL^{-1} NSAID separated on (A) Ascentis Express RP-Amide column and (B) Ascentis Express F5 column at a wavelength 270 nm. 1—ketoprofen, 2—naproxen, 3—sodium diclofenac, 4—indomethacin.

3.2. Optimization of On-Line Extraction

We reported elsewhere [25] that the removal of proteins from nanofibrous sorbent using an aqueous mobile phase occurs within the first minute. This finding was confirmed by monitoring elution at a wavelength of 280 nm that is the absorption maximum for proteins. Human serum contains in addition to proteins also more lipophilic ballast substances, such as, for example, lipoproteins and vitamins, that are difficult to remove from the system since aqueous washing is insufficient. In contrast, a higher concentration of organic solvents in the washing mobile phase and excessive washing lead to undesirable losses of desired analytes. Therefore, the composition of the washing mobile phase and the duration of extraction step had to be optimized to remove most of the lipophilic ballast molecules without any loss of the analytes. Removal of these compounds using the mobile phases containing ACN concentration of 5%, 10%, 15%, and 20% were tested. The recovery of adsorbed analytes on nanofibers decreased with the rising ACN concentration. However, the effect of this process on the removal of ballast substances was negligible. This is why we used a mixture comprising 10% ACN in 0.085% aqueous phosphoric acid. This washing mobile phase was the best compromise between analyte loss and ballast removal. The duration of the extraction step was studied applying the similar approach. The peak of matrix impurities was observed in the chromatogram at 270 nm after 1 min washing. Extension of the washing time by 30 s resulted in a reduction in the area of the matrix ballast peaks without reducing the peak areas of analytes. No further improvement was observed after further increasing the extraction time to 2 min. Thus, the extraction time was finally held at 1.5 min. Chromatogram of standard solution and spiked serum under the optimized conditions of extraction and separation is shown in Figure 3.

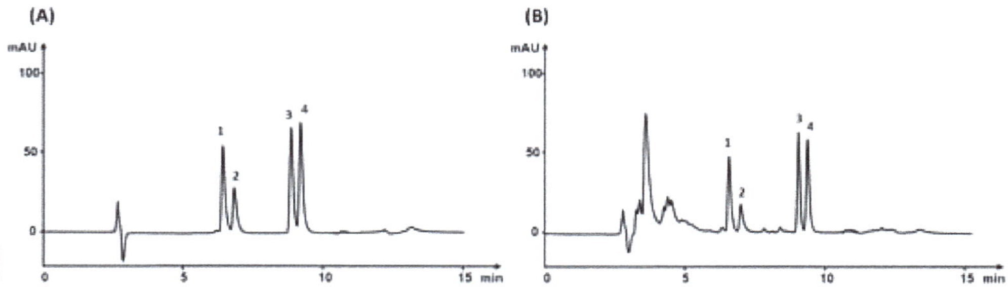

Figure 3. Chromatograms of (A) standard solution and (B) matrix solution spiked with 1000 ng mL^{-1} NSAID at a wavelength 270 nm. 1—ketoprofen, 2—naproxen, 3—sodium diclofenac, 4—indomethacin.

3.3. Extraction Efficiency

The analyte recovery from the human serum matrix was compared to the recovery obtained using extraction from standard solutions and expressed as a percentage value considering the standard solution being 100%. Figure 4 demonstrates that at levels up to 100 ng mL^{-1}, the analytes are extracted from the serum with an efficiency comparable to that found for the standard solutions. KET is an exception since its recovery exceeded 100% at almost all concentration levels. It can be caused by coelution with a matrix interference that increased the total peak area. This finding reduces the accuracy of the method for KET. However, the RSD values ranging from 1.16% to 0.017% and the precision is not affected.

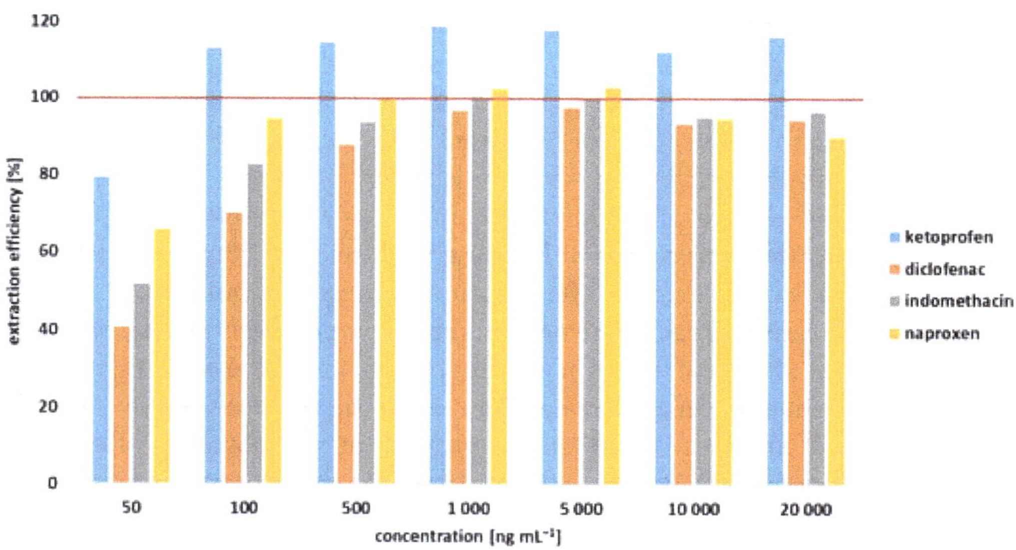

Figure 4. Extraction efficiency of micro/nano PCL sorbent expressed as a percentual recovery of NSAID in human serum.

3.4. Optimization of Injected Sample Volume

Different injection volumes were tested to achieve better sensitivity using UV detection and to simultaneously determine the protein removing capacity of the extraction cartridge. The proteins were completely removed in less than 1 min after the injection volumes 10, 25, and 50 µL. The cleaning of up of 100 µL centrifuged serum was achieved in 1.5 min. A further increase in the injection volume and simultaneous extension of the washing phase resulted in larger analyte loss. Therefore, 100 µL serum was designated as the maximal injection volume for the 10 × 4.6 mm i.d. extraction cartridge.

3.5. Validation

The method was validated with respect to the linearity, precision, accuracy, selectivity, and sensitivity following the ICH Q2 R1 guideline to evaluate the reliability of the results [28]. The validation parameters were chosen to primarily confirm micro/nanofibrous PCL applicability for the NSAID extraction since this sorbent is not standardized and commonly used. The ICH Q2 R1 protocol meets these requirements better than the M10 guideline that is usually used for the validation of bioanalytical methods. The chromatographic system suitability test was carried out confirm the suitability of the HPLC instrument for NSAIDs analysis. Mean values and standard deviations of the retention time, peak capacity, symmetry factor, resolution, and repeatability of the analytical run were calculated from results of six injections of standard solutions and evaluated according to the European Pharmacopoeia recommendations. The results are presented in the Table 1.

Table 1. HPLC system suitability parameters. Sample injection was performed in six replicates.

Retention time [min]	6.41	6.82	9.18	10.49
Retention time repeatability, RSD [%]	0.03	0.03	0.02	0.02
Repeatability of peak areas, RSD [%]	0.85	0.49	0.21	0.56
Peak resolution	1.82	10.32	5.99	
Peak symmetry	1.80	1.63	2.00	1.60
Peak capacity	14.03	13.18	21.65	18.80

Linearity, Accuracy, Precision, and Limits of Detection and Quantification

All samples were measured in triplicate using the optimized conditions. The calibration curves were established for standard and serum matrix solutions at concentration levels of 50; 100; 500; 1000; 2000; 10,000; and 20,000 ng mL^{-1}. The linear relationship between the NSAID quantity and peak area was confirmed for KET, DCF, and IND in the concentration range 50 to 20,000 ng mL^{-1}. The calibration curve for NAP was linear for a range 50 to 10,000 ng mL^{-1}. The correlation coefficient for each drug was exceeded 0.996.

The accuracy was determined via the recovery study carried out for all the drugs in the human serum matrix. The recovery was calculated as a comparison of peak areas of NSAID in standard and serum matrix solutions at three concentration levels of 100; 1000 and 10,000 ng mL^{-1} measured in triplicate. The results of recovery also determined the extraction efficiency for each drug.

The serum solutions were measured six times at three concentration levels of 100; 1000; and 10,000 ng mL^{-1} for intra-day precision determination. The results expressed as RSD (%) were determined in a range of %7.89 to %10.24 for a concentration level of 100 ng mL^{-1}, 0.17% to 0.90% for a concentration level of 1000 ng mL^{-1}, and 0.09% to 0.46% for a concentration level of 10,000 ng mL^{-1}. The inter-day precision was calculated for three consecutive measurements of matrix spiked with 1000 ng mL^{-1} NSAID in three days. The results expressed as RSD (%) were in a value range of 3.62 and %7.88. The lowest concentration of the calibration curve equaled to 50 ng mL^{-1} (10σ) was established as a limit of quantification (LOQ). The limit of detection 15 ng mL^{-1} (LOD) was calculated from the LOQ value as a tree-folds (3σ) variation. The results of the validation are summarized in the Table 2. These results are comparable to those obtained using other methods described for diclofenac determination, including column-switching [18], protein precipitation [29], and SPE via commercial sorbent [30] applied as sample preparation procedures.

Table 2. HPLC method validation results.

Standard calibration range [ng mL-1] 1		50–20,000	50–10,000	50–20,000	50–20,000
Correlation coefficient		0.9997	0.9996	0.9999	0.9998
Matrix calibration range [ng mL-1] 2		50–20,000 1	50–10,000 2	50–20,000 1	50–20,000 1
Correlation coefficient		0.9999	0.9960	0.9999	0.9999
Intra-day precision [%] 3	c1	8.54	7.89	10.24	8.37
	c2	0.60	0.90	0.38	0.17
	c3	0.46	0.29	0.09	0.21
Inter-day precision [%] 4		6.79	3.62	7.44	7.88
Accuracy [%] 5	c1	112.91 ± 4.50	94.31 ± 4.09	70.07 ± 6.20	82.76 ± 6.96
	c2	118.66 ± 0.84	102.28 ± 0.46	96.42 ± 0.15	100.06 ± 0.22
	c3	111.87 ± 0.70	94.24 ± 0.23	93.12 ± 0.83	94.81 ± 0.20
LOD [ng mL-1] 6		2.49	24.40	3.58	26.57
LOQ [ng mL-1] 7		8.29	81.34	11.92	88.57

3.6. Reusability of the Extraction Cartridge

The manually packed extraction cartridge was used during the entire experiments corresponding to the analysis of more than 100 serum samples. Neither extraction efficiency nor the back pressure of PCL composite sorbent had changed during these experiments. Thus, our composite micro/nanofibrous sorbent is better suited for the extensive use in high-pressure liquid chromatography systems than nanofibers reported elsewhere [31]. The microfibers represent a stable scaffold resistant to the high pressure while, simultaneously, the large surface area to volume ratio of the nanofibers contributes to the high extraction efficiency. Moreover, the cotton-like texture of micro/nano composite material is easier to manually fill into the cartridge. The durability of our cartridge paralleled our previous study [20] where we applied 200 injections of 200 μL serum in commercial RAM LiChrospher RP-18 ADS column without observing a decrease in extraction efficiency.

3.7. Analysis of Real Sample

The real patient serum samples after continual intravenous infusion containing 75 mg sodium diclofenac were handled as a serum matrix and the content of diclofenac determined using our method. The diclofenac peak was recognized based on its retention time and UV spectrum. The calculated concentration of diclofenac was 32.57 ng mL^{-1} in ten times diluted serum corresponding to 325.7 ng mL^{-1} in the original patient serum. This result confirmed that the extraction using nanofibrous sorbent is enough sensitive to handle the real samples. The chromatogram of the patient serum is shown in Figure 5.

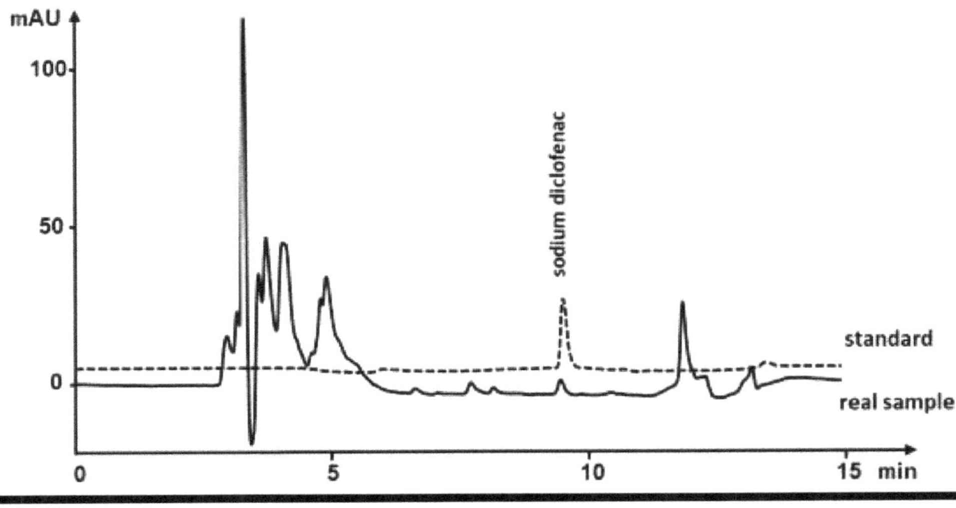

Figure 5. Standard solution spiked with 50 ng mL^{-1} diclofenac and real patient human serum containing the same drug. The concentration of diclofenac in real sample was equal to 32.57 ng mL^{-1} in ten times diluted serum.

4. CONCLUSIONS

Our study built on previous extraction experiments with micro/nano PCL that had promising properties enabling the direct extraction of analytes from proteinaceous matrix. We tested the micro/nano PCL fibers as the extraction sorbent for on-line solid-phase extraction of NSAID. After the injection of 100 μL human serum matrix, the sorbent enabled removal of the proteins and the majority of other macromolecular interferences within 1.5 min. The analyte recovery from the matrix was comparable to that obtained with the extraction from solutions of standards. This demonstrated that micro/nano PCL is a promising sorbent for KET, NAP, DCF, and IND even from the proteinaceous matrixes. The applicability of micro/nano PCL sorbent for the therapeutic drug monitoring was explored using a real sample containing diclofenac. The extraction provided good sensitivity even at low concentrations typical of biological samples and purified them sufficiently. Based on our results, we expect that the nano/micro fibrous PCL sorbent has a potential to improve the bioanalytical methods for sample pretreatment in terms of higher effectiveness at a lower costs compared to commercial RAM. This makes it a valuable tool for the on-line extraction/ chromatography methods developed for the therapeutic drug monitoring enabling fast, safe, and precise analysis.

5. CONCISE SUMMARY

Application of the poly-ε-caprolactone composite sorbent consisting of the micro- and nanometer fibers for the on-line extraction of non-steroidal anti-inflammatory drugs from a biological matrix has been introduced. A 100 μL human serum sample spiked with ketoprofen, naproxen, sodium diclofenac, and indomethacin was directly injected in the extraction cartridge filled with the poly-ε-caprolactone composite sorbent. This cartridge was coupled with a chromatographic instrument via a six-port switching valve allowing the analyte extraction and separation within a single analytical run. The 1.5 min long extraction step isolated the analytes from the proteinaceous matrix was followed by their 13 min HPLC separation using Ascentis Express RP-Amide (100 × 4.6 mm, 5 μm) column. The recovery of all analytes from human serum tested at three concentration levels ranged from 70.1% to 118.7%. The matrix calibrations were carried out in the range 50 to 20,000 ng mL^{-1} with correlation coefficients exceeding 0.996. The detection limit was 15 ng mL^{-1}, and the limit of quantification corresponded to 50 ng mL^{-1}. The developed method was validated and successfully applied for the sodium diclofenac determination in real patient serum. Our study confirmed the ability of the poly-ε-caprolactone composite sorbent to remove the proteins from the biological matrix, thus serving as an alternative to the application of restricted-access media.

REFERENCES

1. Sun, Y.; Takaba, K.; Kido, H.; Nakashima, M.N.; Nakashima, K. Simultaneous determination of arylpropionic acidic nonsteroidal anti-inflammatory drugs in pharmaceutical formulations and human plasma by HPLC with UV detection. J. Pharm. Biomed. Anal. 2003, 30, 1611–1619. [Google Scholar] [CrossRef]
2. Castilhos, T.S.; Barreto, F.; Meneghini, L.; Bergold, A.M. Development of high-throughput multi-residue method for non-steroidal anti-inflammatory drugs monitoring in swine muscle by LC-MS/MS. Food Addit. Contam. Part A 2016, 33, 1166–1174. [Google Scholar] [CrossRef]
3. Martin, M.J.; Pablos, F.; Gonzalez, A.G. Simultaneous determination of caffeine and non-steroidal anti-inflammatory drugs in pharmaceutical formulations and blood plasma by reversed-phase HPLC from linear gradient elution. Talanta 1999, 49, 453–459. [Google Scholar] [CrossRef]
4. El Haj, B.M.; Al Ainri, A.M.; Hassan, M.H.; Bin Khadem, R.K.; Marzouq, M.S. The GC/MS analysis of some commonly used non-steriodal anti-inflammatory drugs (NSAIDs) in pharmaceutical dosage forms and in urine. Forensic Sci. Int. 1999, 105, 141–153. [Google Scholar] [CrossRef]
5. Kim, K.R.; Yoon, H.R. Rapid screening for acidic non-steroidal anti-inflammatory drugs in urine by gas chromatography mass spectrometry in the selected-ion monitoring mode. J. Chromatogr. B Biomed. Appl. 1996, 682, 55–66. [Google Scholar] [CrossRef]
6. Arroyo, D.; Ortiz, M.C.; Sarabia, L.A. Optimization of the derivatization reaction and the solid-phase microextraction conditions using a D-optimal design and three-way calibration in the determination of non-steroidal anti-inflammatory drugs in bovine milk by gas chromatography-mass spectrometry. J. Chromatogr. A 2011, 1218, 4487–4497. [Google Scholar] [CrossRef]
7. Heitmeier, S.; Blaschke, G. Direct assay of nonopioid analgesics and their metabolites in human urine by capillary electrophoresis and capillary electrophoresis-mass spectrometry. J. Chromatogr. B Biomed. Sci. Appl. 1999, 721, 109–125. [Google Scholar] [CrossRef]
8. Bechet, I.; Fillet, M.; Hubert, P.; Crommen, J. Quantitative-analysis of nonsteroidal antiinflammatory drugs by capillary zone electrophoresis. J. Pharm. Biomed. Anal. 1995, 13, 497–503. [Google Scholar] [CrossRef]
9. Maboundou, C.W.; Paintaud, G.; Berard, M.; Bechtel, P.R. Separation of 15 nonsteroidal antiinflammatory drugs using micellar electrokinetic capillary

chromatography. J. Chromatogr. B Biomed. Appl. 1994, 657, 173–183. [Google Scholar] [CrossRef]

10. Navas, N.; Urena, R.; Capitan-Vallvey, L.-F. Determination of celecoxib, rofecoxib, sodium diclofenac and niflumic acid in human serum samples by HPLC with DAD detection. Chromatographia 2008, 67, 55–61. [Google Scholar] [CrossRef]
11. Dadashzadeh, S.; Vali, A.M.; Rezagholi, N. LC determination of piroxicam in human plasma. J. Pharm. Biomed. Anal. 2002, 28, 1201–1204. [Google Scholar] [CrossRef]
12. Kataoka, H.; Saito, K. Recent advances in column switching sample preparation in bioanalysis. Bioanalysis 2012, 4, 809–832. [Google Scholar] [CrossRef]
13. Kataoka, H. New trends in sample preparation for clinical and pharmaceutical analysis. Trends Anal. Chem. 2003, 22, 232–244. [Google Scholar] [CrossRef]
14. Nazario, C.E.D.; Fumes, B.H.; da Silva, M.R.; Lancas, F.M. New materials for sample preparation techniques in bioanalysis. J. Chromatogr. B Anal. Technol. Biomed. Life Sci. 2017, 1043, 81–95. [Google Scholar] [CrossRef] [PubMed]
15. Cassiano, N.M.; Barreiro, J.C.; Moraes, M.C.; Oliveira, R.V.; Cass, Q.B. Restricted access media supports for direct high-throughput analysis of biological fluid samples: Review of recent applications. Bioanalysis 2009, 1, 577–594. [Google Scholar] [CrossRef] [PubMed]
16. Rudolphi, A.; Boos, K.S.; Grimm, C.H. Column switching and restricted access materials for sample preparation. Abstr. Pap. Am. Chem. Soc. 1997, 214, 186. [Google Scholar]
17. Yang, S.H.; Fan, H.; Classon, R.J.; Schug, K.A. Restricted access media as a streamlined approach toward on-line sample preparation: Recent advancements and applications. J. Sep. Sci. 2013, 36, 2922–2938. [Google Scholar] [CrossRef] [PubMed]
18. Lee, H.S.; Jeong, C.K.; Choi, S.J.; Kim, S.B.; Lee, M.H.; Ko, G.I.; Sohn, D.H. Simultaneous determination of aceclofenac and diclofenac in human plasma by narrowbore HPLC using column-switching. J. Pharm. Biomed. Anal. 2000, 23, 775–781. [Google Scholar] [CrossRef]
19. Suenami, K.; Lim, L.W.; Takeuchi, T.; Sasajima, Y.; Sato, K.; Takekoshi, Y.; Kanno, S. Direct determination of non-steroidal anti-inflammatory drugs by column-switching LC-MS. J. Sep. Sci. 2006, 29, 2725–2732. [Google Scholar] [CrossRef]

20. Brabcová, I.; Šatínský, D.; Solich, P. A fast on-line micro-column sample pretreatment with a HPLC column switching technique and large volume sample injection for determination of non-steroidal anti-inflammatory drugs in human serum. Anal. Methods 2013, 5, 6153–6159. [Google Scholar] [CrossRef]

21. Puoci, F.; Iemma, F.; Cirillo, G.; Curcio, M.; Parisi, O.I.; Spizzirri, U.G.; Picci, N. New restricted access materials combined to molecularly imprinted polymers for selective recognition/release in water media. Eur. Polym. J. 2009, 45, 1634–1640. [Google Scholar] [CrossRef]

22. Xiao, D.L.; Liu, S.B.; Liang, L.Y.; Bi, Y.P. Magnetic restricted-access microspheres for extraction of adrenaline, dopamine and noradrenaline from biological samples. Mikrochim. Acta 2016, 183, 1417–1423. [Google Scholar] [CrossRef]

23. de Faria, H.D.; Abrao, L.C.D.; Santos, M.G.; Barbosa, A.F.; Figueiredo, E.C. New advances in restricted access materials for sample preparation: A review. Anal. Chim. Acta 2017, 959, 43–65. [Google Scholar] [CrossRef] [PubMed]

24. Barbosa, A.F.; Barbosa, V.M.P.; Bettini, J.; Luccas, P.O.; Figueiredo, E.C. Restricted access carbon nanotubes for direct extraction of cadmium from human serum samples followed by atomic absorption spectrometry analysis. Talanta 2015, 131, 213–220. [Google Scholar] [CrossRef]

25. Raabová, H.; Háková, M.; Chocholoušová-Havliková, L.; Erben, J.; Chvojka, J.; Solich, P.; Švec, F.; Šatínský, D. Poly-epsilon-caprolactone Nanofibrous Polymers: A Simple Alternative to Restricted Access Media for Extraction of Small Molecules from Biological Matrixes. Anal. Chem. 2020, 92, 6801–6805. [Google Scholar] [CrossRef]

26. Háková, M.; Chocholoušová-Havliková, L.; Chvojka, J.; Erben, J.; Solich, P.; Švec, F.; Šatínský, D. A comparison study of nanofiber, microfiber, and new composite nano/microfiber polymers used as sorbents for on-line solid phase extraction in chromatography system. Anal. Chim. Acta 2018, 1023, 44–52. [Google Scholar] [CrossRef]

27. Erben, J.; Pilařová, K.; Sanetrnik, F.; Chvojka, J.; Jenčová, V.; Blažková, L.; Havlíčeka, J.; Novák, O.; Mikeš, P.; Prosecká, E.; et al. The combination of meltblown and electrospinning for bone tissue engineering. Mater. Lett. 2015, 143, 172–176. [Google Scholar] [CrossRef]

28. European Medicines Association. Note for Guidance on Validation of Analytical Procedures: Text and Methodology; European Medical Agency: Amsterdam, The Netherlands, 1995. [Google Scholar]

29. Emara, L.H.; Taha, N.F.; El-Ashmawy, A.A.; Raslan, H.M.; Mursi, N.M. A rapid and sensitive bioanalytical hplc method for determining diclofenac sodium in human plasma for bioequivalence studies. J. Liq. Chromatogr. Relat. Technol. 2012, 35, 2203–2216. [Google Scholar] [CrossRef]

30. Sahoo, N.K.; Sahu, M.; Rao, P.S.; Ghosh, G. Solid phase extraction and quantification of diclofenac sodium in human plasma by liquid chromatography-tandem mass spectrometry. J. Anal. Chem. 2015, 70, 424–430. [Google Scholar] [CrossRef]

31. Háková, M.; Chocholoušová-Havliková, L.; Chvojka, J.; Solich, P.; Šatínský, D. An on-line coupling of nanofibrous extraction with column-switching high performance liquid chromatography—A case study on the determination of bisphenol A in environmental water samples. Talanta 2018, 178, 141–146. [Google Scholar] [CrossRef]

12 IMPROVING THE MAGNETIC HARVESTING EFFICIENCY OF MICROALGA CHLORELLA VULGARIS

1. INTRODUCTION: AN OVERVIEW

The magnetic harvesting of microalgae, particularly Chlorella vulgaris, has gained significant attention in recent years as a promising technique for optimizing biomass recovery. Microalgae play a crucial role in various industries, ranging from biofuel production to pharmaceuticals, owing to their rich nutritional content and potential applications. However, the efficient harvesting of microalgal biomass remains a challenge due to its small size and low settling velocity. Magnetic harvesting presents an innovative solution by leveraging the unique properties of magnetic materials to streamline the separation process. In this context, this study focuses on the optimization of magnetic harvesting methods for Chlorella vulgaris, aiming to enhance the efficiency and scalability of microalgal biomass recovery. Through systematic experimentation and analysis, this research aims to contribute valuable insights into advancing sustainable practices in microalgal cultivation and harvesting processes. Microalgae have been extensively investigated over the past decades as a source for biofuel production due to their high lipid and carbohydrate yields, as well as being a natural source of high value-added bioactive compounds such as polyphenols, carotenoids, fatty acids, and antibiotics [1,2,3]. Bioactive chemicals derived from natural sources present higher biological activity and acceptance by consumers compared to the synthetic alternatives [4].

Nowadays, microalgae are of significant importance in the fields of human health, cosmetics, food, and animal feed. In comparison to terrestrial crop plants, microalgae can provide higher productivity and photosynthetic performance, and since they can be cultivated on infertile land, they do not compete with existing food production methods [5,6]. Of the several major production steps of microalgae components, harvesting is both energy and time demanding. It is estimated that microalgae biomass harvesting is responsible for 20–30% of the total biomass production cost [7,8]. Moreover, the harvesting step is crucial for the downstream process, since it leads to a slurry of highly concentrated solid matter [9].

Various harvesting techniques have been introduced over the years such as centrifugation, sedimentation, flocculation, filtration, flotation, and their various combinations. These methods provide (variable) sufficient harvesting efficiencies, however, they are relatively energy and time consuming. Furthermore, none of them can be used as an efficient method for all microalgal cell types, and their application depends on the desired end-product of microalgae cells [10,11]. The reduced lipid levels regarding flocculation, as well as clogging and fouling in filtration, are also considerable disadvantages [12,13]. Moreover, the added cost of the chemicals used in chemical-flocculation or the microorganisms in bio-flocculation, along with the potential contamination of the harvested biomass, are further drawbacks [13,14,15,16].

Magnetic separation, which has been extensively used in the food industry, wastewater treatment, and the steel industry, Refs. [3,17,18] has gathered popularity throughout the past decade as a method for harvesting microalgal cells. Magnetic biomass harvesting, which uses micro- or nano-sized magnetic particles and an applied external magnetic field, is a highly efficient process for cell separation; the method is fast and efficient [17,19,20,21,22]. The process is based on the interactions between the magnetic particles and the surface of the microalgae cells. The magnetic particles can attach to the cells via electrostatic forces, hydrogen bonds, acid-base interactions or van der Waals forces, forming magnetic particle–cell bonding [23]. Key factors affecting the magnetic separation efficiency include cell type and their growth stage, magnetic particle dosage, pH, ions in the cultivation medium, temperature, and gradient of the magnetic field [3]. Both naked and surface-functionalized (using polyacrylamides, chitosan, poly diallyldimethyl-ammonium chloride (PDDA), aminoclay, polyethylenimine (PEI), 3-aminopropyl triethoxysilane (APTES)) magnetic nanoparticles have been applied for the harvesting of *C. reinhardtii*, *C. vulgaris*, *N. salina*, *N. maritime*, *S. dimorfus*, and *S. ovalternus* microalgae, among others [2,3,24,25,26,27,28,29]. Nonetheless, successful upscaling of the magnetic harvesting method depends on some important characteristics of the magnetic particles i.e., their biocompatibility, reusability, and ease of use, as well as the optimization of the parameters affecting the harvesting efficiency [30].

Chlorella sp. is a microalga rich in vitamins, polysaccharides, minerals, lipids, and other high-value products [31]. Lutein, a compound with proven anti-cataract properties, is also found in *Chlorella sp.*, while its extracts exhibit anti-tumor, anti-inflammatory, antioxidant, and anti-microbial properties, and reduce blood pressure and cholesterol levels [32,33,34]. As a result, *Chlorella sp.* is one of the most extensively used microalga in cosmetics, pharmaceuticals, food, and animal feed industries [31,35].

Although *Chlorella sp.* holds second place in the global "algal products" market [36], its harvesting process is still a major challenge. That is why the harvesting of freshwater microalgae *Chlorella vulgaris* (using microwave-synthesized Fe_3O_4 magnetic particles) was examined in this work. The purpose of this study is to develop an efficient *C. vulgaris* harvesting method by examining and optimizing the most important parameters. Specifically, three harvesting parameters have been investigated, namely pH, mass ratio of magnetite particles to biomass, and agitation speed; these were characterized using response surface methodology (RSM). Reusability of the magnetic particles, as well as integrity of the microalgae cells, were also examined. Scanning electron microscopy (SEM) was used to explore the interactions between the microalgae cells and the magnetic particles, while the adsorption mechanism of the *C. vulgaris* cells on the magnetite particles was deduced by measuring the adsorption isotherm at ambient conditions. Finally, zeta potential measurements were performed in order to examine the separation mechanism of *C. vulgaris* microalgae with the aid of the synthesized iron oxide magnetic particles.

2. MATERIALS AND METHODS

2.1. Chemical Reagents

For the synthesis of the magnetite (Fe_3O_4) particles, iron (II) sulfate ($FeSO_4{\cdot}7H_2O$) salt (Chem-Lab NV, Zedelgem, Belgium) and sodium hydroxide pellets (NaOH) (Panreac Quimica, SA) were employed. The cultivation medium of the *Chlorella vulgaris* strain was composed of sodium nitrate ($NaNO_3$), calcium chloride ($CaCl_2{\cdot}2H_2O$), magnesium sulfate ($MgSO_4{\cdot}7H_2O$), potassium hydrogen phosphate ($K_2HPO_4{\cdot}3H_2O$), sodium chloride (NaCl), EDTA disodium (Na_2EDTA), iron (III) chloride ($FeCl_3$), manganese (II) chloride ($MnCl_2$) and zinc chloride ($ZnCl_2$) and cobalt chloride ($CoCl_2$) (Sigma-Aldrich, St. Louis, MO, USA). Potassium dihydrogen phosphate (KH_2PO_4), vitamin B1, and vitamin B12 provided from Merck (KGaA, Darmstadt, Germany). Deionized water was used for the solubilization of the chemical reagents. Magnetite nano powder (50–100 nm) was purchased from Sigma-Aldrich (St. Louis, MO, USA) for comparison with the microwave-synthesized Fe_3O_4 particles used in this work. A trypan blue assay was performed for examining cell membrane integrity (Merck, KGaA, Darmstadt, Germany). All the reagents used were of analytical grade.

2.2. Microwave Synthesis of Magnetite (Fe_3O_4) Particles

Magnetic particles were prepared using a simple, rapid, and low-cost precipitation method that has been previously reported [20,27]. Briefly, 1 g of $FeSO_4{\cdot}7H_2O$ was diluted in 100 mL of deionized water at ambient temperature. Gradual addition of 1 M NaOH (with continuous magnetic stirring) was used to adjust the pH of the solution to 12. As a result, a black precipitate of $Fe(OH)_2$ was formed. By adding deionized water, the final volume of the solution was fixed at 200 mL. The solution was then placed in a common microwave oven where it was radiated for 10 min at 700 W. As a consequence, the $Fe(OH)_2$ precipitate was oxidized to Fe_3O_4. Subsequently, the magnetite particles were collected by employing a NdFeB magnet (50.8 mm × 50.8 mm × 25.4 mm, Supermagnete, Gottmadingen, Germany) with 12,600–12,900 G (N40) magnetic induction intensity. Thereafter, the Fe_3O_4 particles were washed several times with deionized water in order to remove any residual ions, dried in an oven at 60 °C under vacuum, pulverized to a fine powder using a mortar and pestle, and stored in dry conditions for further use.

2.3. Microalgal Strain and Cultivation

Chlorella vulgaris strain (UTEX 1809) was purchased from Culture Collection of Algae at the University of Texas (Austin, TX, USA). The strain was grown in a medium with the following composition (per liter): $NaNO_3$, 250 mg; $CaCl_2{\cdot}2H_2O$, 25 mg; MgSO4·$7H_2O$, 75 mg; $K_2HPO_4{\cdot}3H_2O$; 75 mg; KH_2PO_4, 175 mg; NaCl, 25 mg; trace element solution, 6 mL from stock solution; vitamin B1, 0.12 µg; vitamin B12, 0.10 µg. One liter of trace element solution contained Na_2EDTA, 750 mg; $FeCl_3$, 97.0 mg; $MnCl_2$, 41.0 mg; $ZnCl_2$, 5.0 mg; $CoCl_2$, 2.0 mg; and $Na_2MoO_4{\cdot}2H_2O$, 4.0 mg. The pH was adjusted to 6.8–7.0 and the temperature was controlled at 24 °C. Cell cultures of *C. vulgaris* were grown in 1000 mL Erlenmeyer flasks containing a culture volume of 600 mL at 150 rpm. The injected concentration was adapted to an optical density of 0.1 at 600 nm (OD 600nm). Periodic purity assessment was performed by microscopic examination. The photo-bioreactor was illuminated at 50 µmol photon m^{-2} s^{-1} with 7/24 white LED lamps at the cell culture surface. *C. vulgaris* growth was monitored by measuring the OD at 600 nm with a UV-VIS S22- BOECO (Shimadzu, Kyoto, Japan) for up to 15 days. All experiments were conducted in duplicate.

2.4. Characterization of Fe_3O_4 Particles and Microalgae

2.4.1. X-ray Diffraction of Magnetite Particles and Magnetic Properties

In order to determine the crystallographic structure and phase composition of the synthesized magnetic particles, X-ray diffraction was employed by using a Brooker X-ray D8 advance diffractometer equipped with a Cu-Kα radiation source. The radiation of the specimen was generated at 40 kV and 40 mA at room temperature, with a scanning rate of 0.1° per minute, from 2θ 28° up to 89°. The phenomenon mean diameter size of the magnetite crystallites was calculated via the Debye–Scherrer formula method [37] by exploiting the obtained XRD raw data using the Diffrac. Suite Eva software. Furthermore, the same XRD study was performed for the high purity Fe3O4 nanoparticles purchased from Sigma-Aldrich, in order to compare the two spectra and deduce on the purity and micro-structural characteristics of the microwave-synthesized magnetic particles. The magnetic properties of Fe_3O_4 were assessed at room temperature using a vibrating sample magnetometer.

2.4.2. Observation with Scanning Electron Microscopy Coupled with Energy Dispersive X-ray Spectroscopy (SEM-EDS)

The morphological characteristics of treated and untreated *C. vulgaris* algal cells, were studied via scanning electron microscopy (SEM) using an FEI Quanta 200 scanning electron microscope.

Using the same technique, the C. vulgaris—Fe_3O_4 interaction was also evaluated. For this purpose, magnetically harvested C. vulgaris cells were fixed in 2.5% (v/v) paraformaldehyde at 4 °C for 1 h, washed in 0.05 M phosphate buffer (pH 7.4), and consecutively dehydrated in 20%, 35%, 50%, 75%, 90%, and 100% (v/v) ethanol solution [22].

All the aforementioned samples were sputter-coated with Au prior to SEM examination. Moreover, EDS technique was employed (using an EDAX analyzer) in order to obtain the elemental compositional analysis of the microalgal cells and the magnetite particles.

2.4.3. Microalgae Analysis

Algae concentration (g L^{-1}) was calculated using a calibration curve of known optical density at 600 nm and respective dry weights that were determined gravimetrically after drying the algae cells at 60 °C. The optical density was measured with a UV-VIS S-22 BOECO spectrophotometer.

2.4.4. Zeta Potential Measurements

The zeta potential values of the prepared Fe_3O_4 particles and the *C. vulgaris* cells were measured with the Zetasizer Nano-ZS (Malvern, UK) at ambient temperature and calculated according to the Smoluchowski equation using the Zetasizer software. In order to perform these measurements, Milli-Q water was used as the dispersion medium for the magnetite particles (3 mg L^{-1}) and the microalgae cells (50 mg L^{-1}). The pH of the mixtures was fixed to 3.0, 5.0, and 7.0 by adding small quantities of aqueous NaOH and HCl. In each case, at least three zeta potential measurements were recorded, and the reported values correspond to their arithmetic mean.

2.5. C. vulgaris Magnetic Harvesting Using Fe_3O_4 Particles

2.5.1. Magnetic Harvesting Experimental Procedure

The magnetic harvesting experiments of *C. vulgaris* biomass were carried out under ambient conditions in batch mode. The experimental procedure included the following steps: (a) The pH of the microalgal cultivation was fixed at the desired value through the addition of aqueous NaOH or HCl solutions, (b) in order to facilitate the flocculation between the algal cells and the particles, 10 mL of the microalgae cultivation was mixed with a given amount of magnetic particles for 10 min using a mechanical stirrer operating at desired rpm, and (c) the flocs were separated from the cultivation medium with the application of a strong magnetic field, imposed using the same permanent magnet which was employed during the synthesis of the magnetite particles, for 3 min.

Finally, the harvesting efficiency (*HE*%) of the process was calculated from Equation (1):

$$HE\% = \frac{OD_0 - OD_1}{OD_0} \cdot 100 \quad (1)$$

where, OD_0 is the initial absorbance of the microalgae cultivation at a wavelength of 600 nm, and OD_1 is the absorbance at the same wavelength of the supernatant liquid that separates from the microalgae-particles flocs after the application of the magnetic field. The OD_0 and OD_1 values were measured with the UV-VIS S22-BOECO spectrophotometer.

2.5.2. Effect of Harvesting Process Parameters on Microalgae Magnetic Harvesting Using Fe_3O_4 Particles

In order to maximize the harvesting efficiency, the effect of the pH of the mixture, the mass ratio (m_r) of magnetite particles (g)/dry biomass (g) in the algal broth, and the agitation speed (rpm) of the *C. vulgaris* cultivation-Fe_3O_4 particles mixture, were examined. Three values of pH (5.0 ,3.0, and 7.0), three m_r ratios (12:1 ,10:1, and 14:1) and three agitation speeds (350 ,250, and 450 rpm) were compared. Each separate magnetic harvesting experiment was performed three times, and the reported *HE*% results are the mean values.

Preliminary harvesting experiments aiming to evaluate the effect of the duration of flocs' magnetic separation were also performed. According to the results, the strong attachment of Fe_3O_4 on the cell membrane, along with the application of a powerful magnetic field, results in the rapid harvesting of the flocculated biomass, therefore the study of this parameter was excluded.

2.5.3. Experimental Design for Optimization of *C. vulgaris* Harvesting Using the Response Surface Methodology

The optimum conditions for magnetic harvesting with Fe_3O_4 particles were established using central composite design (CCD) under the application of response surface methodology (RSM). The RSM is a combination of mathematical and statistical techniques applicable when several interactive parameters are examined and is considered useful in order to evaluate the relative significance of the operating factors on magnetic harvesting. In the present work, based on preliminary experiments, the independent parameters affecting magnetic harvesting were found to be pH, mass ratio, and agitation speed. Thus, three-level full factorial design (3^k) was used, considering these operational parameters. The experimental design was devised based on the central level (0) between the minimum (-1) and the maximum levels (+1) of the normalized values. Therefore, twenty experiments were carried out using different values of the examined factors (Table 1): pH (3.0–7.0), mass ratio (10:1–14:1), and agitation speed (250–450 rpm). For a complex system, such as the one under consideration, the response cannot be mathematically expressed through a first-order or a second-order polynomial equation. Therefore, the results of the harvesting experiments were fitted to a modified model, and the quality of the fitted model was quantitatively assessed by the ANOVA method to characterize the interaction effect between independent factors and the microalgae-harvesting efficiency.

Table 1. Independent variables and the distribution of their level.

pH	3	5	7
mass ratio (gmagn./gbiomass)	10	12	14
agitation speed (rpm)	250	350	450

Regression analysis and estimation of the coefficients were performed using Design Expert® trial version 12 software.

2.5.4. Adsorption Isotherms

The elucidation of the adsorption capacity and mechanism of the microwave-synthesized magnetite particles on the *C. vulgaris* cells is provided by the adsorption isotherms. In the present study, two different models—Langmuir and Freundlich—were employed in order to analyze the experimental data. These models are frequently reported in related literature as sufficiently accurate to describe the adsorption of magnetic particles on microalgae cells [19,21].

According to the Langmuir model, the amount of dry algae cells adsorbed per unit weight of magnetic particles (Qe [g g^{-1}]), is given by Equation (2):

$$Q_e = Q_m \cdot \frac{C_e \cdot K_l}{1 + C_e \cdot K_l} \quad (2)$$

which can be transformed into the following linear form:

$$\frac{C_e}{Q_e} = \frac{1}{Q_m \cdot K_l} + \frac{C_e}{Q_m} \quad (3)$$

$$Q_e = K_f \cdot C_e^{\frac{1}{n_f}} \quad (4)$$

$$lnQ_e = lnK_f + \frac{1}{n_f} \cdot lnC_e \quad (5)$$

In Equations (2)–(5), Q_m [g g^{-1}] is the maximum adsorption capacity, C_e [g L^{-1}] is the concentration of the microalgae cells in the supernatant after completion of the harvesting process, K_l [L g^{-1}] is the Langmuir adsorption constant, K_f [g g^{-1}] is the Freundlich constant (which is related to the adsorption capacity), and n_f [dimensionless] is the Freundlich factor of the heterogeneity of the adsorption sites.

The adsorption experiments were conducted at ambient temperature in a batch mode in duplicates. For each experiment, an amount of Fe_3O_4 was added to 10 mL of algae cultivation of known initial concentration (C_0 [g L^{-1}]) in order to realize the magnetic harvesting procedure that is described in Section 2.5.1.

Consequently, the amount of dry algae adsorbed per unit weight of magnetite particles (Q_e [g g^{-1}]) was calculated according to Equation (6).

$$Q_e = \frac{C_0 - C_e}{m} \cdot V \quad (6)$$

where, m [g] is the mass of the magnetic particles used and V [L] is the volume of microalgae culture used, i.e., 0.01 L.

2.5.5. Reusability of Magnetic Particles

The study on the reusability of the magnetic particles was realized by carrying out tests with the same (used) particles for cycles, in accordance with the procedure outlined by Markeb et al. [38]. The magnetic particles and microalgae cells obtained from magnetic harvesting were mixed with 10 mL of NaOH (0.5 M) and agitated at 200 rpm for 10 min. The suspension was then sonicated for 10 min. Afterwards, 3 mL of methanol and 3 mL of chloroform were added, and the solution was sonicated for another 10 min. Following sonication, the magnetite particles were collected with a permanent magnet, washed three times, and dried overnight. The detached magnetic particles were used to evaluate the harvesting efficiency tests under the same conditions in order to investigate their reusability.

2.6. Statistical Analysis

Tukey's method, based on one factor ANOVA at the 5% confidence level, was used for the statistical analysis, which was performed with SPSS 15.0.1 software (SPSS

Inc., Chicago, IL, USA). Statistically significant differences were reported when the probability of the results (p) value is less than 0.05 assuming the null hypothesis.

2.7. Cell Membrane Integrity of C. vulgaris Cells

The cell integrity of *C. vulgaris* was determined by the trypan blue staining method. Total of 10 uL of cells were harvested and after addition of 10 uL 1% trypan blue solution the cells were incubated for 10 min at room temperature. The intact cells (viable) remained green (no penetration of the trypan blue solution) while the broken cells appeared blue (stain diffused in the protoplasm).

3. RESULTS AND DISCUSSION

3.1. Characterization of Fe_3O_4 Particles

Figure 1 demonstrates the X-ray diffraction spectra of the magnetite particles purchased from Sigma-Aldrich (Figure 1a) and the magnetic particles synthesized in this work (Figure 1b). According to the analysis conducted via XRD software Diffrac. Suite Eva, one predominant phase was identified, that of magnetite (PDF 02-1035 Fe_3O_4 Magnetite), for both materials.

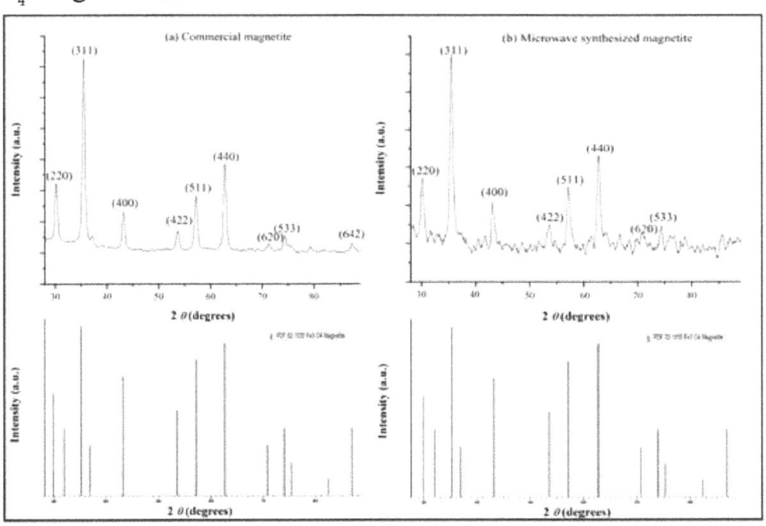

Figure 1. X-ray diffraction spectra of (a) commercially available magnetite particles and (b) the microwave-synthesized magnetite particles.

The noteworthy similarity between the two spectra suggests that they have close microstructural characteristics. Particularly, regarding the microwave-synthesized magnetite (Figure 1b), eight different relatively strong diffraction peaks are displayed in the XRD spectrum at 2 θ ,°53.71 ,°42.96 ,°35.48 ,°30.10 °70.80 ,°62.77 ,°57,17, and °74.27 which correspond to the crystallographic planes (620) ,(440) ,(511) ,(422) ,(400) ,(311) (220), and (533) respectively, indicating a cubic inverse spinel structure. Likewise, the same crystallographic planes, can be observed for the commercial magnetite powder (Figure 1a) at 2 θ ,°35.55 ,°30.18 °71.20 ,°62.82 ,°57,20 ,°53.64 ,°43.20, and °74.34. One more crystallographic plane is present in the spectrum of commercial magnetite at 2 θ °87.36, namely the (642). This last plane cannot be identified with certainty in the XRD pattern of the synthesized magnetite, due to the relatively higher noise encountered. In both cases, it is clearly observed that five diffraction peaks, namely (,(400) ,(311) ,(220 440) ,(511)) are the main ones. The aforementioned analysis of the Bragg reflections for both magnetite particles was realized by identifying each diffraction peak from the raw XRD data using the Diffrac. Suite Eva software.

Using the tools provided by the same software, the phenomenon mean diameter size of magnetite crystallites was calculated according to the Debye–Scherrer equation [37] for each diffraction peak that was identified. So, an average crystallite diameter size equal to 20 ± 4 nm and 20 ± 3 nm was calculated for the synthesized magnetite and the commercial one respectively.

For the determination of the magnetic properties of the microwave-synthesized magnetite particles, the saturated mass magnetization was determined by vibrating the sample in a magnetometer. The magnetization loop is illustrated in Figure 2. From these results, it is evident that the particles follow a unhysteretic loop, illustrating super-paramagnetic behaviour at room temperature, thus indicating that their size is below 35 nm [39], which is verified in our case as the particles demonstrate an average diameter of 20 nm. The synthesized magnetic particles (Fe_3O_4) exhibit saturation magnetization of 60 emu g^{-1}.

Thus, it is concluded that the simple, rapid, and energy-efficient synthesis route followed in this work leads to the formation of high purity nanocrystalline magnetite, which is similar in microstructure to the commercially available that was employed as a means of comparison.

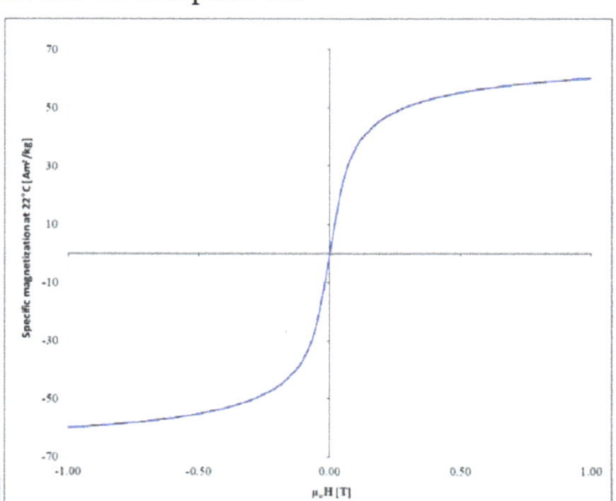

Figure 2. Magnetization curve for the synthesized Fe_3O_4.

3.2. Algae Harvesting Efficiency Optimization

3.2.1. Characterization of *C. vulgaris* and Magnetic Particles Interaction Using SEM

Evaluation of the interactions between the microalgae and the (synthesized) magnetic particles was tested by characterizing *C. vulgaris* before (control) and after mixing with the magnetic particles using the scanning electron microscopy (SEM). The morphology of *C. vulgaris* cells alone and with magnetic particles, are illustrated in Figure 3. As it can be seen in Figure 3a, the surface of the untreated cells is smooth and spherical, whereas after the magnetic harvesting at pH 3.0 (Figure 3c), it can be observed that all the cell surfaces are fully covered with magnetic particles. The cell surfaces appear rougher and the patterns that are visible in Figure 3a are not discerned in Figure 3c. This confirms that the amount of magnetic particles we chose to use is enough to fully cover every single cell. Moreover, energy dispersive X-ray spectroscopy (EDX) reports the iron peak in Figure 3d, verifying the presence of magnetite on the cells in contrast to the untreated cells (Figure 3b).

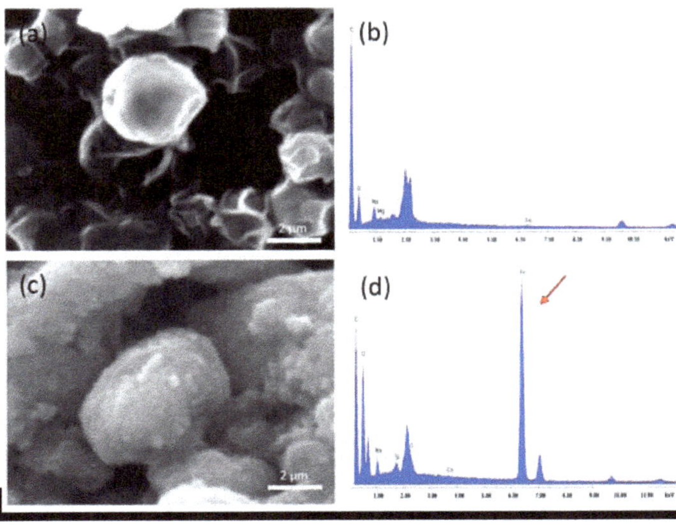

Figure 3. Scanning electron microscopy (SEM) images performed on (a) Chlorella vulgaris cells before magnetic harvesting, (b) EDX analysis for the elemental composition of the cells where the absence of iron peak is reported, (c) harvested Chlorella vulgaris cells at pH 3.0, (d) EDX analysis for the elemental composition of the harvested Chlorella vulgaris cells where the iron peak (arrow) is shown.

3.2.2. Magnetic Harvesting of *C. vulgaris*

A face-centered (distance of each axial point from the center: alpha = 1) composite design was developed for modelling and optimizing (statistically) the magnetic harvesting of *C. vulgaris*. Twenty experiments were carried out examining the effect of pH, mass ratio (m_r), and agitation speed on harvesting efficiency. The range of the examined parameters was chosen based on preliminary experiments and corresponding literature. Specifically, studies on *Chlorella vulgaris* and related species [2,12,40,41] have been conducted in a pH range between 2 and 12, while maximum harvesting efficiency was achieved in most cases at acidic or neutral pH. The *Chlorella vulgaris* cell walls are very robust and remain resistant at acidic pH, in comparison to other microalgal species that do not attain such a resilient cell wall and might be affected. Although harsh conditions can affect the viability of the cells, the harvesting process followed in this study is very quick, without significantly affecting the biomass composition. Regarding stirring, low agitation speed values [19,42] of around 250 rpm [43], or even higher values at 800 rpm [44] have been used for harvesting *Chlorella* species. This data led us to choose a pH range from 3 to 7 and an agitation speed range between 250 and 450 rpm.

The experimental results are presented in Table 2. It is observed that the decrease of pH, and the increase of the other two parameters, improve the harvesting efficiency. Thus, the best experimental harvesting efficiencies higher than 99% were obtained for pH 3.0 and mixing speed greater or equal to 350 rpm. However, even at a pH similar to that of the cell culture (pH 7.0), the harvesting efficiency is proved to be satisfactory (>85%) for m_r = 14:1 and an agitation speed of 450 rpm. These results are in agreement with those reported by Zhu et al. [9] and Bharte and Desai [12], for different *Chlorella* species that demonstrated maximum harvesting efficiency in acidic pH, thus verifying our findings. Analogous studies on *C. vulgaris* harvesting have shown that harvesting based on a flocculation method can lead to high harvesting efficiency, but under different optimum conditions compared to this study. For example, Tork et al. [42] reported a 90% efficiency at basic pH (pH = 11.7) values using cationic starch nanoparticles. Similarly, in the study of Leite et al. [40] for *Chlorella Sorokiniana*, a harvesting efficiency greater than 97% (combined with the appropriate velocity gradient and agitation time) was achieved under highly alkaline conditions (pH = 12). Moreover, a lower optimal agitation rate of 150 rpm was reported by Almomani [19] and Razack et al. [45] (using as a flocculant iron oxide nanoparticles and seed powder of clearing nut respectively). However, the above differences concerning optimal harvesting conditions may be due to the different biomass and magnetic materials, absence or presence of surface modification etc.

Table 2. Design table formed by RSM—CCD, presenting the experimental conditions and the experimental values of harvesting efficiency.

1	7	10:1	450	64.7
2	3	14:1	450	99.6
3	3	10:1	250	83.3
4	7	14:1	250	72.3
5	7	14:1	450	87.2
6	7	10:1	250	51.7
7	3	10:1	450	99.4
8	3	14:1	250	87.8
9	5	12:1	350	76.9
10	3	12:1	350	99.2
11	5	12:1	450	88.8
12	5	12:1	250	66.5
13	7	12:1	350	65.0
14	5	12:1	350	76.7
15	5	10:1	350	70.3
16	5	12:1	350	77.9
17	5	14:1	350	92.5
18	5	12:1	350	80.8
19	5	12:1	350	77.6
20	5	12:1	350	77.1

Analysis of variance (ANOVA) was used to determine the relationship between the dependent (harvesting efficiency %) and independent variables (pH, mass ratio, and agitation). Transformation of the response was considered necessary in order to bound harvest efficiency within reasonable limits. The transformation chosen is presented below:

$$y' = \ln\left(\frac{y - lower}{upper - y}\right) \tag{7}$$

where, y stands for the harvesting efficiency (%), lower and upper for the boundaries (40 and 100 respectively), and y′ for the transformed harvesting efficiency.

Table 3 shows the statistical results of harvesting efficiency for RSM using ANOVA, while a linear model with respect to the examined factors and their interactions is obtained as follows:

$$y' = -2.473 - 2.022 * A + 0.299 * B + 0.027 * C - 0.003 * A * C + 0.229 * A^2 \quad (8)$$

where, y′ stands for the transformed harvesting efficiency, A for pH, B for the mass ratio (gmagn. par./gbiomass), and C for the agitation speed (rpm). It is clarified that through Equation (8) the relation between the independent variables and the transformed dependent variable (y′) is expressed. The calculated harvest efficiency (y) is finally determined by solving Equation (7) with respect to y:

$$y = \frac{100 * e^{y'} + 40}{1 + e^{y'}} \quad (9)$$

Table 3. Analysis of variance results for RSM model regarding harvesting process, where A is pH, B is mass ratio, and C is agitation speed.

Model	51.99	5	10.40	47.58	<0.0001
A-pH	28.47	1	28.47	130.29	<0.0001
B-mass ratio	3.58	1	3.58	16.37	0.0012
C-agitation speed	12.54	1	12.54	57.39	<0.0001
AC	3.21	1	3.21	14.69	0.0018
A^2	4.19	1	4.19	19.15	0.0006
Residual	3.06	14	0.22		
Std. Dev.	0.47		**R^2**	0.94	
Mean	1.1		**Adj R^2**	0.93	
C.V. %	42.52		**Pred R^2**	0.86	
PRESS	7.96		**Adeq Precision**	26.60	

The model was evaluated based on the F- and *p*-values. For an F-value > 1 and *p*-value < 0.01, the model under consideration is deemed valid. In the present work, both the high F-value (47.58) and low *p*-value (<0.0001) of the model prove its significance and accuracy. All the factors considered were significant (high F-value and *p*-value < 0.05). Specifically, the most important variant is shown to be pH (higher F-value). The obtained value of R^2 (0.94) implies the high correlation between the dependent and independent variables, while the adjusted R^2 is only slightly lower (0.93) and in reasonable agreement with the predicted R^2 (0.86). Finally, the adequate precision (26.60) is higher than 4 indicating that the model can be used to navigate the design space.

The juxtaposition of the predicted against the experimental values of harvesting efficiency (shown in Figure 4) proves that most of the data points are close to the 45-degree line, proving the reliability of the model.

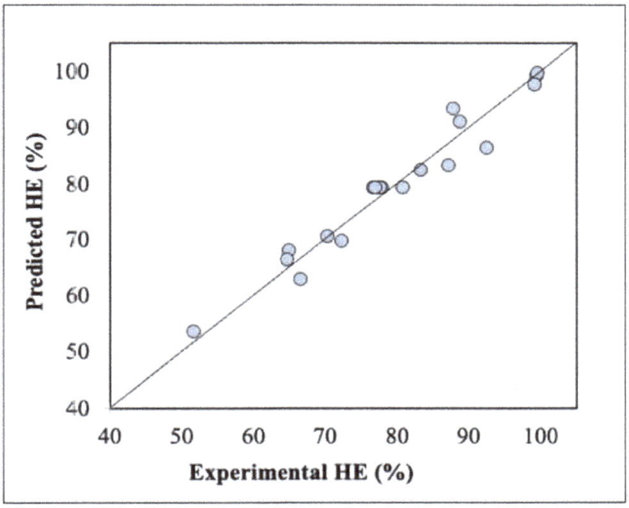

Figure 4. Predicted versus experimental values of harvesting efficiency.

3.3. Separation Mechanism

In order to study the possible electrostatic interactions between the algal biomass and the microwave-synthesized magnetic particles, their zeta potential (ζ) values were measured within the pH range of the magnetic harvesting experiments (pH = 3 to 7). It was found that both the magnetic particles and the biomass cells attain a negative charge. In particular, measurements show ζ values in the range of -17.4 to -33.2, and of -16.6 to -24.9, for the Fe_3O_4 particles and the *C. vulgaris* cells respectively, in the specific dispersing medium (Milli-Q water).

As shown in Figure 5, the zeta potential of the magnetite particles increases with decreasing pH, whereas the *C. vulgaris* cell ζ values remain practically the same as the pH decreases from 7 to 5, and then increases with a further decrease of the pH. Negative surface charge of microalgae biomass is expected in acidic solutions due to the presence of negatively charged groups on the surface of the cells. Concerning the bare Fe_3O_4 particles, a net negative surface charge has been previously reported [20,46].

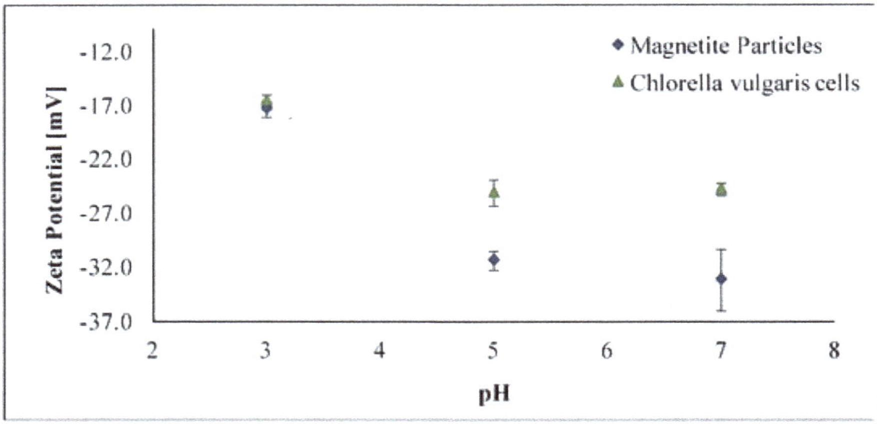

Figure 5. Zeta Potential values of microwave-synthesized Fe_3O_4 particles and *C. vulgaris* cells at different pH values of the dispersing medium.

Since in the present study the magnetic particles and the *C. vulgaris* cells are both negatively charged, it could be assumed that the interactions between them are not dependent on a charge neutralization mechanism. Nevertheless, in any aqueous solution, metal oxide (MeOx) particle surfaces can undergo the following protonation/deprotonation reaction:

$$MeOxO^- + H_2O \overset{H_2O}{\leftrightarrow} MeOxOH \overset{H_2O}{\leftrightarrow} MeOxOH_2^+ \quad (10)$$

Consequently, under acidic conditions and as the pH decreases, an increase of the positively charged sites on the surface of the metal oxide particles is expected, resulting in an increase in the zeta potential values [28] without necessarily altering the net charge of the particles. This is in agreement with the ζ values obtained here for the microwave-synthesized Fe_3O_4 particles. The maximum harvesting efficiency values measured appear at the lowest applied pH values, i.e., at pH = 3. Hence, this behaviour may be attributed to local electrostatic interactions between positively charged sites on the surface of the magnetite particles and the negatively charged surfaces of the *C. vulgaris* cells. This assumption is further confirmed by the fact that the harvesting efficiency increases with decreasing pH.

Nevertheless, electrostatic interactions may not be the primary binding mechanism of the synthesized magnetite nanoparticles with *C. vulgaris* cells. Particularly, Fe3O4 particles can be attached to the microalgal cells through hydrogen bonding; due to the protonation of the magnetite particles under acidic conditions, hydrogen bond donor chemical species OH2+ formed on Fe_3O_4 can interact with hydrogen bond acceptor groups present on C. vulgaris cells, such as amino or carboxy groups [47,48]. Furthermore, due to the nano-size of the Fe_3O_4 particles, a large specific surface area and high surface energy is expected, characteristics that imply strong adsorption of the particles to the cells [49,50].

3.4. Adsorption Isotherms

The adsorption isotherms of the synthesized magnetite particles on the *C. vulgaris* cells are illustrated in Figure 6a,b along with the fitted curves using the Langmuir and Freundlich models respectively. The parameters of both models were obtained using the least squares linear fitting and are presented in Table 4.

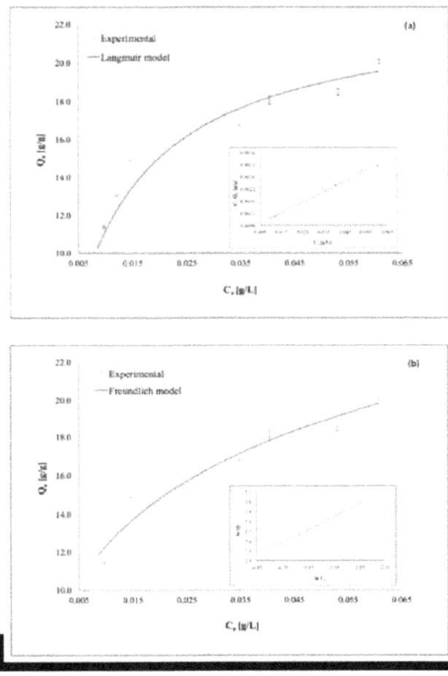

Figure 6. Experimental adsorption isotherm of microwave-synthesized magnetite particles on the C. vulgaris cells at 25 °C, fitted with (a) Langmuir model and (b) Freundlich model; pH = 7, rpm = 450, C. vulgaris cell concentration 0.22 g L−1.

Table 4. Parameters estimated using the Langmuir and Freundlich models.

22.95	95.10	0.99	41.30	3.81	0.94

The results indicate that the adsorption isotherm of Fe_3O_4 particles on the *C. vulgaris* cells is better described by the Langmuir model of adsorption, according to which, a full coverage of the microalgae cells by the magnetic particles occurs. The maximum adsorption capacity, Qm, is predicted to be equal to 22.95 g of dry biomass per gram of magnetic particles. Moreover, the high Kl value given by the Langmuir model denotes a strong attraction between the C. vulgaris cells and the microwave-synthesized magnetic particles [28]. The adsorption mechanism of Fe3O4 particles on C. vulgaris presented here has been previously reported for the same microalgae species both on naked and surface functionalized magnetite particles [19,21].

3.5. Regeneration and Reusability of Fe_3O_4 Particles

Both microalgae harvesting and the downstream processing of harvested biomass (e.g., to obtain biofuels or bio-products) call for recovery and regeneration of the magnetic particles, alongside testing the cell integrity to ensure the intracellular abidance of bioactive compounds and thereby their efficient and easy extraction. Our data indicate that the magnetic particles can be reused for at least five cycles, with less than 20% decrease of the algae harvesting efficiency—even during the 5th cycle, (Figure 7). This is similar to the results reported by Almomani [19]. These findings indicate that the regeneration procedure followed is an acceptable method, since an efficient reusability of the magnetic particles is observed.

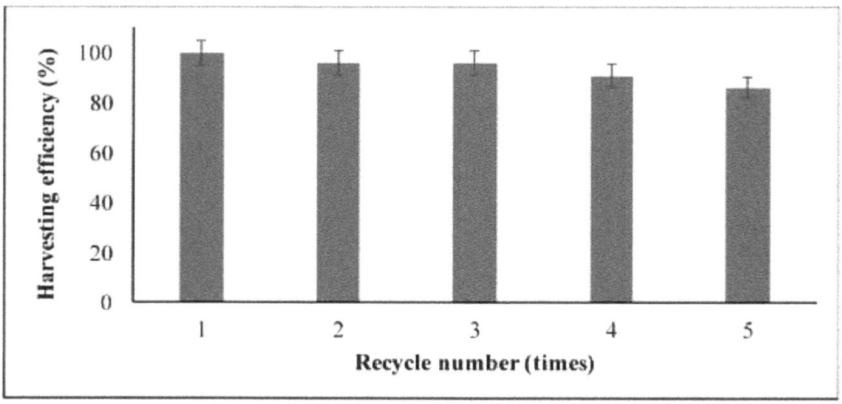

Figure 7. *Chlorella vulgaris* harvesting efficiency (initial concentration of 0.22 g L^{-1}) at five different consecutive cycles.

Analogous studies demonstrated a reusability efficiency of 80% after three cycles with chloroform: methanol treatments and ultrasonication [51], and at least 85% after five cycles of acid-base treatment combined with ultrasonication [52]. Markeb et al. [38], using NaOH, methanol, and chloroform as organic solvents, and ultrasonication, reported a minimum 80% recovery of the magnetic nanoparticles. Treatment with NaOH at pH = 12 resulted in 95% recovery for polypropylene/iron oxide nanoparticles, with the recovered magnetic particles retained almost the same microalgae biomass harvesting efficiency as per the newly synthesized ones [53]. An 80% reusability efficiency of cationic surfactant-decorated iron oxide nanoparticles after microalgae detachment using SDS and sonication has also been reported [54].

Staining with trypan blue (confirming compromised plasma membranes) verified the integrity of the bounded cells on the magnetic particles (Figure 8) and confirmed that no precious intracellular macromolecules were released in the harvesting medium. The trypan blue staining was performed at the most efficient pH for harvesting efficiency (pH = 3). Thus, the cells maintain their ability to produce high value-added products suitable for further use in various biotechnological applications and/or biofuels production.

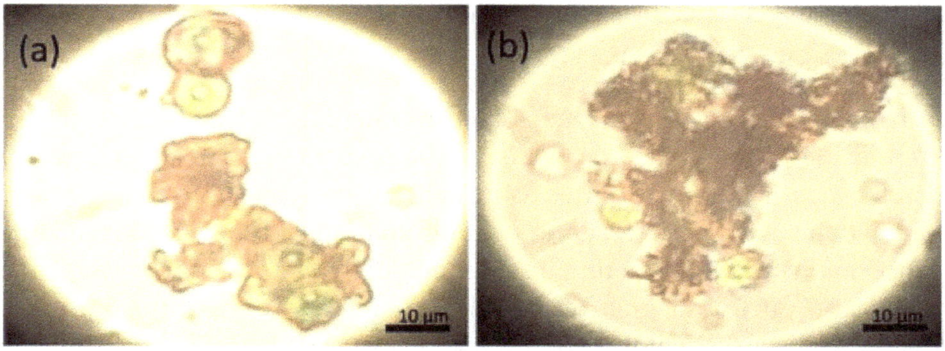

Figure 8. Trypan blue staining (a,b) on *Chlorella vulgaris* cells bound with magnetic particles (at pH 3.0) indicates the integrity of the cell membrane. Intact cells (viable) remained green, without the penetration of the trypan blue solution, while the broken cells appeared blue as stain diffused in the protoplasm.

4. CONCLUSIONS

Chlorella vulgaris harvesting was assessed using iron oxide magnetic particles synthesized using a Fe (II) precursor with the aid of microwave irradiation. The magnetic particles were characterized by X-ray diffraction, identifying one predominant phase, namely that of magnetite, with a mean diameter size of 20 nm. The simple and rapid production route followed in this work leads to the formation of high purity nanocrystalline magnetite with a super-paramagnetic behaviour at room temperature. The adsorption isotherm of the Fe3O4 particles on the C. vulgaris cells at ambient temperature, demonstrated a full coverage of the microalgae cells by the magnetic particles in accordance with the Langmuir model. Furthermore, the magnetite particles do not impose loss of microalgae cells' integrity. Response surface methodology verified the experimental optimum operational harvesting conditions, namely pH = 3, mass ratio = 14:1, and agitation speed = 450 rpm, under which a harvesting efficiency equal to 99.6% was achieved. pH was proved as the most crucial variant. The magnetic particles were successfully detached from the microalgal cells and reused for five cycles maintaining at least 80% of their initial harvesting efficiency. The separation mechanism was primarily attributed to the formation of hydrogen bonds between the magnetite particles and the microalgae cells under acidic conditions and to the expression of nano-size effects related to the high surface energy and large specific surface area of the particles. The high magnetic harvesting efficiencies (greater than 99%) obtained at pH 3.0 and mixing speed greater or equal to 350 rpm using microwave-synthesized naked magnetite (Fe3O4) particles, the cell integrity after the harvesting procedure and the ability to reuse the synthesized magnetic particles for at least five cycles of harvesting, contribute to the novelty of the present work and indicate that the proposed process is efficient and very promising.

5. CONCISE SUMMARY

Harvesting of microalgae is a crucial step in microalgae-based mass production of different high value-added products. In the present work, magnetic harvesting of *Chlorella vulgaris* was investigated using microwave-synthesized naked magnetite (Fe_3O_4) particles with an average crystallite diameter of 20 nm. Optimization of the most important parameters of the magnetic harvesting process, namely pH, mass ratio (mr) of magnetite particles to biomass (g/g), and agitation speed (rpm) of the *C. vulgaris* biomass–Fe_3O_4 particles mixture, was performed using the response surface methodology (RSM) statistical tool. Harvesting efficiencies higher than 99% were obtained for pH 3.0 and mixing speed greater or equal to 350 rpm. Recovery of magnetic particles via detachment was shown to be feasible and the recovery particles could be reused at least five times with high harvesting efficiency. Consequently, the described harvesting approach of *C. vulgaris* cells leads to an efficient, simple, and quick process that does not impair the quality of the harvested biomass.

REFERENCES

1. Khan, M.I.; Shin, J.H.; Kim, J.D. The promising future of microalgae: Current status, challenges and optimization of a sustainable and renewable industry for biofuels, feed, and other products. *Microb. Cell Fact.* 2018, *17*, 36. [Google Scholar] [CrossRef] [PubMed]
2. Sharma, R.K.; Saharia, M.; Srivstava, R.; Kumar, S.; Sahoo, L. Tailoring microalgae for efficient biofuel production. *Front. Mar. Sci.* 2018, *5*, 382. [Google Scholar] [CrossRef]
3. Hossain, N.; Mahlia, T.M.I.; Saidur, R. Latest development in microalgae-biofuel production with nano-additives. *Biotechnol. Biofuels* 2019, *12*, 125. [Google Scholar] [CrossRef] [PubMed][Green Version]
4. Novoveská, L.; Ross, M.E.; Stanley, M.S.; Pradelles, R.; Wasiolek, V.; Sassi, J.-F. Microalgal Carotenoids: A Review of Production, Current Markets, Regulations, and Future Direction. *Mar. Drugs* 2019, *17*, 640. [Google Scholar] [CrossRef] [PubMed][Green Version]
5. Gouveia, L.; Oliveira, A.C. Microalgae as a raw material for biofuels production. *J. Ind. Microbiol. Biotechnol.* 2009, *36*, 269–274. [Google Scholar] [CrossRef]

6. Clarens, A.F.; Resurreccion, E.P.; White, M.A.; Colost, L.M. Environmental life cycle comparison of algae to other bioenergy feedstocks. *Environ. Sci. Technol.* 2010, *44*, 1813–1819. [Google Scholar] [CrossRef]

7. Coons, J.E.; Kalb, D.M.; Dale, T.; Marrone, B.L. Getting to low-cost algal biofuels: A monograph on conventional and cutting-edge harvesting and extraction technologies. *Algal Res.* 2014, *6 Pt B*, 250–270. [Google Scholar] [CrossRef] [Green Version]

8. Sharma, K.K.; Garg, S.; Li, Y.; Malekizadeh, A.; Schenk, P.M. Critical analysis of current microalgae dewatering techniques. *Biofuels* 2013, *4*, 397–407. [Google Scholar] [CrossRef]

9. Zhu, L.-D.; Hiltunen, E.; Li, Z. Using magnetic materials to harvest microalgal biomass: Evaluation of harvesting and detachment efficiency. *Environ. Technol.* 2019, *40*, 1006–1012. [Google Scholar] [CrossRef]

10. Milledge, J.J.; Heaven, S. A review of the harvesting of micro-algae for biofuel production. *Rev. Environ. Sci. Biotechnol.* 2013, *12*, 165–178. [Google Scholar] [CrossRef][Green Version]

11. Molina Grima, E.; Belarbi, E.H.; Acién Fernández, F.G.; Robles Medina, A.; Chisti, Y. Recovery of microalgal biomass and metabolites: Process options and economics. *Biotechnol. Adv.* 2003, *20*, 491–515. [Google Scholar] [CrossRef]

12. Bharte, S.; Desai, K. Harvesting Chlorella species using magnetic iron oxide nanoparticles. *Psychol. Res.* 2019, *67*, 128–133. [Google Scholar] [CrossRef]

13. Papazi, A.; Makridis, P.; Divanach, P. Harvesting Chlorella minutissima using cell coagulants. *J. Appl. Phycol.* 2010, *22*, 349–355. [Google Scholar] [CrossRef]

14. Chen, L.; Wang, C.; Wang, W.; Wei, J. Optimal conditions of different flocculation methods for harvesting *Scenedesmus* sp. cultivated in an open-pond system. *Bioresour. Technol.* 2013, *133*, 9–15. [Google Scholar] [CrossRef] [PubMed]

15. Xu, Y.; Purton, S.; Baganz, F. Chitosan flocculation to aid the harvesting of the microalga Chlorella sorokiniana. *Bioresour. Technol.* 2013, *129*, 296–301. [Google Scholar] [CrossRef] [PubMed]

16. Matter, A.I.; Bui, V.K.H.; Jung, M.; Seo, J.Y.; Kim, Y.-E.; Lee, Y.-C.; Oh, Y.-K. Flocculation Harvesting Techniques for Microalgae: A Review. *Appl. Sci.* 2019, *9*, 3069. [Google Scholar] [CrossRef][Green Version]

17. Borlido, L.; Azevedo, A.M.; Roque, A.C.A.; Aires-Barros, M.R. Magnetic separations in biotechnology. *Biotechnol. Adv.* 2013, *31*, 1374–1385. [Google Scholar] [CrossRef]

18. Yavuz, C.T.; Prakash, A.; Mayo, J.T.; Colvin, V.L. Magnetic separations: From steel plants to biotechnology. *Chem. Eng. Sci.* 2009, *64*, 2510–2521. [Google Scholar] [CrossRef]

19. Almomani, F. Algal cells harvesting using cost-effective magnetic nanoparticles. *Sci. Total Environ.* 2020, *720*, 137621. [Google Scholar] [CrossRef]

20. Boli, E.; Savvidou, M.; Logothetis, D.; Louli, V.; Pappa, G.; Voutsas, E.; Kolisis, F.; Magoulas, K. Magnetic harvesting of marine algae *Nannochloropsis oceanica*. *Sep. Sci. Technol.* 2017, *6395*, 1–8. [Google Scholar] [CrossRef]

21. Gerulová, K.; Bartošová, A.; Blinová, L.; Bártová, K.; Dománková, M.; Garaiová, Z.; Palcut, M. Magnetic Fe_3O_4-polyethyleneimine nanocomposites for efficient harvesting of *Chlorella zofingiensis*, *Chlorella vulgaris*, *Chlorella sorokiniana*, *Chlorella ellipsoidea* and *Botryococcus braunii*. *Algal Res.* 2018, *33*, 165–172. [Google Scholar] [CrossRef]

22. Savvidou, M.G.; Boli, E.; Logothetis, D.; Lymperopoulou, T.; Ferraro, A.; Louli, V.; Mamma, D.; Kekos, D.; Magoulas, K.; Kolisis, F.N. A Study on the Effect of Macro-and Micro-Nutrients on Nannochloropsis oceanica Growth, Fatty Acid Composition and Magnetic Harvesting Efficiency. *Plants* 2020, *9*, 660. [Google Scholar] [CrossRef] [PubMed]

23. Taghizadeh, S.M.; Berenjian, A.; Chew, K.W.; Show, P.L.; Mohd Zaid, H.F.; Ramezani, H.; Ghasemi, Y.; Raee, M.J.; Ebrahiminezhad, A. Impact of magnetic immobilization on the cell physiology of green unicellular algae *Chlorella vulgaris*. *Bioengineered* 2020, *11*, 141–153. [Google Scholar] [CrossRef][Green Version]

24. Eroglu, E.; Agarwal, V.; Bradshaw, M.; Chen, X.; Smith, S.M.; Raston, C.L.; Swaminathan Iyer, K. Nitrate removal from liquid effluents using microalgae immobilized on chitosan nanofiber mats. *Green Chem.* 2012, *14*, 2682–2685. [Google Scholar] [CrossRef]

25. Hu, Y.R.; Guo, C.; Wang, F.; Wang, S.K.; Pan, F.; Liu, C.Z. Improvement of microalgae harvesting by magnetic nanocomposites coated with polyethylenimine. *Chem. Eng. J.* 2014, *242*, 341–347. [Google Scholar] [CrossRef]

26. Prochazkova, G.; Podolova, N.; Safarik, I.; Zachleder, V.; Branyik, T. Physicochemical approach to freshwater microalgae harvesting with magnetic particles. *Colloids Surf. B Biointerfaces* 2013, *112*, 213–218. [Google Scholar] [CrossRef] [PubMed]

27. Prochazkova, G.; Safarik, I.; Branyik, T. Harvesting microalgae with microwave synthesized magnetic microparticles. *Bioresour. Technol.* 2013, *130*, 472–477. [Google Scholar] [CrossRef]

28. Xu, L.; Guo, C.; Wang, F.; Zheng, S.; Liu, C.Z. A simple and rapid harvesting method for microalgae by in situ magnetic separation. *Bioresour. Technol.* 2011, *102*, 10047–10051. [Google Scholar] [CrossRef]
29. Fraga-Garcia, P.; Kubbutat, P.; Brammen, M.; Schwaminger, S.; Berensmeier, S. Bare iron oxide nanoparticles for magnetic harvesting of microlagae: From interaction behavior to process realization. *Nanomaterials* 2018, *8*, 292. [Google Scholar] [CrossRef][Green Version]
30. Franzreb, M.; Siemann-Herzberg, M.; Hobley, T.J.; Thomas, O.R.T. Protein purification using magnetic adsorbent particles. *Appl. Microbiol. Biotechnol.* 2006, *70*, 505–516. [Google Scholar] [CrossRef]
31. Sharma, R. Effects of Culture Conditions on Growth and Biochemical Profile of Chlorella Vulgaris. *J. Plant Pathol. Microbiol.* 2012, *3*, 1000131. [Google Scholar] [CrossRef][Green Version]
32. Zhang, R.; Chen, J.; Mao, X.; Qi, P.; Zhang, X. Anti-inflammatory and anti-aging evaluation of pigment-protein complex extracted from *Chlorella Pyrenoidosa*. *Mar. Drugs* 2019, *17*, 586. [Google Scholar] [CrossRef][Green Version]
33. Shibata, S.; Natori, Y.; Nishihara, T.; Tomisaka, K.; Matsumoto, K.; Sansawa, H. Antioxidant and Anti-Cataract Effects of Chlorella on Rats with Streptozotocin-Induced Diabetes. *J. Nutr. Sci. Vitaminol.* 2003, *49*, 334–339. [Google Scholar] [CrossRef]
34. Ishiguro, S.; Robben, N.; Burghart, R.; Cote, P.; Greenway, S.; Thakkar, R.; Upreti, D.; Nakashima, A.; Suzuki, K.; Comer, J.; et al. Cell wall membrane fraction of *Chlorella sorokiniana* enhances hot antitumor immunity and inhbits colon carcinoma growth in mice. *Integr. Cell Ther.* 2020, *19*, 1–10. [Google Scholar]
35. Belasco, W. Algae burgers for a hungry world? The rise and fall of chlorella cuisine. *Technol. Cult.* 1997, *38*, 608–634. [Google Scholar] [CrossRef]
36. Rani, K.; Sandal, N.; Sahoo, P.K. A comprehensive review on chlorella-its composition, health benefits, market and regulatory scenario. J. Pharm. *Innov.* 2018, 7, 584–589. [Google Scholar]
37. Dardavila, M.M.; Hamilakis, S.; Loizos, Z.; Kollia, C. Ni/ZrO_2 composite electrodeposition in the presence of coumarin: Textural modifications and properties. *J. Appl. Electrochem.* 2015, *45*, 503–514. [Google Scholar] [CrossRef]
38. Markeb, A.A.; Llimós-Turet, J.; Ferrer, I.; Blánquez, P.; Alonso, A.; Sánchez, A.; Moral-Vico, J.; Font, X. The use of magnetic iron oxide based nanoparticles to improve microalgae harvesting in real wastewater. *Water Res.* 2019, *159*, 490–500. [Google Scholar] [CrossRef]

39. Aivazoglou, E.; Metaxa, E.; Hristoforou, E. Microwave-assisted synthesis of iron oxide nanoparticles in biocompatible organic environment. *AIP Adv.* 2018, *8*, 048201. [Google Scholar] [CrossRef][Green Version]

40. Leite, L.D.S.; Daniel, L.A. Optimization of microalgae harvesting by sedimentation induced by high pH. *Water Sci. Technol.* 2020, *82*, 1227–1236. [Google Scholar] [CrossRef]

41. Zhao, Y.; Liang, W.; Liu, L.; Li, F.; Fan, Q.; Sun, X. Harvesting Chlorella vulgaris by magnetic flocculation using Fe_3O_4 coating with polyaluminium chloride and polyacrylamide. *Bioresour. Technol.* 2015, *198*, 789–796. [Google Scholar] [CrossRef]

42. Tork, M.B.; Khalilzadeh, R.; Kouchakzadeh, H. Efficient harvesting of marine Chlorella vulgaris microalgae utilizing cationic starch nanoparticles by response surface methodology. *Bioresour. Technol.* 2017, *243*, 583–588. [Google Scholar] [CrossRef] [PubMed]

43. Fayad, N.; Yehya, T.; Audonnet, F.; Vial, C. Harvesting of microalgae Chlorella vulgaris using electro-coagulation-flocculation in the batch mode. *Algal Res.* 2017, *25*, 1–11. [Google Scholar] [CrossRef]

44. Liu, Y.; Jin, W.; Zhou, X.; Han, S.F.; Tu, R.; Feng, X.; Wang, Q. Efficient harvesting of Chlorella pyrenoidosa and Scenedesmus obliquus cultivated in urban sewage by magnetic flocculation using nano-Fe_3O_4 coated with polyethyleneimine. *Bioresour. Technol.* 2019, *290*, 121771. [Google Scholar] [CrossRef]

45. Razack, S.A.; Duraiarasan, S.; Shellomith, A.S.; Muralikrishnan, K. Statistical optimization of harvesting Chlorella vulgaris using a novel bio-source, Strychnos potatorum. *Biotechnol. Rep.* 2015, *7*, 150–156. [Google Scholar] [CrossRef][Green Version]

46. Zhao, Y.; Fan, Q.; Wang, X.; Jiang, X.; Jiao, L.; Liang, W. Application of Fe_3O_4 coated with modified plant polyphenol to harvest oleaginous microalgae. *Algal Res.* 2019, *38*, 101417. [Google Scholar] [CrossRef]

47. Wang, T.; Yang, W.L.; Hong, Y.; Hou, Y.L. Magnetic nanoparticles grafted with amino-riched dendrimer as magnetic flocculant for efficient harvesting of oleaginous microalgae. *Chem. Eng. J.* 2016, *297*, 304–314. [Google Scholar] [CrossRef]

48. Wang, X.; Zhao, Y.; Jiang, X.; Liu, L.; Li, X.; Li, H.; Liang, W. In-situ self-assembly of plant polyphenol-coated Fe_3O_4 particles for oleaginous microalgae harvesting. *J. Environ. Manag.* 2018, *214*, 335–345. [Google Scholar] [CrossRef]

49. Hu, Y.R.; Wang, F.; Wang, S.K.; Liu, C.Z.; Guo, C. Efficient harvesting of marine microalgaeNannochloropsismaritimausingmagneticnanoparticles. *Bioresour. Technol.* 2013, *138*, 387–390. [Google Scholar] [CrossRef]

50. Li, Y.G.; Gao, H.S.; Li, W.L.; Xing, J.M.; Liu, H.Z. In situ magnetic separation and immobilization of dibenzothiophene-desulfurizing bacteria. *Bioresour. Technol.* 2009, *100*, 5092–5096. [Google Scholar] [CrossRef]

51. Lin, Z.; Xu, Y.; Zhen, Z.; Fu, Y.; Liu, Y.; Li, W.; Zhang, D. Application and reactivation of magnetic nanoparticles in Microcystis aeruginosa harvesting. *Bioresour. Technol.* 2015, *190*, 82–88. [Google Scholar] [CrossRef] [PubMed]

52. Xu, Y.; Wang, X.; Fu, Y.; Hu, F.; Qian, G.; Liu, Q.; Sun, Y. Interaction energy and detachment of magnetic nanoparticles-algae. *Environ. Sci. Technol.* 2020, *41*, 2618–2624. [Google Scholar] [CrossRef] [PubMed]

53. Hena, S.; Fatihah, N.; Tabassum, S.; Lalung, J.; Jing, S.Y. Magnetophoretic harvesting of freshwater microalgae using polypyrrole/Fe_3O_4 nanocomposite and its reusability. *J. Appl. Phycol.* 2016, *28*, 1597–1609. [Google Scholar] [CrossRef]

54. Seo, J.Y.; Praveenkumar, R.; Kim, B.; Seo, J.-C.; Na, J.-G.; Seo, J.C.; Park, J.-Y.; Na, J.-G.; Jeon, S.G.; Park, S.B.; et al. Downstream integration of microalgae harvesting and cell disruption by means of cationic surfactant-decorated Fe_3O_4 nanoparticles. *Green Chem.* 2016, *18*, 3981–3989. [Google Scholar] [CrossRef]

13

COVALENT IMMOBILIZATION OF PAPAIN ON FILMS COMPRISING CHITOSAN–CLAY NANOCOMPOSITES: APPLICATION IN SYNTHETIC AND REAL WHITE WINE

1. INTRODUCTION: AN OVERVIEW

The application of enzymes as powerful catalysts in a large number of industrial fields, including pharmaceuticals and food, allows the development of efficient processes with low environmental impact [1]. In the wine-making industry, some enzymes have been traditionally used whereas others have been only recently introduced. Proteases, able to catalyze the hydrolysis of proteins, have been proposed as a selective alternative to the conventional bentonite fining for removing the haze-forming proteins, which are responsible for haze formation in white wine under post-bottling conditions.

In recent decades, immobilization techniques have been successfully applied to enhance enzyme properties, in particular stability and reusability [2]. The material employed for the support preparation is of fundamental importance in the immobilization process, affecting the catalytic properties of the produced biocatalyst. A wide range of inorganic and organic materials, as well as composites, may be applied as carriers for enzyme immobilization [3,4] in different shapes (e.g., films [5], beads [6,7] and fibers [8]). Recently, increasing attention has been paid to organic–inorganic composite materials for food application. Regarding composite films as supports for enzyme immobilization, Yang et al. [9] reported that the chemical and thermal stability of organic membranes could be enhanced by an inorganic phase. The latter usually allows the stabilization of enzyme–support interactions [10].

Among biopolymers, chitosan (CS, a linear polysaccharide made of N-acetyl-D-glucosamine and D-glucosamine units [11]) is the most commonly used for the production of immobilization supports as it is, as well as in composite/nanocomposite forms [2], to be applied in the food and pharmaceutical industries.

Several studies have been aimed at improving the mechanical and thermal stability of neat CS films [12,13,14] by seeking CS-based nanocomposite, with the addition of specific nanoparticles to enhance the aforementioned properties [3,13,15]. The nanomaterials dispersed within the chitosan matrix, e.g., nanoclays, carbon structures, and metal/metal oxides, via physical or chemical interaction [16], allow not only to enhance the physical, mechanical, and thermal stability but also to endow the composite with their intrinsic features, e.g., high specific surface area [11].

Among the nanoparticles, clay is a fine-grained soil material that contains metal oxides or hydroxides with traces of organic matters. Due to its smallest dimension, excellent colloidal properties [11], and ability to interact with CS by electrostatic force, clay has been used for the preparation of CS–clay nanocomposites through adsorption, gelation, or intercalation [13,17].

Recently, CS–clay nanocomposite films have been prepared by using different nanoclays (i.e., bentonite, sepiolite and montmorillonite (MMT)) and have been applied as supports for the covalent immobilization of stem bromelain in wine production [13,15]. In our previous works, we proposed nanocomposites based on CS and nanoclays in very low concentrations (1–5% w/w) [12] and systems based on CS and nanoclays in high concentrations (70–80% w/w) [14], maintaining, in both cases, the total weight of the composite constant. Papain from Carica papaya L. latex (EC 3.4.22.2), a cysteine protease applied in brewing for the removal of chill haze [18] and for the clarification treatment of pomegranate juice [19], has been covalently immobilized on CS. Moreover, Benucci et al. [20] immobilized papain on commercial chitosan beads by direct linkage for the protein stabilization of white wines.

This study was focused on the production of novel CS–clay nanocomposite films, using a fixed amount of CS (1% *w/v*) and a food-grade activated MMT (Optigel, OPT) or a high-purity unmodified MMT (SMP), in four different weight percentages with respect to CS (i.e., 20, 30, 50, 70% *w/w*), as carriers for the covalent immobilization of papain.

The physical properties, in terms of the morphology, thermal stability and the mechanical behaviour, of the produced nanocomposites were investigated. Finally, the catalytic properties of immobilized papain were tested in synthetic wine and its effectiveness in protein stabilization was investigated in real white wines.

2. MATERIALS AND METHODS

2.1. Enzyme, Chemicals and Wines

Papain from Carica Papaya latex (EC 3.4.22.2), shellfish-derived chitosan (CS) powder (low molecular weight 50–190 kDa; percentage of deacetylation 75%), glutaraldehyde

(GDH, 25% v/v), and glycerol (≥99.5%) were obtained from Sigma-Aldrich (Milan, Italy). Two different nanoclays, i.e., activated food-grade montmorillonite (Optigel, OPT) and high-purity unmodified montmorillonite (SMP), were kindly provided by BYK Additives GmbH (Wesel, Germany) and Zhejiang Fenghong New Material Co., Ltd. (Huzhou, China), respectively. The selected synthetic tripeptide chromogenic substrate (Bz–Phe–Val–Arg–p-nitroaniline (pNA)), applied for the kinetic characterization of immobilized papain, was purchased from Bachem (Bubendorf, Switzerland). All the other chemicals were of analytical grade (Sigma Aldrich, Milan, Italy). Manzoni and Sauvignon Blanc unfined white wines (vintage 2018) were kindly provided by Casale del Giglio winery (Le Ferriere, LT, Italy) and their oenological parameters are summarized in Table 1.

Table 1. Oenological parameters of the unfined Manzoni and Sauvignon Blanc wines.

pH	3.37 ± 0.01	3.34 ± 0.01
Total acidity (g/L tartaric acid)	6.00 ± 0.05	6.00 ± 0.05
Alcohol level (% v/v)	11.2 ± 0.3	13.2 ± 0.2
Free SO2 (mg/L)	6 ± 1	10 ± 1
Total SO2 (mg/L)	30 ± 2	48 ± 3
Total phenols (mg/L catechin)	217 ± 1	219 ± 4
Total protein (mg/L)	465 ± 43	120 ± 7
ΔNTU Index0	600 ± 11	116 ± 8

2.2. Preparation of CS–Clay Nanocomposite Films by Solvent Casting

CS–clay supports were obtained by solvent casting technique, using low molecular weight CS (1% w/v) blended with glycerol (CS:glycerol 75:25, in %w/w) [5], and adding two different nanoclay types, i.e., OPT and SMP, in four different weight concentrations (i.e., 20, 30, 50, 70% w/w). The set-up procedure was reported elsewhere [5]. Briefly, nanoclays aqueous suspensions were prepared by ultrasonication using Sonics Vibracell CV33 (Sonics & Materials Inc., Newtown, CT, USA) in the following conditions: power 750 W, frequency 20 kHz, amplitude 30%, time 30 min. Afterwards, acetic acid (2% v/v), glycerol and CS were added and the mixtures were maintained under continuous magnetic stirring overnight. The final suspensions were solvent cast onto plastic Petri dishes, in a fume hood, and dried at room temperature for 48 h. Furthermore, as a reference, a system composed of only CS and glycerol was produced. The obtained samples were designed as Clay-free for the CS/glycerol-based film and OPT-x and SMP-x in the case of clay-loaded composites (where x is the w/w percentage of the used nanoclay with respect to the CS, i.e., 50 ,30 ,20, and %70 w/w).

2.3. Physical Characterization of CS–Clay Nanocomposite Films

The produced films were morphologically, thermally and mechanically analyzed by Field-Emission Gun Scanning Electron Microscope (FEG-SEM, Leo Supra 35, Carl Zeiss SMT Ltd., Cambridge, UK), differential scanning calorimetry (DSC, TAInstruments Q2000, New Castle, DE, USA) and uniaxial tensile tests (Lloyd LRX, Lloyd Instruments Ltd., West Sussex, UK), respectively. In detail, the SEM micrographs were acquired on gold-coated samples, applying a voltage of 3–5 kV. DSC measurements were carried out between 25 and 400 °C (heating rate 10 $°C·min^{-1}$), in a nitrogen atmosphere (N_2 flow rate 50 $cc·min^{-1}$), using a sample weight of ~5 mg. For the mechanical investigation, the ASTM D1708 and ASTM D882 standards were followed, and the nominal specimen cross-section was considered for the measurement of all mechanical features. Dog-bone specimens (width 4.8 mm, length 22.25 mm) were obtained from the produced films. The tests were performed at 1.2 $mm·min^{-1}$, using a 50 N load cell. All measurements were taken in triplicate.

2.4. Papain Immobilization on CS–Clay Nanocomposite Films

Before enzyme immobilization, the CS-film samples were neutralized by overnight shaking in 26% (v/v) ethanol–NaOH 2 M solution and then cut into squares (10 mm × 10 mm) with a razor blade. The immobilization procedure was performed as briefly reported in the following [13]. The film surface was activated by adding 1 mL of 3% (v/v) GDH as a cross-linker, and keeping at room temperature under constant agitation (120 rpm) for 2 h. The activated films were thoroughly washed with distilled water; thereafter, 1 mL of papain preparation (0.12 $mg_{protein}$/mL solubilized in the tartaric buffer) was added. After the overnight incubation (150 rpm at 20 °C), the obtained biocatalysts were carefully washed with tartaric buffer, and then left to stand for 20 min in 0.1 M glycine solution. At the end of the immobilization procedure, the biocatalysts were washed three times with 2 M NaCl solution to remove all non-covalently bound proteins. The immobilization yield (IY, %) was calculated as the difference between the protein concentration in the enzyme solution, before and after immobilization. The protein concentration was evaluated by the Bradford method [21] using bovine serum albumin (BSA) as a standard protein.

2.5. Proteolytic Activity Assay

Proteolytic activity toward the tripeptide chromogenic substrate (Bz–Phe–Val–Arg–pNA) was tested in model wine (0.03 M, tartaric acid/sodium tartrate solution pH 3.2, containing 12% v/v of ethanol) at 20 °C, as reported by Benucci et al. [13].

2.6. Kinetic Characterization of Immobilized Papain

A kinetic study of papain immobilized on CS–clay nanocomposite films was carried out in model wine, fortified with Bz–Phe–Val–Arg–pNA substrate (0–750 μM). Kinetic parameters (V_{max} and K_M (Michaelis–Menten constant)) were determined according to the Michaelis–Menten equation using a non-linear regression procedure (GraphPad Prism 5.01, GraphPad Software, Inc., San Diego, CA, USA). Moreover, k_{cat} (turnover number) and K_a (affinity constant) were calculated as described by Benucci et al. [13].

2.7. Wine Stabilization Treatment in the Batch-Scale Stirred Reactor

Ten milliliters of each white wine were treated using the immobilized biocatalyst in a laboratory-scale stirred reactor (120 rpm, 20°C) for 24 h. Trials were performed in triplicate, in three different stirred reactors, in order to obtain three independent replicates. To investigate the potential absorption effect of the CS–clay nanocomposite films, a blank correction was made using the supports covered with protease firstly immobilized and then deactivated with NaOH 16 N.

2.8. Wine Protein Content Determination

The total protein content of both white wines, before and after the enzymatic treatment in the laboratory-scale stirred reactor, was determined using the potassium dodecyl sulphate method, according to Gaspar et al. [22]. All measurements were made in triplicate.

2.9. Heat Test

The protein stability of both white wines, before and after the enzymatic treatment in laboratory-scale stirred reactor, was investigated by heat test, incubating the wines at 80 °C for 6 h, and then keeping them at 4 °C for 16 h [23]. Haze formation was measured using HD 25.2 turbidimeter (Delta Ohm, Padua, Italy), and the turbidity was expressed in nephelometric turbidity units (NTU). Wine stability was calculated as the difference in the wine NTU after and before the heat test (haze potential). The turbidity removal yield (TRY (%)) was expressed as the percentage of haze potential decrease due to the proteolytic treatment. All measurements were made in triplicate.

2.10. Statistical Analysis

One-way completely randomized analysis of variance (ANOVA), with an EXCEL® Add-in macro, was employed to analyze the data, obtained from the average of three replicate measurements. Then, a Tukey honestly significant difference (Tukey HSD) post-hoc test ($p = 0.05$) for the multiple comparisons of samples was applied.

3. RESULTS AND DISCUSSION

3.1. Physical Properties of CS–Clay Nanocomposite Films

All the samples were morphologically analyzed. As an example, the SEM micrographs of SMP70 and OPT70 are compared in Figure 1. Two very different filler distribution behaviours were observed. OPT clays were better dispersed, promoting a more uniform and homogeneous surface. On the contrary, the SMP addition induced the development of several agglomerates, leading to a very rough surface, characterized by several defects (Figure 1).

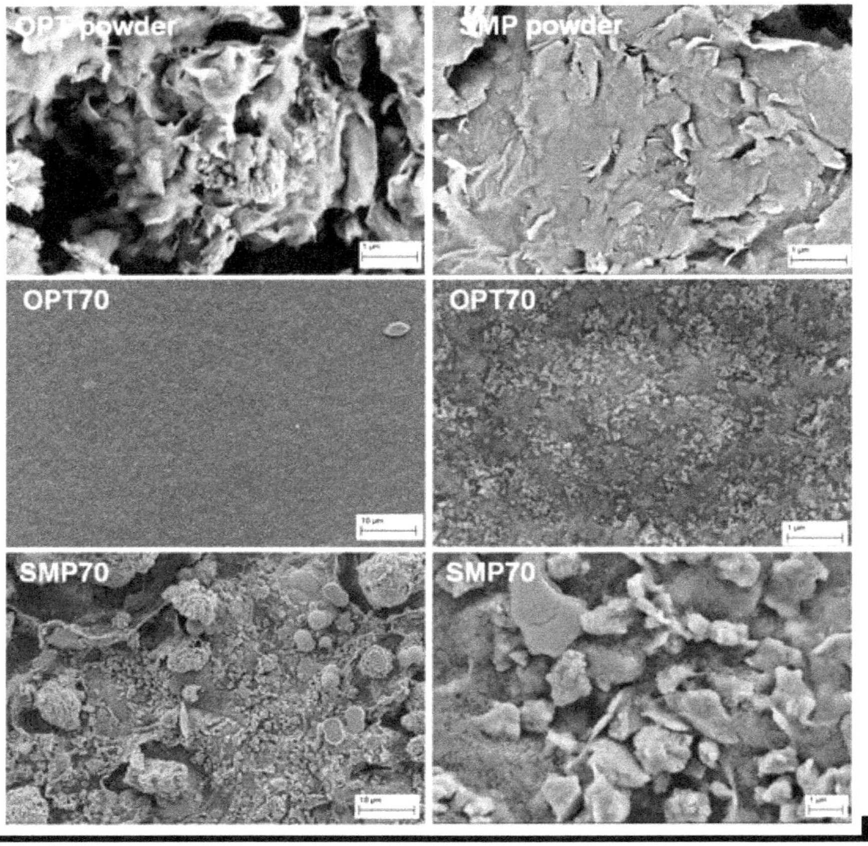

Figure 1. SEM micrographs of the food-grade activated montmorillonite (Optigel, OPT) and high-purity unmodified montmorillonite (SMP) powders, and the OPT- and SMP-based films.

The influence of OPT and SMP on the thermal properties was explored by DSC measurements. In Table 2, the acquired thermal properties are collected and the DSC thermograms of OPT20 and SMP20, as an example, are compared to that of the clay-free sample in Figure 2.

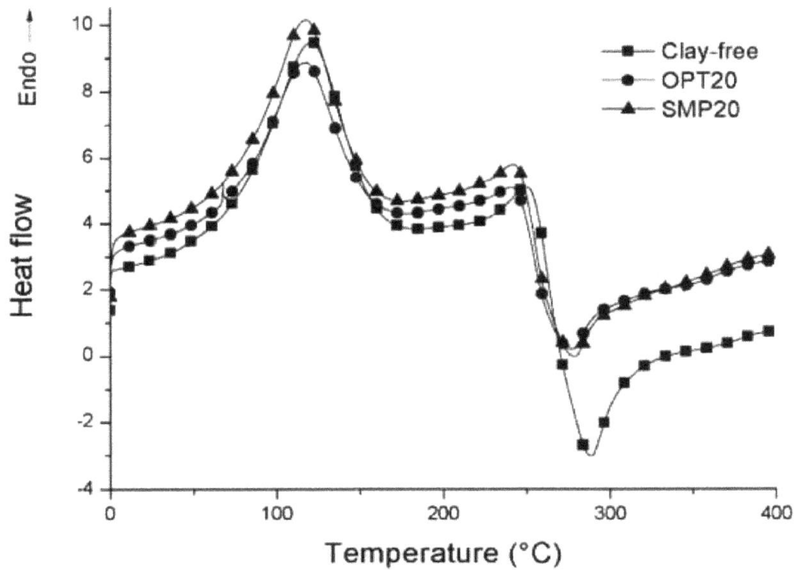

Figure 2. Differential scanning calorimetry (DSC) thermograms of Clay-free, OPT20 and SMP20 samples.

Table 2. Thermal properties of the chitosan and nanoclay-based supports.

TIHB (°C)	121 ± 0.5	118 ± 0.4	124 ± 0.7	111 ± 0.6	118 ± 0.4	115 ± 0.4	105 ± 0.3
Td (°C)	287 ± 1.2	274 ± 1.3	272 ± 0.9	260 ± 1.0	279 ± 1.1	280 ± 1.4	284 ± 1.5
ΔHmI (J/g)	259 ± 1.5	259 ± 1.4	238 ± 1.2	142 ± 0.9	274 ± 1.1	296 ± 1.3	335 ± 2.0

In all cases, two main peaks, one endothermic and the other one exothermic, were detected at approximately 105–120 °C and 235–335 °C (Table 2), respectively. The first one was ascribed to the dissociation process of interchain hydrogen bonding of CS (T_{IHB}) [24,25], whereas the second one to CS decomposition (T_d) [26,27,28]. In all the composite systems, a decrement of T_{IHB} temperature was revealed, compared to Clay-free (121 °C). Particularly, a progressive decrease increasing the nanoclay amount was evidenced in the case of SMP-based films. For the related enthalpies, different trends were recorded for OPT- and SMP-based films. For the former, a decrement of the enthalpy was evidenced increasing OPT amount, whereas, an opposite trend was revealed for SMP-based composites. This experimental evidence indicates that higher energy is necessary in order to induce the dissociation process of interchain hydrogen bonding of CS in the case of SMP-based systems. Concerning the chitosan degradation temperature, a decrease was evidenced in the case of composite systems with respect to Clay-free film, suggesting a not good filler dispersion within the polymeric matrix, with the consequent development of many agglomerates.

The effect of nanoclays on the mechanical behaviour was analyzed performing uniaxial tensile tests (Figure 3, Table 3).

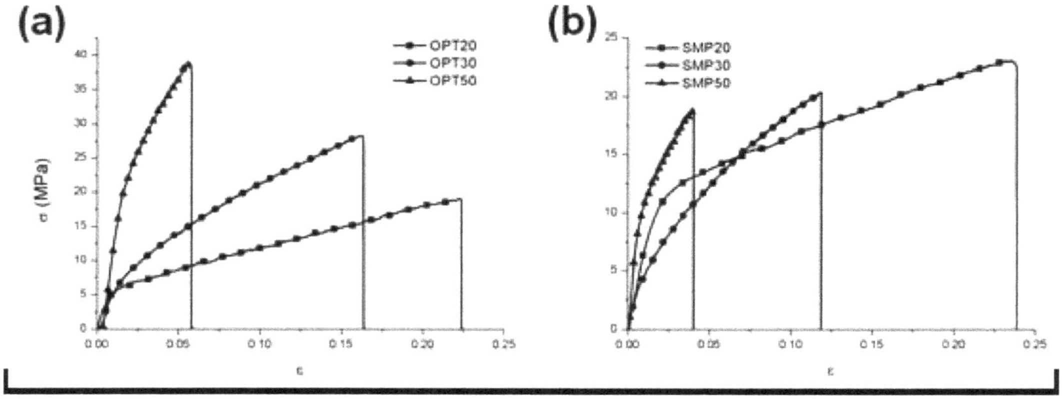

Figure 3. σ–ε curves of OPT (a) and SMP (b) based films.

Table 3. Mechanical properties of the OPT- and SMP-based films, compared to the neat one (Clay-free).

Clay-Free	579 ± 23	16 ± 1	0.30 ± 0.01
OPT-20	738 ± 20	21± 6	0.22 ± 0.08
OPT-30	653 ± 71	23 ± 4	0.13 ± 0.05
OPT-50	1670 ± 202	36 ± 7	0.06 ± 0.03
SMP-20	580 ± 55	24 ± 4	0.28 ± 0.05
SMP-30	528 ± 4	22 ± 3	0.12 ± 0.03
SMP-50	1582 ± 210	19 ± 4	0.04 ± 0.01

An improvement in the mechanical responsiveness, in terms of the Young modulus and maximum stress value (σ_{max}), was evidenced in the case of all composite systems, particularly those based on OPT, due to its better distribution within the polymeric matrix, as testified by the reported SEM micrographs (Figure 1) and DSC data (Table 2). The OPT and SMP-based films presented different mechanical behaviour. In detail, a progressive increment of the ultimate tensile strength (σ_{max}) was evidenced for OPT-based samples with the OPT amount, whereas an opposite trend was observed in the case of SMP-based ones (Table 3), probably due to the presence of several agglomerates whose amount tended to increase with filler percentage (Figure 1), as expected.

Additionally, a remarkable increase in the Young Modulus value was observed for a nanoclay amount of 50% w/w, independently of the considered nanoclay. On the contrary, up to 30% w/w), comparable values were recorded (Table 3). As expected, by increasing the nanoclay amount, a progressive decrease in the elongation at break (ε_{max}) was revealed.

Taking into account all the collected morphological, thermal and mechanical properties, it is evident that OPT presented a higher affinity towards CS, as testified by its better distribution within the polymeric matrix, promoting a good filler-polymeric chains interaction.

Indeed, SMP and OPT led to distinct behaviour due to their different chemical nature and microstructure. In details, even if both SMP and OPT powders were composed of platelet-like particles (Figure 1), a higher tendency to agglomeration was evidenced in the case of SMP powder, whereas OPT nanoclays were more homogeneously dispersed within the polymeric matrix. These experimental evidences could be ascribed to OPT good wettability and compatibility, able to promote a significant improvement of the mechanical behaviour (Table 3).

3.2. Kinetic Properties of Papain Immobilized on CS–Clay Nanocomposite Films

CS–clay films, produced by the addition of two nanoclays (OPT and SMP) in different amounts (20, 30, 50, 70% w/w with respect to CS), were applied as supports for the papain covalent immobilization.

Data in Table 4 show that the IY was significantly higher for SMP%60) 70-) with respect to Clay-free sample (%45), thus proving that the addition of such nanoclay remarkably increased the amount of covalently bound protein, probably due to its higher surface roughness and better affinity between SMP clay and the papain. Similarly, Basak et al. [29] demonstrated that the addition of bentonite into chitosan beads (ratio 50:50% w/w) significantly increased the catalase immobilization yield. Furthermore, the addition of OPT nanoclay significantly decreased the IY, with the only exception of OPT-70. These data appeared consistent with those described by Cacciotti et al. [15], who found that the IY slightly decreased when the bromelain from pineapple stem was immobilized on nanocomposites obtained combining MMT (i.e., SMP and OPT) with low amounts of CS in different relative ratios (MMT:CS 70:30, 75:25 and 80:20, % w/w), compared to Clay-free support. Moreover, the lowest IY was found for OPT-30 and OPT-50 films (26 and 29%, respectively), whereas no relevant differences were revealed comparing the other samples.

Table 4. Immobilization yield and kinetic properties of papain immobilized on the CS–clay nanocomposite systems.

Clay-free	45 ± 6 b,c	6.9 ± 0.2 e	88 ± 9 a	3183 ± 0 e	36 ± 1 d	0.99
OPT-20	41 ± 4 c,d	15.3 ± 0.9 c	75 ± 16 a,b	6535 ± 1 c	87 ± 4 c	0.97
OPT-30	26 ± 5 e	21.4 ± 1.0 b	106 ± 23 a	12507 ± 1 b	118 ± 6 b	0.98
OPT-50	29 ± 2 de	38.9 ± 2.0 a	102 ± 14 a	20417 ± 2 a	199 ± 3 a	0.99
OPT-70	55 ± 3 b	11.6 ± 0.5 d	50 ± 8 b,c	4225 ± 0 d	85 ± 3 c	0.98
SMP-20	54 ± 7 a,b,c	2.6 ± 0.1 f	34 ± 8 c	1017 ± 0 h	30 ± 2 d,e	0.97
SMP-30	41 ± 5 c,d	3.5 ± 0.2 f	46 ± 11 b,c	642 ± 0 i	14 ± 1 f	0.97
SMP-50	56 ± 2 b	4.5 ± 0.1 f	49 ± 2 b,c	1221 ± 0 g	25 ± 0 e	1.00
SMP-70	60 ± 4 a	7.3 ±0.1 e	29 ± 3 c	2503 ± 0 f	86 ± 1 c	1.00

The kinetic characterization of the biocatalysts was performed in order to explore the effect of the nanoclays' inclusion within the polymeric matrix on the behaviour of immobilized papain. All the biocatalysts followed the hyperbolic trend described by the Michaelis–Menten equation (Figure 4) and the related catalytic properties are reported in Table 4. Papain on OPT films showed higher V_{max} and k_{cat} values with respect to the enzyme linked to all the other supports (Clay-free and SMP-based films). Notably, the highest V_{max} and k_{cat} values were found for papain immobilized on OPT38.9) 50- mIU/mg_{IP} and 20,417 min^{-1}, respectively), thus indicating a greater product release velocity. However, both parameters significantly increased as the percentage of clay in the nanocomposite increased, except for the ratio 30:70% w/w (Table 4). Basak et al. [29] described similar results, finding an increased V_{max} value (about %20+) for lipase immobilized on CS–clay (hydrophilic bentonite) with respect to the neat CS support. As widely described in the literature for α-amylase and α-amylase immobilized on clay-CS carriers, an increased V_{max} value may arise from the conformational changes of biocatalysts, which usually occur after the linkage to the support [30].

In contrast, papain immobilized on SMP-20, SMP-30 and SMP-50 films showed significantly lower V_{max} and k_{cat} values with respect to the enzyme linked to Clay-free support, whereas no significant differences were revealed for SMP70- (Table 4). These results are in agreement with what Cacciotti et al. [14] reported for another protease belonging to the papain-superfamily (bromelain from pineapple stem).

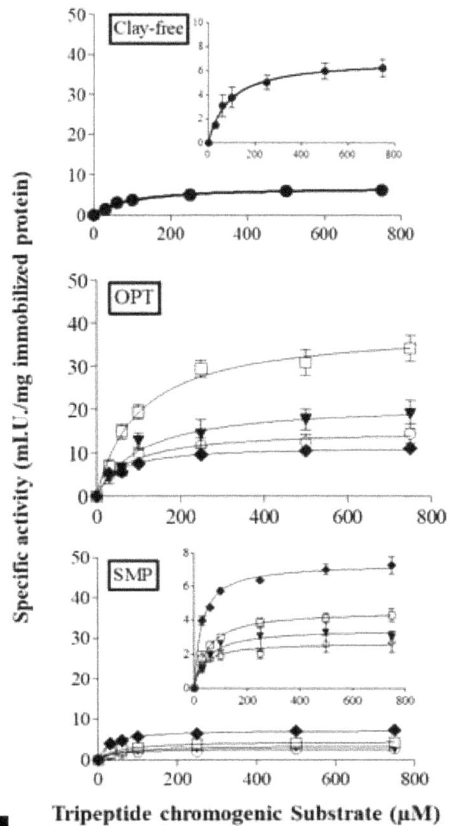

Figure 4. Kinetic curves of papain immobilized on CS–clay composite systems loaded with OPT or SMP at various weight percentages:♣ 30 ,(○) 20 (▼), 50 (□) and 70 (♦) % *w/w*.

All SMP biocatalysts and OPT-70 showed a significant decrease in K_M values (on average %55- and %43- for SMP-based samples and OPT70-, respectively) as compared with Clay-free. No remarkable differences were found for OPT20-, OPT-30 and OPT50- (Table 4), thus indicating no changes in the enzyme–substrate complex formation in comparison to the reference biocatalyst. Similar results were described by Basak et al. [29], who found comparable K_M values for the lipase immobilized on CS and on CS–clay. Moreover, the K_a values revealed for papain immobilized on OPT films were 2.4 (OPT20- and OPT70-) to -5.5fold (OPT50-) higher as compared with the enzyme immobilized on Clay-free, suggesting a greater apparent affinity of the protease on the aforementioned nano-composites for the synthetic substrate. Concerning the biocatalysts on the SMP-based supports, their K_a values always appeared lower or not statistically different (i.e., SMP20-) with respect to Clay-free film.

Overall, in spite of the lowest IY observed for papain immobilized on OPT-based films, OPT-30 and OPT-50 exhibited the best catalytic performances, both in terms of V_{max}, k_{cat} and K_a, thus indicating a higher efficiency in the release of the reaction product. As known, the decrease in the biocatalyst activity could be ascribed to a great amount of enzyme molecules immobilized on the carrier, since the latter could limit the accessibility of the substrate to the active sites, due to diffusion limitation [31]. Furthermore, as described in the literature [13], the improved catalytic properties observed for papain bound on OPT films could be likely due to the hydrophilic nature of OPT clay, which contributes to create an environment conducive to the interaction between substrate and enzyme. Otherwise, the hydrophobic clay which constitutes SMP-based films, does not facilitate the aforementioned interaction [13].

Thus, among all the produced biocatalysts, papain immobilized on the OPT-50 film was selected as the most suitable to be applied for the protein stabilization treatment of white wines.

3.3. Wine Stabilization Treatment in Batch-Scale Stirred Reactor

The partial stabilization of white wines via proteolytic treatment has been already described by other authors [1]. Benucci et al. [20] described the efficacy of a laboratory bench-scale packed-bed reactor, containing papain bound on chitosan beads, in stabilizing white wines.

The effectiveness of papain immobilized on OPT-50 film was tested in a laboratory-scale stirred reactor with the aim of reducing the protein instability of two real white wines (i.e., Manzoni and Sauvignon Blanc), characterized by a different haze potential (601 ± 11 ΔNTU and 116 ± 8 ΔNTU, respectively), as well as by a different initial protein content (465 ± 63 mg_{BSAeq}/ L and 124 ± 1 mg_{BSAeq}/L, respectively). Papain covalently immobilized on OPT-50 nanocomposite film significantly reduced both the haze potential and the protein content in the two wine samples (Table 5). The biocatalyst exhibited the greatest stabilization efficiency in Manzoni wine, with a TRY reduction rate of about 83% and a decrease in total protein amount of approximately 73%. The lowest stabilization efficiency revealed in the Sauvignon Blanc wine (TRY reduction: 31%; total protein reduction: 12%) could be ascribed to the higher content of potential inhibitors found in Sauvignon Blanc with respect to Manzoni wine, especially SO_2 free (about -2fold higher in the former compared to the latter wine) and the highest alcohol content (about 2% v/v). Indeed, it has been proven that these substances exhibit a strong inhibiting action against the catalytic activity of papain [20].

Table 5. Protein stability data of Manzoni and Sauvignon blanc white wines, treated in a laboratory-scale stirred reactor with papain immobilized on OPT-50 carrier at 20 °C (120 rpm, 24 h). Reported data are mean of triplicate measurements.

Manzoni				
Untreated wine	601 ± 11 a	-	465 ± 63 a	-
OPT-50	102 ± 41 b	83	124 ± 35 b	73
Sauvignon Blanc				
Untreated wine	116 ± 8 a	-	124 ± 1 a	-
OPT-50	80 ± 1 b	31	109 ± 1 b	12

4. CONCLUSIONS

The CS-based nanocomposite films produced by the addition of two nanoclays (OPT and SMP) showed improved mechanical properties. The better distribution of OPT clay within the polymeric matrix, as proved by the SEM micrographs and DSC data, imparted a higher Young modulus and σ_{max} to the OPT nanocomposites.

Papain was successfully bound on such carriers, showing the highest efficiency in the release of the reaction product after the covalent immobilization on OPT-30 and OPT-50. The latter biocatalyst reliably reduced both the haze potential and the protein content in the two tested white wine samples, namely Manzoni and Sauvignon Blanc, even if the stabilization efficiency was affected by the content of the potential inhibitors in the wine.

5. CONCISE SUMMARY

Increasing attention has been recently paid to the development of nanocomposite materials for food application as new tool to enhance the mechanical and thermal properties of polymers. In this study, novel chitosan–clay nanocomposite films were produced as carriers for the covalent immobilization of papain, by using a fixed amount of chitosan (1% w/v) and a food-grade activated montmorillonite (Optigel, OPT) or a high-purity unmodified montmorillonite (SMP), in four different weight percentages with respect to chitosan (i.e., 20, 30, 50, 70% w/w). Both nanoclays (OPT and SMP) improved the mechanical properties of the obtained nanocomposites, and the OPT films showed the highest Young modulus and mechanical resistance (σ_{max}). The nanocomposites were used as carriers for the covalent immobilization of papain, which was preliminarily characterized in model wine towards a synthetic substrate, showing the highest efficiency in the release of the reaction product when it was bound on OPT-30 and OPT-50 films. Finally, the latter biocatalyst (papain on OPT-50 film) was applied for the protein stabilization of two different unfined white wines, and it efficiently reduced both the haze potential and the protein content.

REFERENCES

1. Ottone, C.; Romero, O.; Aburto, C.; Illanes, A.; Wilson, L. Biocatalysis in the winemaking industry: Challenges and opportunities for immobilized enzymes. Compr. Rev. Food Sci. Food Saf. 2020, 19, 595–621. [Google Scholar] [CrossRef][Green Version]
2. Verma, M.L.; Kumar, S.; Das, A.; Randhawa, J.S.; Chamundeeswari, M. Chitin and chitosan-based support materials for enzyme immobilization and biotechnological applications. Environ. Chem. Lett. 2020, 18, 315–323. [Google Scholar] [CrossRef]
3. Zdarta, J.; Meyer, A.S.; Jesionowski, T.; Pinelo, M. A general overview of support materials for enzyme immobilization: Characteristics, properties, practical utility. Catalysts 2018, 8, 92. [Google Scholar] [CrossRef][Green Version]
4. Liburdi, K.; Straniero, R.; Benucci, I.; Garzillo, A.M.V.; Esti, M. Lysozyme immobilized on micro-sized magnetic particles: Kinetic parameters at wine pH. Appl. Biochem. Biotechnol. 2012, 166, 1736–1746. [Google Scholar] [CrossRef]
5. Zappino, M.; Cacciotti, I.; Benucci, I.; Nanni, F.; Liburdi, K.; Valentini, F.; Esti, M. Bromelain immobilization on microbial and animal source chitosan films, plasticized with glycerol, for application in wine-like medium: Microstructural, mechanical and catalytic characterizations. Food Hydrocoll. 2015, 45, 41–47. [Google Scholar] [CrossRef]
6. Benucci, I.; Lombardelli, C.; Cacciotti, I.; Liburdi, K.; Nanni, F.; Esti, M. Chitosan beads from microbial and animal sources as enzyme supports for wine application. Food Hydrocoll. 2016, 61, 191–200. [Google Scholar] [CrossRef]
7. Benucci, I.; Mazzocchi, C.; Lombardelli, C.; Cacciotti, I.; Esti, M. Multi-enzymatic systems immobilized on chitosan beads for pomegranate juice treatment in fluidized bed reactor: Effect on haze-active molecules and chromatic properties. Food Bioprocess Tech. 2019, 12, 1559–1572. [Google Scholar] [CrossRef]
8. Cacciotti, I.; Pallotto, F.; Scognamiglio, V.; Moscone, D.; Arduini, F. Reusable optical multi-plate sensing system for pesticide detection by using electrospun membranes as smart support for acetylcholinesterase immobilisation. Mater. Sci. Eng. C 2020, 111, 110744. [Google Scholar] [CrossRef]

9. Yang, Y.; Yang, H.; Yang, M.; Shen, G.; Yu, R. Amperometric glucose biosensor based on a surface treated nanoporous ZrO2/chitosan composite film as immobilization matrix. Anal. Chim. Acta 2004, 525, 213–220. [Google Scholar] [CrossRef]

10. Yang, H.C.; Hou, J.; Chen, V.; Xu, Z.K. Surface and interface engineering for organic—Inorganic composite membranes. J. Mater. Chem. A 2016, 4, 9716–9729. [Google Scholar] [CrossRef]

11. Kumar, S.; Ye, F.; Dobretsov, S.; Dutta, J. Chitosan nanocomposite coatings for food, paints, and water treatment applications. Appl. Sci. 2019, 9, 2409. [Google Scholar] [CrossRef][Green Version]

12. Wu, Z.; Huang, X.; Li, Y.C.; Xiao, H.; Wang, X. Novel chitosan films with laponite immobilized Ag nanoparticles for active food packaging. Carbohydr. Polym. 2018, 199, 210–218. [Google Scholar] [CrossRef] [PubMed]

13. Benucci, I.; Liburdi, K.; Cacciotti, I.; Lombardelli, C.; Zappino, M.; Nanni, F.; Esti, M. Chitosan/clay nanocomposite films as supports for enzyme immobilization: An innovative green approach for winemaking applications. Food Hydrocoll. 2018, 74, 124–131. [Google Scholar] [CrossRef]

14. Jamróz, E.; Kulawik, P.; Kopel, P. The effect of nanofillers on the functional properties of biopolymer-based films: A review. Polymers 2019, 11, 675. [Google Scholar] [CrossRef] [PubMed][Green Version]

15. Cacciotti, I.; Lombardelli, C.; Benucci, I.; Esti, M. Clay/chitosan biocomposite systems as novel green carriers for covalent immobilization of food enzymes. J. Mater. Res. Technol. 2019, 8, 3644–3652. [Google Scholar] [CrossRef]

16. Saleh, T.A. Nanomaterials: Classification, properties, and environmental toxicities. Environ. Technol. Inno. 2020, 101067. [Google Scholar] [CrossRef]

17. Islam, M.A.; Tan, Y.L.; Islam, M.A.; Romić, M.; Hameed, B.H. Chitosan–bleaching earth clay composite as an efficient adsorbent for carbon dioxide adsorption: Process optimization. Colloids Surf. A Physicochem. Eng. Asp. 2018, 554, 9–15. [Google Scholar] [CrossRef]

18. Mastanjević, K.; Krstanović, V.; Lukinac, J.; Jukić, M.; Vulin, Z.; Mastanjević, K. Beer—The importance of colloidal stability (non-biological haze). Fermentation 2018, 4, 91. [Google Scholar] [CrossRef][Green Version]

19. Cerreti, M.; Liburdi, K.; Benucci, I.; Spinelli, S.E.; Lombardelli, C.; Esti, M. Optimization of pectinase and protease clarification treatment of pomegranate juice. LWT Food Sci. Technol. 2017, 82, 58–65. [Google Scholar] [CrossRef]

20. Benucci, I.; Lombardelli, C.; Liburdi, K.; Acciaro, G.; Zappino, M.; Esti, M. Immobilised native plant cysteine proteases: Packed-bed reactor for white wine protein stabilisation. J. Food Sci. Tech. 2016, 53, 1130–1139. [Google Scholar] [CrossRef][Green Version]

21. Bradford, M.M. A rapid and sensitive method for the quantification of microgram quantities of protein utilizing the principle of protein-dye binging. Anal. Biochem. 1976, 72, 248–254. [Google Scholar] [CrossRef]

22. Gaspar, L.M.; Machado, A.; Coutinho, R.; Sousa, S.; Santos, R.; Xavier, A.; Simões, J.S. Development of potential yeast protein extracts for red wine clarification and stabilization. Front. Microbial. 2019, 10, 2310. [Google Scholar] [CrossRef] [PubMed]

23. Colangelo, D.; Torchio, F.; De Faveri, D.M.; Lambri, M. The use of chitosan as alternative to bentonite for wine fining: Effects on heat-stability, proteins, organic acids, colour, and volatile compounds in an aromatic white wine. Food Chem. 2018, 264, 301–309. [Google Scholar] [CrossRef] [PubMed]

24. Chuang, W.Y.; Young, T.H.; Yao, C.H.; Chiu, W.Y. Properties of the poly (vinyl alcohol)/chitosan blend and its effect on the culture of fibroblast in vitro. Biomaterials 1999, 20, 1479–1487. [Google Scholar] [CrossRef]

25. Kittur, F.S.; Prashanth, K.H.; Sankar, K.U.; Tharanathan, R.N. Characterization of chitin, chitosan and their carboxymethyl derivatives by differential scanning calorimetry. Carbohydr. Polym. 2002, 49, 185–193. [Google Scholar] [CrossRef]

26. Sakurai, K.; Maegawa, T.; Takahashi, T. Glass transition temperature of chitosan and miscibility of chitosan/poly (N-vinyl pyrrolidone) blends. Polymer 2000, 41, 7051–7056. [Google Scholar] [CrossRef]

27. Zeng, M.; Fang, Z.; Xu, C. Effect of compatibility on the structure of the microporous membrane prepared by selective dissolution of chitosan/ synthetic polymer blend membrane. J. Membr. Sci. 2004, 230, 175–181. [Google Scholar] [CrossRef]

28. Garavand, F.; Rouhi, M.; Razavi, S.H.; Cacciotti, I.; Mohammadi, R. Improving the integrity of natural biopolymer films used in food packaging by crosslinking approach: A review. Int. J. Biol. Macromol. 2017, 104, 687–707. [Google Scholar] [CrossRef]

29. Basak, E.; Aydemir, T.; Dinçer, A.; Becerik, S.Ç. Comperative study of catalase immobilization on chitosan, magnetic chitosan and chitosan-clay composite beads. Artif. Cells Nanomed. Biotechnol. 2013, 41, 408–413. [Google Scholar] [CrossRef]

30. Chang, M.Y.; Juang, R.S. Activities, stabilities, and reaction kinetics of three free and chitosan–clay composite immobilized enzymes. Enzyme Microb. Technol. 2005, 36, 75–82. [Google Scholar] [CrossRef]

31. Jiang, D.S.; Long, S.Y.; Huang, J.; Xiao, H.Y.; Zhou, J.Y. Immobilization of pycnoporus sanguineus laccase on magnetic chitosan microspheres. Biochem. Eng. J. 2005, 25, 15–23. [Google Scholar] [CrossRef]

INDEX

www.ingramcontent.com/pod-product-compliance
Lightning Source LLC
LaVergne TN
LVHW062220200926
848484LV00006B/1441